NEW COLLEGE, SWINDON

20 21002 154

KU-451-571

Lab Dynamics

MANAGEMENT AND LEADERSHIP
SKILLS FOR SCIENTISTS

THIRD EDITION

Carl M. Cohen, Ph.D.

Suzanne L. Cohen, Ed.D.

COLD SPRING HARBOR LABORATORY PRESS
Cold Spring Harbor, New York • www.cshlpress.org

Lab Dynamics: Management and Leadership Skills for Scientists
Third Edition

2021002154

All rights reserved

© 2018 by Carl M. Cohen and Suzanne L. Cohen
Published by Cold Spring Harbor Laboratory Press, Cold Spring Harbor, New York
Printed in the United States of America

Publisher and Acquisition Editor	John Inglis
Director of Editorial Services	Jan Argentine
Project Manager	Inez Sialiano
Director of Publication Services	Linda Sussman
Production Editor	Kathleen Bubbeo
Production Manager	Denise Weiss
Cover Designer	Michael Albano
Illustrations	Michiyo Namura

Front cover artwork: zentilia/Shutterstock.com.

Library of Congress Cataloging-in-Publication Data

Names: Cohen, Carl M., author. | Cohen, Suzanne L., author.
Title: Lab dynamics : management and leadership skills for scientists / Carl M. Cohen, Ph.D., Suzanne L. Cohen, Ed.D.
Description: Third edition. | Cold Spring Harbor, New York : Cold Spring Harbor Laboratory Press, [2018] | Includes bibliographical references and index.
Identifiers: LCCN 2018028371 (print) | LCCN 2018032299 (ebook) | ISBN 9781621823162 (ePub3) | ISBN 9781621823179 (Kindle-Mobi) | ISBN 9781621823155 (printed hardcover)
Subjects: LCSH: Science--Management. | Technology--Management. | Scientists--Relations. | Organizational behavior.
Classification: LCC Q180.55.M3 (ebook) | LCC Q180.55.M3 C63 2018 (print) | DDC 506.8--dc23
LC record available at https://lccn.loc.gov/2018028371.

10   9   8   7

All World Wide Web addresses are accurate to the best of our knowledge at the time of printing.

Authorization to photocopy items for internal or personal use, or the internal or personal use of specific clients, is granted by Cold Spring Harbor Laboratory Press, provided that the appropriate fee is paid directly to the Copyright Clearance Center (CCC). Write or call CCC at 222 Rosewood Drive, Danvers, MA 01923 (978-750-8400) for information about fees and regulations. Prior to photocopying items for educational classroom use, contact CCC at the above address. Additional information on CCC can be obtained at CCC Online at www.copyright.com.

For a complete catalog of all Cold Spring Harbor Laboratory Press publications, visit our website www.cshlpress.org.

For Naftali, Nomi, and Wilder

Contents

Preface

This is a book for scientists and technical professionals about surviving and succeeding in the organizations and groups in which they work. It is also for science managers and executives who want to know how best to manage scientists. If you work with scientists, this book can provide you with a better understanding of the world in which they live and the challenges that they face.

Practical advice and exercises show scientists and science managers how to interact with others in ways that improve their effectiveness and increase their productivity. The book also shows how to apply improved self-awareness and interpersonal skills to specific problems that science professionals encounter every day. If you are a scientist, the skills that you learn will enable you to better identify, focus on, and achieve your objectives. You will become more productive in your job and more successful at what you do, whether your field is molecular biology or astrophysics.

Unless you have well-developed self-awareness and interpersonal skills, all of the management tools in the universe will not be of much use to you. If you are a scientist, chances are that your self-awareness and interpersonal skills are not as well developed as your technical skills; this limitation can impede your work. We provide concepts, concrete tools, and exercises that will help you to improve these skills. Our approach is designed to aid you in overcoming the barriers to knowing yourself, what to do, and how to do it.

The third edition of this book comes at a time we frankly never thought we would see. Since the second edition was published in 2012, we have seen a steady increase and interest in management and leadership training in the scientific realm. This has been manifested in a steady growth in requests for us to run workshops at educational institutions, private companies, and government agencies both in the United States and internationally. In addition, more and more organizations (NIH, NSF, and private companies) are now requiring or strongly recommending that their scientific managers and leaders, or scientists themselves, receive annual training in management and leadership skills. We would like to think that this book and the work we have done in this field has played some part in this gradual shift. The degree to which attention has been focused on this realm was brought home to us recently when Carl was referred to as "… the de facto Godfather of teaching EI (emotional intelligence) to scientists" (Kidder 2016).

In this third edition, we have edited and revised nearly every chapter of the book and have added two new substantial chapters. Chapter 4 (Bring Them On! Interviewing, Selecting and Hiring Scientists) and Chapter 6 (How Am I Doing? Setting Goals, Giving Feedback, and Doing Performance Reviews). The genesis of Chapter 4 is interesting. Carl has been running workshops on management and leadership for scientists for almost 20 years. At the end of every workshop Carl hands out feedback forms so that participants can evaluate the workshop. On the form is a place where participants can answer the question "Are there other themes you would like to see in future or new workshops?" In the last two years, but not even once before then, we have received multiple responses asking for a workshop on how to interview and hire scientists. Just recently (in January 2018), Carl participated in a panel discussion on management and leadership for scientists at the Massachusetts Institute of Technology. When the moderator asked the audience (MIT scientists from multiple disciplines) what they would like the panel to address, the very first request (which was heartily seconded by several others in the audience) was about how to select and hire scientists for their groups.

Perhaps this focus on hiring reflects something about the current job market for scientists (is it now harder to hire qualified scientists?) or perhaps it reflects changing demographics in our workshops (maybe we're attracting more junior faculty starting or getting ready to build a lab). Whatever the reason, Chapter 4 will provide useful tools for those who need to staff their labs. We think the approach we advocate in Chapter 4 will be useful to you even if, or perhaps especially if, you've been hiring scientists for years. The new Chapter 6 is a logical extension of the new chapter on hiring. It provides the tools science managers need to help those they hire be as productive as possible. The chapter shows how to set goals and provide clear direction, give useful feedback, and help scientists know how they're doing relative to their own and their mentor's expectations. We provide step-by-step guidance and scripts on which to base these important conversations.

In addition to the new chapters, we have updated most of the others. We added a section on creating a lab culture and another on delegation in Chapter 5. Chapter 7 contains a new section on the importance of "psychological safety" in optimizing team performance, and Chapter 12 has a new section entitled "Ten essential characteristics of scientific team leaders."

We especially want to highlight that we added important sections that relate to issues of gender, racial, and minority bias to Chapters 4 and 13. These sections were informed in part by the excellent report from the National Academies of Sciences, Engineering, and Medicine, *Sexual Harassment of Women: Climate, Culture and Consequences in Academic Science, Engineering, and Medicine* (2018). In Chapter 4 on hiring we discuss a variety of implicit biases (including gender and racial bias) that affect how we interact with others and that can influence hiring or promotion decisions. We specifically draw attention to resources available to our readers who wish to explore and mitigate the impact of their own biases. In Chapter 13, a new section, "Recognizing and acknowledging attitudes towards gender, race, ethnicity, and minority status," emphasizes the pivotal role played by those in leadership positions in shaping the culture and behaviors in their own organizations. The message is that leaders must set examples and actively work to create organizational cultures that recognize and eliminate discriminatory and harassing behaviors of all types.

The book draws heavily on examples and experiences from Carl's 30-year career in science, both in academia and the private sector. The new chapters, in particular, use material that Carl developed for the workshops that he has run during the years since the second edition. The book also relies heavily on Suzanne's long career as a psychologist and clinician, working with individuals and groups, and on her insights into people in general and scientists and technical professionals in particular. Our suggestions and guidelines work. They are all based on techniques that we have tried and used ourselves and that we have helped others use.

REFERENCES

Kidder FK. 2016. For managers, emotional intelligence trumps IQ, *Lab Manager*, Sept. 2016. http://www.labmanager.com/leadership-and-staffing/2016/09/for-managers-emotional-intelligence-trumps-iq#.WpWwvqjwZhE.

National Academies of Sciences, Engineering, and Medicine. 2018. *Sexual harassment of women: Climate, culture, and consequences in academic sciences, engineering, and medicine.* National Academies Press, Washington, DC.

Acknowledgments

We would like to thank the many scientists, administrators, and executives who have attended our workshops on management, negotiation, and leadership over the years. Your anecdotes, questions, and comments have served as inspiration for many of the themes developed in the book, especially in the new chapters on hiring and feedback. Your positive and encouraging comments reinforce our belief in the importance of the topics about which we have written and in the utility of the approaches that we advocate. We especially thank the scientists who have attended all of the Cold Spring Harbor Workshops on Leadership in Bioscience which we have run annually since 2011. The case studies you prepared, the exercises in which you engaged, and the individual discussions you participated in with us provided invaluable background and inspiration for this book. Finally, we thank the Cold Spring Harbor Press and its Executive Director and Publisher John Inglis, for seeing value in this book when we were trying to get it published in 2005 as well as the Cold Spring Harbor Laboratory Meetings and Courses Program, its Executive Director, David Stewart, and Manager for Training, Outreach and Evaluation, Charla Lambert, for their steadfast support of our CSHL workshops.

CARL M. COHEN
SUZANNE L. COHEN
Newton, Massachusetts
February 2018

Introduction

The interactions of scientists in the workplace are often fascinating to the lay public because they expose the human element of endeavors that would otherwise be suitable only for textbooks. What is so riveting, and at the same time paradoxical, about these accounts is the immense impact of human interactions on the process and outcome of a scientific project. James Watson's account (Watson 1968) of his and Francis Crick's rivalry with Linus Pauling, their acquisition of data from and interaction with Rosalind Franklin, and their amusing and fruitful interactions with each other have been eagerly read by scientists and laymen alike. Stories about the effects of rivalries, collaborations, and relationships on the daily conduct of science, which is often imagined to be a purely rational and intellectual endeavor, are endlessly fascinating.

Despite the lively debates and discussion generated by Watson's book, little if any analysis has focused on whether the working relationships among the various protagonists could have been more productive. In this book, we have tried to fill that void and, by doing so, show that there are ways to train scientists and run science organizations that improve the conduct of science, facilitate communications, and maximize productivity.

Science training programs are designed to impart technical skills and scientific knowledge to their trainees. The fact that these same programs provide no training in management nor interpersonal skills sends trainees the implicit message that these skills are largely irrelevant in science. In this sense, educational institutions have two characteristics in common with most science- and technology-based organizations: (1) a single-minded focus on technology and (2) a lack of appreciation for the importance of social and interpersonal skills.

Among scientists in all fields of specialization is a strongly held belief that if you just get the science right, everything else will fall into place or become irrelevant. Yet experience shows that this belief is false. Analyses of the chemical industry (Perrow 1993), the space program (Vaughan 1996), and military campaigns (Cohen and Gooch 1991) highlight the central role that human interactions and group dynamics have in their success or failure. It has been proposed that the principal reason military commanders fail to learn from military disasters is the tendency of analysts to focus exclusively on technical and logistical explanations for failure (Cohen and Gooch 1991). This narrow focus betrays naïve indifference to the importance of human interactions and communication and to the individual and organizational characteristics that foster them.

One could make the same indictment of scientific enterprises. For example, the challenges faced by the pharmaceutical industry to replenish their diminishing drug pipelines have been discussed and reviewed extensively (one might say obsessively) during the past decade or more. Pharmaceutical companies struggle to find new operating models or paradigms to improve the discovery and development of new therapeutics. Most often, these approaches focus on the application of new technologies or tools (e.g., high-throughput screening, genomics, genetic knockouts, RNA interference) or new ways to structure the companies themselves (i.e., Glaxo-SmithKline's Centers of Excellence for Drug Discoveries). Like technical and logistical analysis of military failures, these approaches address only one facet of what is likely a multidimensional problem. Other facets might include improving the communications, management, and leadership skills of those engaged in, or managing, the scientific enterprise. It is an encouraging sign that during the last decade, some of the best and most insightful of the big pharmas have started providing some management training for their scientists. But what kind training are they actually providing?

Efforts to train scientists and science managers to function beyond the lab bench are often limited to project management, running meetings, doing performance reviews, and team building. Although scientists are efficient at learning the nuts and bolts of management—Gantt charts, work plans, etc.—mastery of these skills cannot compensate for poor self-awareness and a paucity of empathy. Without these and other personal and interpersonal skills, managerial functions will be implemented in a mechanical fashion, without heed to individual, interpersonal, or group nuances. Managing in this way makes no more sense than driving a car blindfolded: You may know how to manipulate all the mechanical controls and levers, but you are dangerously blind to context and feedback.

Interpersonal skills can be taught. For some, the abilities to relate productively, to notice how others respond to you, and to forge both personal and professional bonds come naturally. For others, and this includes many science and technical professionals, these skills are not an integral part of their personalities. Fortunately, you do not need to change your personality to become interpersonally savvy.

Interpersonal skills comprise a set of behaviors and responses that can be learned. Such learning can and should be part of the professional education of scientists. In the chapters that follow, Carl provides examples from his own career, illustrating how adopting some of these skills has influenced his development as a scientist, mentor, and leader.

Our experience with the workshops that we have run shows that scientists are eager to learn the skills presented in this book. We have also found that the best way to learn them is to try them and use them. Despite the fact that many scientists are initially reluctant to test new skills in role-playing exercises, their experimental nature quickly takes over. Most are ultimately convinced by the data—their own improvement as negotiators during the course of the workshop. For scientists, data rule!

It takes practice and skill to be able to observe yourself, capture what you experience, and view your behavior, body language, and facial expressions as others do. It also takes skill to attend to how others act and react to you. These skills include new ways of listening—in a manner of speaking, listening between the lines. The first section of this book describes how to acquire these skills or to improve on those you already have.

The first three chapters of the book provide a set of core skills and concepts that serve as the foundation for much of what follows. Chapter 1 provides you with the opportunity to examine your behavior in the scientific workplace in light of what is known about scientists in general. Two self-assessment exercises help you to discover which facets of self-awareness and interpersonal skills you need to work on. Chapter 2 offers guidelines and exercises for developing or improving these skills. We teach you to become an active observer of yourself and others and how that informs you of your feelings, helps you to choose appropriate behaviors, and enables you to assess the effectiveness of new behaviors. Chapter 3 shows how to apply your skills of self-observation, self-management, and observation in the context of negotiation. This chapter presents a framework for using these skills to guide you through the difficult situations you encounter every day. We also show you that learning and practicing negotiation is one of the best ways to acquire and improve your interpersonal skills and put them into practice.

The second section of the book teaches you to apply your new skills and powers of observation to three interpersonal domains—with employees, with peers, and with bosses. Chapter 4, written by popular demand, is new to the third edition. It introduces a data-driven approach to identifying, selecting, and hiring scientists and managers of scientists and provides guidance to help recognize and mitigate the impact of bias (including gender and racial) in hiring decisions. The chapter offers step-by-step guidance and editable (and downloadable) forms that you can use to help you evaluate and rank applicants. We think that this chapter will make a significant difference in improving how you go about hiring scientists.

Chapters 5 and 6 answer the question "Now that I've hired my group of scientists, what's the best way to manage them and keep them productive?" Chapter 5 offers methods for improving your effectiveness as a manager or leader of other scientists and alerts you to the most common problems when managing teams of scientists. Chapter 6, another new chapter in this edition, provides guidelines and tools for keeping your team productive and on track. This includes how to set realistic but challenging goals for scientists, how to give feedback that improves performance, and how to let scientists know whether they're performing up to your expectations. Chapter 7 extends the theme of managing scientists into the realm of groups and teams. In this chapter, we specifically focus on the all-important theme of running team or group meetings, the pitfalls and roadblocks that leaders may encounter, and how to recognize and manage them. Chapter 8 shows you how to use your new management skills when dealing with your boss. There, we illustrate the most common problems that scientists have with their bosses and the most effective ways of handling them. Building on the previous chapters, in Chapter 9, we show that improving your ability to recognize and deal with conflict, along with practicing the skills you learned in previous chapters, will improve your ability to interact productively with peers.

The final section of the book addresses special management problems associated with the organizations in which scientists work and study. Chapter 10 describes how trainees and mentors can use self-awareness, observation, and interpersonal skills to survive in and improve the academic training experience. Chapter 11 shows how these same skills, as well as an understanding of the unique challenges that accompany the transition from academia, can improve the productivity of scientists in the private sector. Chapter 12 addresses the concept of leadership. We explore what we mean by leadership in a scientific setting and provide concrete steps you can take and behaviors you can adopt to hone and improve your leadership skills. Finally, Chapter 13 provides a review of the skills that we have presented using an extended case study. This last chapter also suggests ways in which you can use the concepts and tools that we have introduced to improve the practice of science and productivity in your own organization and concludes with recommendations on how leaders can help make their organizations more inclusive.

At the end of most chapters, we provide exercises and experiments designed to help you acquire and practice the skills that we present. In some cases, the exercises are in the form of experiments that allow you to use and evaluate the effectiveness of new behaviors. Use the ones that work for you.

If you believe that there is more to leading in science than intimidation, more to interacting with your peers than jockeying for advantage or withdrawing when others do, and more to working with your boss than being defensive or resentful, you will find this book valuable. If you believe this but haven't a clue as to how to change your behavior, read on. The skills that you learn will enable you to better identify, focus on, and achieve your objectives, making you more productive and effective. Finally, and perhaps most importantly, you will learn to accomplish all of this in a way that promotes openness and trust in your interactions with others.

REFERENCES

Cohen EA, Gooch J. 1991. *Military misfortunes. The anatomy of failure in war.* Vintage/Random House, New York.

Perrow C. 1993. *Normal accidents: Living with high-risk technologies,* 3rd ed. Princeton University Press, Princeton, NJ.

Vaughan D. 1996. *The Challenger launch decision: Risky technology, culture, and deviance at NASA.* University of Chicago Press, Chicago.

Watson JD. 1968. *The double helix: A personal account of the discovery of the structure of DNA.* Atheneum, New York.

CHAPTER 1

People Who Do Science: Who They Are and Who They Can Be

In terms of behavior patterns, affect and even some intellectual matters, we know more about alcoholics, Christians, and criminals than we do about the psychology of the scientist.

MAHONEY (1979)

We are all familiar with the stereotypes of scientists portrayed in movies, television, and paperback thrillers as aloof, arrogant, intense, and distracted. Each of us is, of course, much more complex and nuanced than such simplistic characterizations, but like most stereotypes, these all have a kernel of truth in them. Scientists as a group do have personality characteristics that distinguish them from, say, social workers as a group. Although you may not share every one of these characteristics or the others discussed below, you likely share some

▸ Technical professionals are different
  The Tea Bag Company exercise
  What research shows about the personalities of scientists
▸ Why you should pay attention to your personality
▸ The good news: Scientists are psychologically flexible and quick learners
▸ Summary
▸ References
▸ Exercises and experiments
  1. Self-assessment: Who you are
  2. Self-assessment: Dealing with others
  3. Identifying themes

of them. Much of this book focuses on helping you to discover which of these characteristics you share and anticipating how they may affect your behavior as well as your effectiveness as a scientist.

Many of the exercises that we present at the end of this and subsequent chapters focus on helping you to improve your interpersonal awareness and self-awareness. We focus on these two characteristics in particular because we know from personal experience that these represent areas in which many scientists are weak. We have also supplemented our personal experiences with a review of the psychological literature pertaining to the personality characteristics of scientists. Our objective in presenting this information is to help you notice and identify in yourself some of the traits that have been noted in others. As you read the following sections, take note of those characteristics that sound or feel familiar, or that others may have noted that you display. At the end of the chapter, we provide a brief questionnaire to help you to identify some of these characteristics of which you may not already be aware.

TECHNICAL PROFESSIONALS ARE DIFFERENT

The Tea Bag Company exercise

Applying the stereotypes mentioned above to all scientists seems, and certainly is, both crude and extreme. But before we dismiss such categorization out of hand, perhaps we should ask just how much truth there is to these and other popular notions of scientists as a group. The following story recounts how Carl first became convinced of the presence of many of these characteristics in himself.

A number of years ago, when Carl first became interested in improving his management skills, he took a course at the Harvard University Extension School. The course was entitled, "Understanding Your Management Style," and it was taught by Robert Benfari, who had written a book of the same title (Benfari 1991; now updated as Benfari 2013). The course was intended to give students some insight, from a psychological perspective, into how they approached the various tasks associated with management. Carl enrolled in this particular course for the excellent reason that no other course in management was being offered in a time slot that was convenient for him.

During the first couple of classes, Dr. Benfari spoke at some length about "personality types" and the utility of assigning people into one of 16 categories using the Myers–Briggs Type Indicator (MBTI) inventory, a widely used and much studied psychological instrument (Briggs Myers and Myers 1995). In fact, the students had taken this inventory during the first class but had not seen the results yet. Being a scientist and a skeptic, Carl spent a lot of time arguing that such categorization was artificial, simplistic, and without validity.

The third meeting of the course consisted of an in-class exercise. Dr. Benfari read students' names from a list that he had prepared, and in so doing divided them into five groups. He told the class that within the groups they were to come up with a plan to address and solve an organizational problem that he would pose. He had also assigned one student from each group to observe how the group went about its task, without participating. He referred to the exercise as the "Tea Bag Company" exercise.

Each group was to put itself in the role of senior managers of Tea Bag Company, Inc. As managers, they had just learned from their sales and marketing department that sales were down catastrophically over the last two quarters. The task was to determine a solution to this problem. The group would have 45 minutes to work on the problem and then each group would report to the class on what they had decided to do.

Within Carl's group, several members immediately suggested that they convene near the white board so that they could use it to organize their strategy. They did so, effectively preventing any other group from accessing the board. Within a few minutes, Carl's group was intensely involved. Members were interrupting each other, talking in loud voices, and grabbing the marker from one another to write on the board. They all agreed that they needed to take an analytical approach to the problem. They mapped out a marketing survey to determine whether consumers' tastes had changed. They allocated resources for analytical testing of the tea bags to see if quality had slipped. They crafted a backup plan to move into coffee, if that seemed prudent. They were really very efficient and logical and completed the task easily within the allotted time.

After the 45 minutes, Dr. Benfari reconvened the class, instructing them to report on their plans, one group at a time. Carl's group was asked to report first. One of them outlined the series of logical steps they took and how they focused on objective measures of success and economic outcomes. The observer accurately described their deliberations as being lively and competitive, and noted that they all jockeyed for board time, interrupted one another, and spoke more than they listened. This was not surprising to the group; it was how they behaved all the time.

Dr. Benfari then asked the second group to report. The spokesperson for that group said that the group's primary concern was the welfare of the employees of the Tea Bag Company.

The group believed that because the problem was so acute, a plan should be in place to ensure that if the company were to go under, the employees would be provided for; they would have adequate outplacement services, and health benefits would continue for as long as possible. They also scheduled an emergency stockholders meeting to allay the concerns of the company's investors. They did start to deal with how to address and fix the sales problem, but had not progressed very far when their time ran out.

Carl recalls listening to this presentation and thinking that these people must have landed in corporate America from the Moon. He was baffled by their approach. He was further baffled when the observer assigned to that group reported that the discussion had been quiet and respectful. The observer said that group members waited for one another to finish speaking before speaking themselves, and that one member of the group had gone out and brought back sodas for the whole group during their discussion.

After the other groups reported, it became clear that a very wide spectrum of approaches had been taken. But none was so remarkably different from Carl's group as that of the second group that had reported.

When all of the reports had been delivered, Professor Benfari told the class that he had composed the groups using their personality types as determined from the MBTI inventory that they had taken during the first class. He said that Carl's group was dominated by "NTJ" personalities, which in Myers–Briggs jargon stands for intuitive, thinking, and judging. We do not need to go into the arcana of the Myers–Briggs categories (for a more detailed discussion of MBTI types and how they relate to a chosen profession, see Tieger and Barron-Tieger 1992); suffice it to say that NTJ people have a tendency to be highly intuitive (N), analytical and logical (thinking: T), and can be very judgmental (J). NTJs are good at facts and relationships among facts and tend to be "visionaries." "NTJs are 'commandant' types who insist on ruling by their particular vision" (Benfari 1991).

The second group in the class was composed largely of members determined to be "NFPs." NFP stands for intuitive (N), feeling (F), and perceiving (P). NFP people are highly relational and react to the world in a feeling mode, rather than in a thinking or analytical one. As Benfari says, "When presented with a task such as developing a more marketable product, they consider their real task to be developing their own potential and that of their colleagues" (Benfari 1991). In other words, they are in many ways the exact opposite of the NTJs.

Carl recalls being completely dumbstruck by this revelation. None of the students had known the basis on which they were placed into their respective groups. They all went about working on their task in ways that came naturally to them and behaved precisely as the MBTI would have predicted!

People really are different, and they are different in ways that can be described and measured. As much as we hate generalities and categorization, we know that many of our scientist colleagues are NTJs or STJs. The "S" ("sensing") suggests that some of us have a more data-driven way of coming to conclusions, compared with the Ns, who are more intuitive. And we also know that we work in ways that are different from the way NFPs and many others work. Carl became a believer in the MBTI, not as a diagnostic or classification tool, which is how it typically is used, but as a tool for insight into himself. Do not be fooled into thinking that just because the MBTI can identify people who share behavioral characteristics that it should be relied on to choose a profession or direct others into a profession. As has been amply noted, most recently by Annie Murphy Paul in her book *The Cult of Personality: How Personality Tests Are Leading Us to Miseducate Our Children, Mismanage Our Companies, and Misunderstand Ourselves* (Paul 2004), the predictive value of these tests is overrated and the tests themselves overused.

Moreover, don't be fooled into thinking that by rote application of the Myers–Briggs classification system you will be able to "psych out" your dysfunctional colleagues or employees. Carl once spent an entire day in a workshop designed to teach a method by which people you interact with could be quickly categorized into MBTI types. The objective was laudable—to help people communicate better in the workplace—in this case a mid-sized biotechnology company. The idea makes sense in principle. If someone is a "thinker" rather than a "feeler" (i.e., a T instead of an F in MBTI parlance), you might be better off trying to convince them of something by appealing to data rather than to human consequences. In the Tea Bag exercise discussed above, Carl's group was clearly much more focused on the bottom line, collecting and analyzing data that would help the group make decisions and keep the company running. If you were making a proposition to that group, you might want to show up with graphs, spreadsheets, and financials. The other group, the NFPs, were more focused on the human consequences of their actions and decisions. In trying to sway this group you might do well to appeal to the consequences of their decisions on the company's employees and its customers.

From this perspective, the MBTI scheme might be useful if it attunes you to the possibility that scientists as a group may be superb at focusing on tasks, but may be less attuned to the interpersonal. To go about managing scientists without taking into account who they are as people, and how their personalities might differ from other types of people, is like trying to train a pack of tigers using a training manual meant for parakeets.

However, there are problems with the use of schemes like the MBTI on a day to day basis. First, actually remembering how to categorize people, doing it in real time, and translating those categorizations into action turns out to be a lot of work. And in our experience, those of us who do not do it full time, or for a living, do a lousy job of it. Second and more important, we are all multidimensional. The NFPs will be swayed by data and financials and the NTJs by feelings and emotion. Both axes are important, and both need to be addressed. Finally, human behavior is complex and is powerfully influenced by culture, background, environment and more. So the next time you get the urge to use any of the popular psychological scales or dimensions to predict how someone will respond or behave read *Behave. The Biology of Humans at our Best and Worst* by Robert Sapolsky (2017) and *Five Constraints on Predicting Behavior* by Jerome Kagan (2017). By highlighting the myriad factors that influence human behavior, these two books should be sufficient to warn you away from any scheme that purports to make understanding or managing other people "simple."

The best that could be said of the day-long workshop Carl attended was that it stressed the importance of lifting his mind out of the science and data and routinely paying attention to the people he was working with. In essence, that's what the rest of this book is about.

What research shows about the personalities of scientists

The Tea Bag story might make you ask whether scientists and technical professionals share certain personality characteristics. In the following section we review a few of the studies that have attempted to answer this question and to identify shared characteristics. Despite the fact that the quote at the beginning of this chapter suggests that there is a paucity of such data, the data that exist are revealing.

The opening quote was cited in a comprehensive review of the psychology of science and scientists by Feist and Gorman (1998). This review contains references to more than 150 scholarly publications relating in one way or another to the psychological characteristics of scientific and technical professionals. The following is a list adapted from that article, and compiled from the literature of experimental psychology, that compares the personalities of scientists to those of nonscientists. Compared with nonscientists, scientists are

- more conscientious and orderly
- more dominant, driven, or achievement oriented
- more independent and less sociable
- more emotionally stable or impulse controlled

A bit more intriguing is the summary in the same article of the differences in personality between "eminent" and "less eminent" and "creative" and "less creative" scientists (let us not obsess here over how eminence and creativity were quantified). According to the article, compared with less eminent and less creative scientists, eminent and creative scientists are more

- dominant, arrogant, self-confident, or hostile
- autonomous, independent, or introverted
- driven, ambitious, or achievement oriented
- open and flexible in thought and behavior

Beginning to get the picture? Of course, there is always the issue of cause and effect. Does a career in science promote arrogant, antisocial behavior, or does science attract those who already have a tendency to show these characteristics? Feist and Gorman take the safe middle road and suggest a bidirectional interaction between personality and science.

In another publication, Greene (1976) reported that, "The psychological problems most frequently encountered with...scientists stem from (a) communications difficulties, (b) confusion about the role of the expert, (c) emotional and interpersonal needs, and (d) failure experiences."

In a study of 99 academic researchers (all full professors), Feist (1994) concluded that "[eminent scientists]...were rated by observers as more exploitative, more fastidious, more deceitful, less giving, and less sensitive to the demands of others.... In sum, complex thinkers about research are influential in their discipline and are well cited, but are considered by observers to be neither warm nor sociable."

Finally, in a study of 100 technical project team leaders (in unspecified technical areas), Gemmill and Wilemon (1997) listed the top ways in which scientific and technical project team leaders misread events within project teams. These leaders

- were unaware of interpersonal conflict among members of the team
- were unaware of hidden agendas on the part of team members
- did not understand the motivation and needs of team members
- were unaware of expectations of team members
- did not listen carefully to team discussion
- misread lack of argument as agreement
- interpreted conflict as unhealthy when it was actually constructive
- misread team members' ability to work together as a team

So, at the considerable risk of overgeneralization, the data suggest that as a group, science and technical professionals are poorly attuned to the dynamics of their interactions with others and to the needs and feelings of those around them.

WHY YOU SHOULD PAY ATTENTION TO YOUR PERSONALITY

If the only consequences of exhibiting some or all of the traits mentioned above were that you might be viewed as being aloof and uncaring, you might be excused for having little or no motivation to take note of them in yourself. However, the consequences of such traits and the behaviors they engender can be far more profound, even to the point of being dire. Let us examine some hypothetical consequences in the science workplace of a few of the personality traits identified in the studies cited above. These brief vignettes outlining the consequences of behaviors, which were found to be common among scientists, are based on actual cases.

String together enough outcomes like those listed below and before you know it, the people in your group, company, or organization are confused and alienated, projects are foundering more often than they should, and decisions are being made for other than scientific reasons.

Trait	Consequence
Dominant, driven, or achievement-oriented	You forge ahead on projects with your own ideas, listening politely but usually failing to take into account the suggestions or objections of colleagues or employees. Most of the time this works well because you know more than they do. But the one time that you do not, you continue to follow your own agenda and end up spending millions of dollars on a failed project that you should have abandoned two years ago.
Arrogant or hostile	Of course you are arrogant: You are the smartest, most accomplished scientist in the company. But when it comes time to seek the support of others for a controversial new technology that you want to acquire, you find yourself isolated and without support. The technology is actually just what your company needs, but because you have created enemies with your arrogance, you cannot get anyone else to agree with you. The company suffers and so do you.
Introverted	Paying attention to other people is a distraction. It takes you away from your work. Moreover, it is hard, and you figure that people are complicated and unpredictable. You fail to notice that, over time, people are excluding you from their informal discussions because you would rather be in your office analyzing data. The result is that you do not hear about the new project until the formal announcement is made, by which time all the team leaders have been selected. You wonder why you were left out.
Less sensitive to the needs or demands of others	You figure that just like you, everyone has their own agenda, and it is pretty hard to know what that agenda is. You have always felt that people complain all the time—it is only natural. One day Carol, your senior and most productive postdoc, announces that she is leaving in three weeks for a job in industry where the salary is higher and the advancement prospects are better. You recall that over the past several months, she has been asking you about a salary increase and whether you would support her for a faculty position, but you kept putting her off. Now you are faced with the prospect of a major setback in your most important research project.
Unaware of interpersonal conflict among members of the team	You hired Alice and gave her the same project as Hans because you thought that competition would drive them both to work harder. The result was that Hans hoards reagents and signs up for equipment time that he does not need to prevent Alice from getting the better of him. Others in the lab mention the brewing conflict, but you shrug it off with the comment that the competition will make each of them stronger. Three months later, Alice goes to Human Resources and files a sexual harassment complaint against Hans. The resulting turmoil sets both of them, and the lab, back a year.

(Continued)

Trait	Consequence
Unaware of hidden agendas on the part of team members	You are trying to decide on the appropriate version of a recombinant protein with which to go into production. Eric, the director of protein expression has been arguing vehemently for version 2C, whereas others in the group believe that several other variants are more appropriate. You believe that everyone is arguing the case for each variant on its merits. Months later, you learn that the reason the director was arguing for 2C had nothing to do with its scientific merits: He had prematurely anticipated the use of 2C and had his group produce several grams of it. Had you known this, you might have reassured him that you would have been happy to sacrifice the produced 2C in favor of making the best choice.
Unaware of expectations of team members	As head of a task force on in-licensing a technology for fabricating a new type of solar cell, you ask Sandra, one of your scientists, to review what is known about the physics behind the technique. She spends a week on the project with the expectation that she will present her conclusions to the executive committee that makes the final acquisition decision. When she delivers her findings to you, you indicate that they will simply be attached as an appendix to the final report. She accuses you of misleading her, and you counter that she spent way too much time on the report. She loses trust in you, and you are annoyed by what you see as her unreasonable expectations.
Do not listen carefully to team discussion	During a group meeting, Ed, one of your postdocs, comes up with an idea for a new project. Others in the group are enthusiastic, but you are preoccupied because you are answering e-mails on your phone. A month later, you suggest this same project to a new postdoc. When Ed finds out about this, he assumes that you stole his idea intentionally and complains to the department chair. Several days of everyone's valuable time are wasted sorting out the misunderstanding.
Interpret conflict as unhealthy when it can be used productively	You are managing a project team of engineers and biologists developing a new brain imaging technology. The engineers keep insisting that the biologists have not collected enough data on glucose metabolism in the brain to ensure that the instrument will have adequate dynamic range. The biologists insist that the engineers are being overly compulsive and are demanding data impossible to obtain. The groups have reached an impasse. You believe that the project will self-destruct unless you can defuse the situation. You finally insist that the groups cease arguing and that the engineers move ahead with the design despite the disagreement. When tested in the clinic, the prototype instrument is not sensitive enough to changes in glucose metabolism to be useful.
Misread team members' ability to work together	Your group will move into a new lab in six months and you are finalizing its layout. Lab benches and office space need to be assigned. You are about to leave for a meeting in Italy and tell your group that they can work out the assignments themselves and you will review their recommendations when you return. Several members of the group feel disenfranchised by the process, which is dominated by a few highly aggressive group members. They accept space assignments that do not meet their needs because they refuse to argue with the others. They never mention this to you, but their morale deteriorates and they minimize their interactions with the others in the lab. You never notice.

What is so insidious about the behaviors in the above examples is that in each case, the protagonist was behaving in a "reasonable" manner. No overt hostility was evident and, with one exception, there were no actions that might be cause for allegations of misconduct or mistreatment. Moreover, in many of the examples the actions of the protagonist were the result of considerable thought and deliberation. If all of this is true, what went wrong? Quite simply, the thought and deliberation were all focused on scientific and technical matters and not at all on interpersonal consequences.

Whereas the above examples were heavily weighted toward team leaders as protagonists, bench scientists and technicians are no less prone to encountering such problems. Indeed, even the most mundane interactions during a typical day in the lab can have unintended consequences for scientists who fail to anticipate the effects of their behavior on others. A typical day for a working scientist might include the following interactions.

- Meet with a technician to discuss plans for the day. Because the technician also works for a colleague, it is necessary to negotiate the technician's schedule with the colleague. The colleague is uncooperative.

- Talk to a different colleague and request some of their confocal microscope time, because the microscope schedule is fully booked for the next two weeks. The colleague refuses. Stomp back to the bench and wonder how to finish experiment on schedule.

- Start weighing out reagents for an experiment. Discover that a previous user spilled an unidentified chemical on the electronic balance. Instantly identify the culprit and resolve to confront them.

- Schedule time to meet with the lab director to discuss attending the annual meeting of the American Society for Cell Biology, although the director had already said that travel funds were exhausted. You are angry because two of your peers are going.

- Start an experiment, try to put irritation with colleagues out of mind, and concentrate on work. Become distracted by loud rock music coming from the next lab bench. The longer you hear it, the angrier you become.

These are fairly typical, perhaps even understated, examples of interactions with which scientists deal on a daily basis. Although they take place within the context of professional activities, how you deal with them will depend on your personality. Some avoid conflict, whereas others exacerbate it; some become aggressive and others withdrawn or reticent; some act in ways that establish productive alliances with colleagues, whereas others work in isolation; and some have satisfying relationships with supervisors and others spend their careers feeling manipulated and unappreciated.

Each of these contrasting behaviors has very different interpersonal and professional consequences. The success of your work and the progression of your career are strongly influenced by your behavior and whether you interact with others in a productive or an antagonistic manner. For example, take the examples of typical laboratory interactions that were listed above.

- If you become accusatory and confrontational when trying to negotiate with your technician about time on the confocal microscope, not only do you risk not getting the time you need, you may create animosities that make these negotiations more difficult in the future.

- If you angrily accuse a colleague of spilling chemicals, you may increase the likelihood that they will deny it even if they were responsible. Alternatively, you may be blaming the wrong person.

- If you go to your lab director contending that you are being unfairly treated and become accusatory or act insulted, you may actually reduce your chances of changing their mind.

- If you continue to bottle up your annoyance at the loud music, you may end up losing your temper with someone else over an unrelated and trivial matter.

Of course, everyone has bad days, snaps at a colleague, or says something thoughtlessly that they later regret. But it just may be that scientists do this more often than others.

At one of his workshops for scientists, Carl asked participants to answer several questions about ways in which their interactions with others in the lab affected their work. Here are some of their responses to three of the questions.

- More than three-quarters reported spending between 10% and 25% of their time at work on "people problems."

- More than two-thirds reported having between one and five "uncomfortable interactions" with people at work each week.

- Nearly two-thirds reported that interpersonal conflict had hampered progress on a scientific project between one and five times during their career.

If these figures are even close to being representative of the science community as a whole, we are wasting a lot of manpower, resources, and time because of interpersonal problems. Many of the scientists in Carl's workshops routinely ignore these problems or try to resolve them in ways that create more animosity than was present to begin with. It is not uncommon for us to avoid problems in the workplace because we lack the skills to resolve them diplomatically. But scientists as a group, and science organizations as a whole, may be more prone to such avoidance than others. Why is this so?

During years of biomedical research, Carl spent countless hours regularly reading the scholarly scientific literature in his field. He also read journals dedicated to promulgating the latest time-saving and clever techniques. He even read advertisements for products that promised to accelerate his research with the latest technological innovations. He spent several weeks each year at scientific meetings to learn of the latest ideas, discoveries, and techniques that might help him in his work. At these meetings, he even went so far as to speak with the dreaded salesperson about new instrumentation, reagents, and gadgets for his lab. He did all this religiously for almost 15 years before he thought to seek out any information about managing his laboratory and its research personnel more effectively.

If scientists are willing to invest time and effort in learning a new technical skill, why not do the same for management and interpersonal skills? And why, when they do decide to do so, do they wait until the need is acute? Recently Carl was engaged by a research organization to help resolve what it referred to as a crisis among some of its senior members. Many of them were not on speaking terms with others, and the leader of the organization was at his wits' end about how to solve the problem. When Carl asked how long this "crisis" had been going on, he was stunned by the answer: 10 years!

Not everyone waits 10 years, of course. Many people attend Carl's workshops to proactively gain the skills that we teach, and many more attend in response to a particular problem that they are having at that time. When Carl asks participants for questions at various points in his workshops, he is always struck by how specific they are. "What if you had a person in your lab who always lied about whether they had made the mess in the fume hood?" Or "Let's say you worked for a group vice president who was always trying to micromanage everything you did?" It is clear that these are not hypothetical examples; they represent issues regularly dealt with at work.

THE GOOD NEWS: SCIENTISTS ARE PSYCHOLOGICALLY FLEXIBLE AND QUICK LEARNERS

Fortunately, data from the scholarly studies that we cited above are not all bad news for scientists. The studies also found that scientists were emotionally stable, impulse controlled, and open and flexible in thought and behavior. What this suggests is that despite less-than-optimal interpersonal skills, technical professionals have a high capacity, motivation, and willingness to learn and improve. What they need is data showing the utility of improvement, as well as the opportunity to learn.

Technical professionals may spend as many as 10 years in college and professional school and never experience a single hour of training to help them manage themselves and others. The training in working with and managing others that they do receive comes from observing the behavior of their mentors, many of whom are themselves untrained and often poor managers (see Chapter 10 for more on this topic). If you are a scientist working for a company, chances are that you have been sent to one or more management training seminars over the course of your career. On the other hand, if you work in academia, it is likely that no one has even suggested that you attend a management training seminar.

Whether run by your company or an outside agency, these seminars typically focus on the nuts and bolts of managing: budgeting, time management, goal setting, and project management. These are all important skills and worth learning. However, your success at applying these skills will not be determined by how well you learn them or even how long you use them. Your success will be determined by how well you understand, relate to, and respond to the people you manage and with whom you work.

Standard management training may provide concrete guidance for dealing with overt behaviors, but it does not help you to see beneath the surface to the underlying motivations or needs that drive those behaviors. Thus, you may have learned how to fill out the annual performance review form for your employees, but if you present your feedback in an aloof or devaluing manner, you may do more harm than good. You may have learned how to create and implement a project plan, but if you fail to notice and deal with conflict among members of the project team, your plan may founder. You may have learned how to organize and run group meetings, but if you fail to notice and address the fact that several key participants routinely remain silent during these meetings, you may be running the project at half steam.

If you are oblivious to conflict, insensitive to the needs and aspirations of others, and unaware of the impact of your own behavior on others, you are managing under a handicap. If you are interpreting silence as agreement, repeated absences as laziness, or failure to follow instructions as forgetfulness, you may be missing important underlying dynamics that can hamper or derail an important project.

The good news is that it does not have to be this way. Improving your interactions with others does not require a personality transplant, and learning how to notice and manage your reactions and behaviors in difficult situations does not require years of psychotherapy. The following chapters present steps that you can take to improve your situation, and they outline ways in which scientific organizations can take the lead in promoting and legitimizing the importance of interpersonal expertise as much as they promote technical expertise.

SUMMARY

Research shows that professionals in science and technology are more likely than others to have personality characteristics that lead them to avoid or miss important interpersonal cues. They may also act in ways that show a lack of appreciation of the effects of their own behavior on others. These traits can have unanticipated negative consequences on their careers and scientific progress. Becoming aware of your own personality characteristics is the first step towards becoming a more effective scientist and science manager.

REFERENCES

Benfari R. 1991. *Understanding your management style: Beyond the Myers–Briggs type indicators.* Lexington Books, New York.

Benfari R. 2013. *Understanding and changing your management style. Assessments and tools for self-development.* Josey Bass, San Francisco.

Briggs Myers I, Myers PB. 1995. *Gifts differing: Understanding personality type.* Davies Black, Mountain View, CA.

Feist GJ. 1994. Personality and working style predictors of integrative complexity: A study of scientists' thinking about research and teaching. *J Pers Soc Psychol* **3:** 474–484.

Feist GJ, Gorman ME. 1998. The psychology of science: Review and integration of a nascent discipline. *J Gen Psychol* **2:** 3–47.

Gemmill G, Wilemon D. 1997. The hidden side of leadership in technical team management. In *The human side of managing technological innovation: A collection of readings,* 1st ed. (ed. Katz R). Oxford University Press, New York.

Greene RJ. 1976. Psychotherapy with hard science professionals. *J Contemp Psychother* **8:** 52–56.

Kagan J. 2017 *Five constraints on predicting behavior.* MIT Press, Cambridge, MA.

Mahoney MJ. 1979. Psychology of the scientist: An evaluative review. *Soc Stud Sci* **9:** 349–375.

Paul AM. 2004. *The cult of personality: How personality tests are leading us to miseducate our children, mismanage our companies, and misunderstand ourselves.* Simon & Schuster, New York.

Tieger PD, Barron-Tieger B. 1992. *Do what you are: Discover the perfect career for you through the secrets of personality type.* Little, Brown, Boston.

Sapolsky, Robert M. 2017 *Behave. The biology of humans at our best and worst.* Penguin Press, New York.

EXERCISES AND EXPERIMENTS

1 Self-assessment: Who you are

This, and the following exercise, are designed to help you become more aware of how you interact with others and how your behavior influences others' responses to you. It is a self- assessment in that you are answering the questions about yourself. The goal is not to label yourself as having one character type or another. Rather it is to provide a practical framework within which you can notice or observe your own behavior in the workplace. By the very process of thinking about and answering the questions, you will have taken an important first step in recognizing how your feelings, reactions, and responses can impact your effectiveness in the scientific workplace. The traits or characteristics listed on page 14 were taken from several of the studies described in the section beginning on page 5 of this chapter.

The traits in the self-assessment exercise on the following page ("Identify your traits") are separated into groups that have the following characteristics:

Group 1. May limit your effectiveness in managing or collaborating with others. May result in your being seen as untrusting and uncollaborative.

Group 2. Inherently neutral, but if manifested in an extreme manner, can result in consequences similar to those of group 1.

Group 3. Characteristic of task orientation and achievement; generally thought of as positive attributes. If traits are overly dominant, you may be seen as "cold-blooded" or self-centered.

Group 4. Will contribute to the maintenance of collaborative and productive group interactions.

Read through your responses to the exercises. Does the overall picture describe you? Do you see trends that surprise or dismay you or limit your effectiveness as a leader or team member? If so, make note of them and pay special attention to sections of the following chapters that address those issues.

If you found that you had more than one or two entries checked "do not know" in the "I think I am" group, this is a hint that your self-awareness may need some work. If you do not know how to characterize your attitudes, feelings, or reactions, you are not paying close-enough attention to your own behavior. The next chapter provides exercises to help improve your self-awareness. Take this inventory again in a few months, after you have worked on self-awareness.

If you checked a lot of "do not know" in the "others think I am" category, you may need to improve your skills at paying attention to how others react to you. The section beginning on page 238 of Chapter 9 may help you in this area. Retake this inventory after you have read that section and note whether your answers have changed.

Exercise: Identify your traits						
Group/traits	I think I am			Others think I am		
	Yes	No	Do not know	Yes	No	Do not know
Group 1						
Overly dominant						
Arrogant						
Hostile						
Introverted						
Uncommunicative						
Ungiving						
Insensitive						
Group 2						
Autonomous						
Driven						
Fastidious						
Group 3						
Conscientious						
Orderly						
Emotionally stable						
Impulse-controlled						
Able to listen carefully to discussion						
Self-assured						
Achievement oriented						
Group 4						
Open and flexible						
Aware of conflict						
Aware of hidden agendas						
Able to understand others' motivations and needs						
Able to listen carefully to discussion						
Able to channel conflict to achieve desired results						
Good at identifying people who are compatible with one another						

2 Self-assessment: Dealing with others

The goal of this exercise is to help you to identify situations involving others that you find difficult. Like above, the objective is not to provide a diagnosis or assessment of your problems, but rather to help you notice and remember what you find difficult or uncomfortable when dealing with others. You may find it useful to return to these same questions periodically, every six months or so, to assess whether your views change as you develop the skills presented in the following chapters.

1. Describe, in as much detail as you want, one specific difficult interpersonal interaction that took place in the context of your work as a science professional. Describe the impact this interaction had on your work.

2. List three other difficult situations and specify the nature of your relationship with the person(s) involved (example: unpleasant conversation with another postdoc about author-ship on a paper). In this case, write no more than one sentence for each.

3. How many times per week do you engage in what feels like an uncomfortable interpersonal interaction with someone in your scientific workplace?
   ___never___1–2___3–5___6–10___more than 10___too many to count

4. Rank the following categories of people in order of frequency of difficult or conflictual interactions over the past year (4 = most frequent; 1 = least frequent).
   ___colleague or peer   ___direct superior
   ___employee   ___administrator or clerical worker

5. Approximately how many times in your career has progress of a scientific project been negatively affected by an interpersonal conflict that was not handled well?
   ___never ___once or twice___3–5 times ___more than 5 times ___do not know

6. About what percent of your time do you spend thinking about or dealing with interpersonal or "human" issues in your professional day?
   ___none ___less than 10% ___10%–25% ___26%–50% ___51%–75% ___more than 75%

7. Check the boxes on page 16 that best describe the degree to which you agree or disagree with the statements (scale: 5 = strongly agree; 1 = strongly disagree).

   Like the previous inventory, take note of the frequency with which you answered "do not know." If you answered this way to questions about observations of yourself, pay attention to the exercises for self-awareness at the end of Chapter 2. If you answered this way to questions about how others see you, pay attention to the exercises for observing others at the end of Chapter 9.

3 Identifying themes

How easy was it for you to answer the above questions? List any questions that seemed particularly difficult, because you had a hard time understanding or answering them. Questions that you found difficult or troublesome may refer to interpersonal themes or personal characteristics that you find uncomfortable to think or talk about. Making note of these themes may enable you to anticipate work situations in which you are uncomfortable and in which you may not perform optimally. As we show in the following chapter, such anticipation is often the key to intercepting ineffective behaviors.

Are you able to identify themes in your answers that give you better awareness of our own behavior and reactions? By answering these questions, have you identified some behavioral themes that are common to the various situations? Some examples of themes that you might identify include

- "I have a lot of difficulty dealing with my peers."
- "There is a disconnect between the way I see myself and the way others see me."
- "I was unaware of how much my behavior affects others."

By answering these questions and finding common themes, you are becoming more aware of your interpersonal style, which is the first step toward managing yourself and others in a more informed manner.

	Strongly agree 5	4	3	2	Strongly disagree 1	Do not know
I am very collaborative						
Others think that I am very collaborative						
I think that I am confrontational and argumentative						
Others think that I am confrontational and argumentative						
I am sensitive to others' needs and feelings						
Others think that I am sensitive to their feelings						
I am receptive to suggestions from others						
Others see me as being open to suggestions						
I avoid interacting with colleagues that I do not like						
I manage to interact with colleagues as needed, regardless of whether I like them						
I tend to withdraw in tense or conflictual situations						
I tend to become aggressive in conflictual situations						

CHAPTER 2

The Mote in Your Own Eye: Manage Yourself First

Not only are scientists less likely to be attuned to the dynamics of interpersonal relationships, but they are also less likely to notice and manage their own emotions and reactions. Recognizing these characteristics in yourself is the first step to improving them.

In this chapter, we show how self-awareness can improve your ability to manage yourself and others and improve your effectiveness at what you do. We review case studies that illustrate problems scientists typically encounter and show how improved self-awareness can help you to deal with these problems more effectively.

▶ The importance of being self-aware
▶ Developing and using self-awareness
 Knowing what you are feeling: To solve a problem, you must be able to describe it
 Anticipating how feelings will affect your behavior
 Deciding: Determining the appropriate response
▶ Three common dichotomous reactions to difficult situations
▶ Helping others to improve their self-awareness
▶ Using self-awareness to help guide behavior
▶ References
▶ Exercises and experiments
 1. Notice what you are thinking
 2. Notice what you are feeling
 3. Notice your sensations
 4. Putting it all together
 5. Know your hot buttons
 6. Identifying behavioral polarities

THE IMPORTANCE OF BEING SELF-AWARE

Some of the underlying themes of this book, especially as they relate to self-management and self-awareness, were influenced by the "emotional intelligence" school of thought. Daniel Goleman's book *Emotional Intelligence* (Goleman 1995), as well as the more recent book *Primal Leadership* (Goleman et al. 2002), argue convincingly for the application of the skills associated with emotional intelligence in the workplace and also suggest that curricula need to be developed that are aimed at teaching these skills at every level of education.

One of the key elements of emotional intelligence is self-awareness, the capacity to notice what we feel and think. The ability to do this in "real time"—that is, to be able to notice and name feelings and reactions as you experience them—is the hallmark of self-awareness. Another element of self-awareness is the capacity to anticipate your own behavior—that is, imagining what you will or could do in the moment, before you actually do it. For example, experiencing anger and anticipating that your response might be to berate or insult the person against whom it is directed places you in the position of being able to decide how to behave, instead of simply behaving in accordance with how you feel. Thus, self-awareness allows you to exercise behavioral options and choose the behavior that will be the most effective, rather than the one that comes naturally or that may make you feel good for the moment, but that you will later regret.

Let us start with an example from Carl's career of how this process works.

■ *Case Study: Left Out*

I worked with a team on a drug discovery project that involved personnel from all parts of the company and required multiple team meetings every week. During an especially tense phase of the project, the meeting organizer told me that the weekly meeting with senior managers would not be held that week. However, when the day and time of the meeting arrived, I saw most of the regular participants filing into the conference room, presumably for the very meeting I was told had been canceled. I could not ascertain who knew about the meeting and who did not.

At this point, my sensitivity to rejection took over. My first reaction was to assume that the meeting was being held without me and that I was intentionally being excluded. I had been especially outspoken at the last meeting and worried that my viewpoints were so objectionable that the meeting organizer had decided that it would be best to leave me out of the next meeting. My reaction to this feeling was to resolve never to go to another meeting unless I was given a personal invitation. My second reaction was to mentally compose stinging e-mails of rebuke and invective to the organizer of the meeting.

I spent some time experiencing these feelings of withdrawal and hostility without acting on them. I knew from past experience that I have a particular sensitivity to being excluded, and that I had on other occasions misinterpreted another's actions as being exclusionary when in fact they had nothing whatsoever to do with me. Reminding myself of this helped me to put my current feelings in perspective and made me cautious about saying or doing anything about the situation until I had more information. Even though I could not imagine why they were having the meeting without me, I nonetheless resolved to keep an open mind, refrain from jumping to conclusions and, more importantly, avoid reacting overtly to my feelings.

The next day, I discovered that the meeting really had been canceled but some of the participants had decided to get together at the original meeting time to discuss another project in which they were involved. Because I had no role in that project, there was no need to invite me to the new meeting. This explanation made perfect sense and I was thankful that I had not reacted in accord with how I had initially felt.

To be honest, it took me a long time to get to the point of being able to take a step back and examine my feelings before acting on them (I suspect that some of my colleagues will say that I still have a way to go). I also had to learn that although acting in a hostile or withdrawn manner was the most overt reaction to my feelings, just getting those reactions under control was not always sufficient. Reactions to feelings can take more subtle forms as well. Even behaviors such as being less participatory than usual in a meeting can be a reaction to feeling excluded, and making a seemingly harmless but snide comment to a colleague can be a reaction to feelings of hostility. The behavior may be less extreme, but the consequences can be every bit as negative. The more subtle the reaction, the more practice you will need to recognize it for what it is.

The key is to develop an awareness of your feelings and sensitivities. In my case, a tendency to take things personally and a willingness to make assumptions about others' motivations were also contributing factors. Self-awareness is the first step to managing yourself in tough or uncomfortable situations. Chapter 3 shows us that this ability to monitor your internal thermostat is also a key element to becoming an effective negotiator.

Before we get into how you can develop your self-awareness and use it to your advantage, another illustrative case study follows.

Every organization seems to employ someone who is an interpersonal and managerial disaster, but whose knowledge or skills makes them seemingly indispensable. The situation is often complex: On the one hand, this person may possess unique knowledge and experience, and thus must be kept involved in projects. On the other hand, this person may behave in such a way as to prevent others who have important contributions to make from speaking or participating during team meetings.

How can an organization retain the input and participation of a domineering manager or senior scientist with low self-awareness while limiting the damage he does to others and to the group? In the following case study, a senior scientist behaves in ways that are damaging to his colleagues and company. Because the scientist did not possess good self-awareness, his company had to find a creative way of retaining his expertise while removing him from a position in which he could hinder the work of others.

▪ Case Study: The Clueless Systems Analyst

Dr. Raju is a founding member at Triotech, a leading designer of cutting-edge enterprise software. He knows the Triotech business inside and out and believes that he knows the best ways to adapt Triotech's products to a company's needs. He has been promoted within the organization over the years so that he now manages a group of midlevel employees. The problem is that morale in his group is low. Raju is domineering in team meetings; he belittles others, and provides support only when colleagues agree with him. Alex, a senior systems analyst, feels so put down by Raju that he rarely speaks in team meetings any more. The last time Alex tried to disagree with Raju, Raju told him that he was being "stupid and he needed to go back to school."

The difficulty is that Raju does possess a lot of knowledge and experience, but people will seek his advice only in the most extreme circumstances because they would rather not deal with him. Junior and midlevel staff are dissatisfied with this situation, but management is at a loss about what to do.

Triotech recently hired Elizabeth, a new vice president who was told that she needed to deal with the "Raju problem." Elizabeth spent some time getting to know Raju to understand why he was behaving in this way. Conversations with members of the executive team led her to the hypothesis that Raju had long felt bitter about not being promoted to the position of vice president. He sorely wished to be a member of the executive management team and felt excluded from the inner workings of the company. Management viewed Raju as a "loose cannon" with no executive or strategic capability. Elizabeth suspected that Raju felt underappreciated and his actions in the group environment were a reaction to this. She also thought that Raju might be humiliating others to bolster his own sense of importance, in essence, saying to senior management, "look how knowledgeable I am."

Elizabeth had had some experience with people such as Raju and thought that she might be able to help him gain some insight into his behavior and its consequences. However, after several conversations with Raju, Elizabeth concluded that he had little insight into his own motivations and behavior. When opportunities arose for Elizabeth to discuss Raju's behavior and motives with him, he reacted with resistance, deflection, and hostility. When she suggested that Raju examine how he behaved in a meeting from which they had just emerged, Raju exploded in anger and said that his behavior was not the problem; the problem was the stupidity of the people he had to work with every day.

After a number of such attempts, Elizabeth concluded that in the short term, neither Raju's self-awareness nor his behavior was likely to be improved by mentoring or counseling, either by her or someone else. Therefore, Raju could not remain in his current role.

The solution that Elizabeth crafted was to suggest that Raju be promoted to a slightly higher position, something akin to a senior adviser to the executive team. In this new role, he would be invited to many of the senior staff meetings from which he had previously been excluded. However, he would also be expected to continue to participate in project team meetings as a kind of elder statesman, although he would no longer have line management responsibility. Everyone hoped that Raju would continue to provide all the benefits of his knowledge to the company, feel appreciated and acknowledged by senior management, and cease browbeating team members.

This is what in fact happened, but it took the better part of a year for the transition to stabilize, and not without friction. In the end, members of the junior staff were promoted to line management positions and felt empowered and recognized. Eventually, Raju felt sufficiently secure to treat his former subordinates with respect, and the company continued to benefit by having access to his skills.

Two of the keys to this success were Elizabeth's determination that Raju's knowledge was critical to the continued success of Triotech and her recognition that Raju lacked the insight and willingness to examine his own behavior and its consequences. She spent time with Raju to determine how important his knowledge was to the future of the organization. In this case, the company was about to embark on a fund-raising campaign. Although she felt capable of managing the process without Raju, she suspected that this would be a mistake. Few others in the company possessed Raju's ability to address in-depth questions about the company's methodologies.

This story shows how a manager who is attentive not just to the technical facets of her job, but also to individual and interpersonal issues, can arrive at creative solutions to difficult problems. In this case, Elizabeth helped the company retain the knowledge of a valuable scientist and created opportunities for junior staff to contribute and excel in ways that they could not have had previously.

The solution was designed to enable the company to achieve its objectives and satisfy some of Raju's interests at the same time. It involved elements of "out of the box" thinking in that it created a new kind of position for Raju, rather than just marginalizing him. Not all companies will have a vice president like Elizabeth, who found a creative solution to a tough problem. In other circumstances Raju might have been fired, demoted, or marginalized. In those cases, Raju would have paid dearly for his inability to examine his own behavior and its effects on others, and the company would have lost as well. Most frequently, behavior such as Raju's is ignored because the scientific leadership of the company either is reluctant to confront such behavior (see Chapter 5 for more on conflict avoidance in science professionals) or fails to see its destructive nature.

Many of us have known individuals such as Raju, who manage others in destructive ways and are unable to examine their own behavior and motives. If we manage such people and allow their behavior to continue, we own responsibility for the consequences just as much as do the Rajus of the world. Some people who act destructively will be amenable to coaching. Others will need to be moved into positions with little or no line management responsibilities. Others will need to be separated from the organization.

Often, when we see a difficult scientist getting away with the kinds of behaviors that Raju was exhibiting, you can bet that part of the problem is that management does not know how to handle him. In a sense, management is authorizing that scientist's destructive behavior. This typically happens when management feels that it has few alternatives or does not even recognize the presence of the problem. In these cases, the root of the problem may rest as much with management as it does with the scientist. Managers may feel held hostage by someone who possesses domain expertise in areas in which they are ignorant. Management's approach to such problems must begin with an acceptance of its own complicity in the scientist's behavior. If appropriate, coaching and mentoring can be offered. If the scientist is not amenable to these approaches, other tactics, such as the one devised by Elizabeth, can be used.

From our perspective, Raju's difficulty in examining or changing his behavior was due to his poorly developed self-awareness. Greater self-awareness allows us to monitor our feelings and behavior to ensure that they are appropriate to the circumstance. This increases the likelihood of having productive and appropriate interactions with employees, colleagues, and supervisors.

For managers such as Elizabeth, the ability to notice or assess a scientist's self-awareness helps them to assign appropriate roles within a team or department. Managers with the ability to help scientists develop their self-awareness are in a position to assist promising scientists develop into leaders (Chapter 13 deals with this theme in more detail).

Scientists such as Raju are not uncommon. From our experience and the research cited in the previous chapter, scientists as a group have low self-awareness. Their inclination is to be analytical and rational, and they may believe that their behavior represents a logical response to external events. Despite this belief, scientists are no less prone than others to experiencing strong or troublesome feelings. If they are oblivious to the impact of these feelings on their thoughts and behavior, they will act in ways that cause them and their organizations considerable trouble.

DEVELOPING AND USING SELF-AWARENESS

Developing and using self-awareness, in the sense that we have defined it at the beginning of this chapter, involves the following three elements:

1. **knowing** what you are feeling

2. **anticipating** how this feeling may affect your behavior

3. **deciding** whether the behavior is appropriate to the situation and will help you

Knowing what you are feeling: To solve a problem, you must be able to describe it

At the beginning of the workshops Carl runs for scientists, he asks participants to pair off and take part in a role-playing negotiation exercise. He deliberately chooses topics for negotiation that most scientists dread—for example, negotiating first authorship of a scientific publication. The groans and eye-rolling from the audience when he announces this topic attest to their discomfort. Carl describes the hypothetical scenario for the group, making it as difficult as possible for them to come up with an easy solution. He then gives them 10 minutes to negotiate an agreement.

After the 10 minutes, he asks how the negotiation went. Because the group has not learned the principles of negotiation yet, the outcome is predictable: Many do not reach an agreement.

Carl also asks what they felt during the negotiation because he knows that much of the difficulty people have in tense situations stems from primal behavioral responses to anxiety and feeling threatened. Of course, he could see what they felt as he walked around the room while negotiations were taking place: discomfort and anxiety. Responses to his inquiry of how they felt usually consist of the following:

- I wanted to win.
- I felt that we were stuck.
- I could not get my point across.
- I could not think of any good compromises.
- I felt that my partner was inflexible.

After Carl patiently writes these phrases on a flip chart, he tells the participants that these are not descriptions of feelings at all; they are descriptions of thoughts. He typically tries very hard to get them to tell him what they felt, not what they thought, and he often does not have much success.

After one such workshop, Carl relayed to Suzanne his frustration at not being able to get scientists to describe what they were feeling. Suzanne immediately diagnosed the problem. She said that the scientists could not describe what they felt because they were not accustomed to using the language of feelings, which is somewhat like using a foreign language—if you do not know the vocabulary, you cannot describe what you are experiencing. Moreover, she pointed out something even more profound: If you do not have the vocabulary to describe what you are experiencing, you may not even be aware that you are experiencing it.

What were the words that the workshop participants could not use? They were words such as anxious, scared, threatened, angry, and hurt. These words describe powerful feelings and emotions that have profound effects on behavior. If Carl's workshop participants were not aware of their feelings and could not name them or put them into words, what were the chances that they could recognize the impact of their feelings on their behaviors?

The ability to monitor one's emotions in real time is a key element of self-awareness. Self-awareness is a prerequisite for anticipating your reactions in response to strong feelings, some of which may be inappropriate to the situation. This skill is essential for functioning effectively in tense and emotion-laden situations such as negotiations (discussed in the following chapter) or other difficult interactions.

Now when Carl conducts workshops for scientists, he lists "feeling" and "thinking" words to help them see the distinctions and get used to using such words.

Thinking words	Feeling words
Want to win	Threatened
Determined	Frustrated
Strategizing	Angry
Not getting point across	Hurt
Stuck	Sad
Not being listened to	Happy
Anything that begins with "I felt that..."	Anxious

Thinking phrases often start with "I felt that... ." Although this sounds like a description of a feeling, "that" is the tip-off that it is not. See for yourself how different it feels to say, "I felt that I was in danger" compared with, "I was scared." The first puts you at a distance from the feeling, whereas the second expresses the feeling.

If the ability to notice what you are feeling in real time is the first step to increasing your self-awareness, naming the feeling is the first step to noticing it. The exercises at the end of this chapter will help you to learn how to accomplish this in everyday situations.

Anticipating how feelings will affect your behavior

The second step to becoming more self-aware is anticipating how you will react after you experience a strong feeling. For example, as illustrated in the first case study in this chapter ("Left Out"), Carl knows from past experience that his reaction to feeling rejected is to withdraw and pout. He also reacts by developing imaginary scenarios to get even with the person whom he believes rejected him. Over time, as he learned that this is how he reacts to feeling rejected, he learned to anticipate his reactions as soon as he experienced the feeling of rejection.

Carl took control of his behavior because he was able to notice and identify his feelings and know that his initial reactions to these feelings might be inappropriate to the circumstances. Note that even if he had been intentionally excluded from the meeting, his initial reactions would still have been inappropriate. There are many effective ways in which he could have dealt with being intentionally or inadvertently excluded from a meeting to which he should have been invited. If his goal was to solve the problem of why he was not invited, and to make sure that it does not happen again, he could have met with the meeting organizer to find out why his presence was not needed. Chances are that there would have been a good explanation. Becoming hostile or withdrawn would not have helped him to achieve his goal. It is almost always more effective to ignore insults or slights, real or perceived, and remain focused on your mission and objectives (see Chapter 3).

Anticipating how you will respond to a feeling requires that you study and learn your own pattern of responses. In the case study about being excluded from a meeting, Carl anticipated that his feelings of rejection would lead him to withdraw and become vindictive. Others react differently to feelings of insecurity or fear of being marginalized, and the following case study illustrates how someone else might have responded.

▪ Case Study: Zombie Anxiety

Ted was part of a three-person task force empowered to reorganize the research division of Zombie Biotech in conjunction with a team of external consultants. Ted was anxious that the reorganization would result in his loss of power and influence. This anxiety caused him to control group meetings and meet surreptitiously with the consultants without other team members. Eventually these behaviors came to the attention of the senior vice president of the division, who called Ted to task for his behavior. He told Ted that his behavior was disruptive to the project and promoted mistrust among the other team members. He threatened to remove him from the task force unless he became more of a "team player."

The paradoxical turn of events in this case was that Ted's behavior in response to his fear of losing influence almost resulted in his removal from the task force. Ted was reacting to his feelings of insecurity by trying to overcontrol the situation. Possibly, he had good reason to feel insecure; he may have learned that the senior vice president was unhappy with his performance. It is also possible that Ted's fears were groundless and he may have had a history of feeling insecure in dynamic situations. Whatever the case, Ted's actions were clearly counterproductive.

If Ted wants to change his behavior before it affects his career, he first needs to recognize and name his feelings, which in this case were insecurity and feeling threatened. The second step is to anticipate what these feelings might lead him to do. This is not as hard as it may seem; typically, this is the first action that pops into one's head when experiencing those feelings. Ted can use his past experience to recall how similar situations and feelings have caused him to react.

The self-awareness exercises at the end of the chapter can help you to identify your feelings in specific situations and examine what you are likely to do in response, before you do it.

Deciding: Determining the appropriate response

Once Ted recognizes and names his feelings and anticipates what he might do in response to them, he can decide whether his anticipated behavior is appropriate to the situation. This is perhaps the most important step in the process. Where most of us get into trouble is when we act according to how we feel, often without considering the consequences. Ask yourself the following questions to assess the appropriateness of your reaction to a situation.

- Have I acted this way in the past? Did my behavior have a negative consequence?
- Will my behavior have a negative effect on others (i.e., will it harm, insult, or embarrass)?
- Is what I feel like doing inconsistent with my professional responsibilities?
- Am I about to act out of fear or anger? Is my behavior vindictive?

If you answered "yes," even once, think twice about your behavior. One of the guidelines Carl offers in his negotiation workshops is that if you are imagining how vindicated or self-satisfied you are going to feel after you say or do something, don't do it. Of course, it is necessary to be sufficiently self-aware to recognize the fact that you are anticipating acting this way.

Behavioral rule #1
If you're anticipating how good you're going to feel after you say something critical or sarcastic, DO NOT SAY IT.

It takes some experience and practice to evaluate the appropriateness of your behavior. In the example above, even if Ted had taken the time to examine his behavior, he may not have recognized that it was going to have negative consequences. Had he been working on self-awareness for some time, he might have recalled clues from previous situations in which he reacted similarly and which had caused negative consequences. Or he might simply use this situation as an opportunity to learn how not to react in the future.

THREE COMMON DICHOTOMOUS REACTIONS
TO DIFFICULT SITUATIONS

Just as having the language to name feelings helps you to become aware of them, so too does the ability to name common behavioral patterns help you to become aware of them. The following can assist you in recognizing and naming some common reactions to difficult situations. The reactions are listed in pairs that are behaviorally opposite from one another.

1. **Retreat/attack.** When threatened, do you tend to retreat or attack? Many people behave passively when threatened: They become silent or withdrawn or attempt to leave or avoid the situation altogether. Other passive responses might include ignoring or not speaking to someone who hurt or insulted you. Others become hostile, angry, and threatening. Observe yourself in uncomfortable circumstances and note whether your behavior fits either of these categories.

2. **Verbal/nonverbal.** When you find yourself in an uncomfortable situation do you react verbally or nonverbally? Some people launch verbal defenses or attacks whereas others express their discomfort or hostility nonverbally. Avoiding eye contact, making facial

expressions that show your feelings, or moving or holding your body in either aggressive or submissive postures are all nonverbal modes of behavior. Observe whether you manifest these behaviors and use that information to get a better reading of your feelings as well as how you come across to others.

3. **Internalize/externalize.** Do you have a tendency to blame yourself (internalization) or others (externalization) for problems? When some people are criticized, or when they encounter a problem, they automatically blame themselves. They are ready to take responsibility and feel guilty for everything, regardless of whether it is actually their fault. In contrast, others immediately shift blame and responsibility to someone else, even when most or all of the responsibility is their own. Observe your own behavior to learn if you have a tendency to behave in either of these extreme ways. This can help you to anticipate what might be an inappropriate reaction the next time you find yourself in such a situation.

If you find that you often behave according to one or more predictable patterns (e.g., "I always retreat and blame myself"), it is likely that your reactions have less to do with the specific situations you're reacting to than with your personality. The final exercise at the end of this chapter makes use of these behavioral polar opposites to help you identify and remember your own behavior patterns.

HELPING OTHERS TO IMPROVE THEIR SELF-AWARENESS

Thankfully, not every scientist is like Raju. Many are receptive to feedback and with coaching can use that feedback to change their behavior and improve their self-awareness. In the following case study, a manager uses feedback to help a capable scientist become a more effective team member and eventually a team leader.

■ *Case Study: The Angry Scientist*

Several years ago I was advising Colin, a senior scientist, who frequently lost his temper and lashed out at colleagues during meetings. He just could not seem to control himself. Frequently, during discussions of scientific results at team meetings I could see him begin to squirm in his seat. He would become more and more agitated until he could take it no longer, and then he would release a scathing critique of whatever was being presented. His behavior typically resulted in hostile responses from others, and the meetings would often degenerate into shouting matches.

This behavior was, of course, hurting Colin's career. Discussions about appointing leaders for new project teams often included Colin, but he would never be chosen. He had made too many enemies and no one was confident that he could lead a team in a constructive manner.

After an especially confrontational outburst, I privately asked Colin why during the meeting he had shouted that a colleague's work was sloppy and useless. His response was, "Shoddy work makes me furious! I can't stand it and I can't lie about it! I have to be honest!" Colin knew that his outbursts were hurting him professionally, but he refused to compromise what he saw as his scientific and personal integrity by keeping his opinions bottled up. He believed that he owed it to himself and the company to be honest.

I explained to Colin that as a senior scientist he was expected to use his knowledge to solve problems, and that shouting at colleagues was not the way to do this. I asked Colin whether he could express his scientific concerns in a helpful manner, as opposed to a hostile one. He was astonished by my question. He said that by expressing his hostility he was behaving honestly and could not see anything wrong with that. He felt that not expressing what he felt, or keeping his fury bottled up, would be duplicitous.

After he said this, I saw a look of confusion settle on his face. I suspected that he was making the connection between his insistence on being honest about his feelings and the disastrous outcomes for him personally and for the group as well.

I asked Colin why he thought that it was his responsibility to express his every feeling. I suggested to him that the issue was not honesty but rather one of deciding whether to act on every feeling that he happened to have. I reminded him that it was one thing to feel angry and quite another to act angry; the decision was his. It had nothing to do with honesty or duplicity and everything to do with choosing a behavior that was most likely to get the job done.

Somehow this line of reasoning had never occurred to Colin, yet I could see that it made a lot of sense to him. I also told him to think about what he hoped to accomplish by criticizing the presentation in his aggressive way. I reminded him again that his role in the company was that of senior scientist, not senior arguer. His responsibility was to ensure that good science was done. To accomplish this, he needed to provide his expertise to his colleagues in a way that they would hear and accept. Behaving in a hostile, argumentative manner would not accomplish this.

Colin thought it over and agreed that choosing his behavior, rather than simply acting on his feelings, allowed more control over himself and the situation. I suggested that the next time he felt himself getting furious with someone's presentation he just sit there, remain quiet, and pay attention to what he was feeling. Once he experienced his anger consciously, he could then have an internal dialogue with himself about the consequences of expressing that anger. I also suggested that at his next meeting, he try to make just one comment that reflected his scientific concerns in an emotionally neutral manner.

Once Colin took control of his feelings, his behavior improved, so much so that he was eventually promoted to group leader. Over time, Colin became far more aware of both his own feelings and their effect on others, which in turn helped him to control both. Colin was one of the most rigorous scientists I have ever known, yet he was also very open to feedback about his behavior and to discussing it and his feelings.

Had Colin not been so open, the manager's suggestions might have fallen on deaf ears or been met with hostility and denial, such as Elizabeth's discussions with Raju. There is no foolproof way to predict how people will react when presented with an opportunity to examine themselves. But if you ease into the discussion slowly, you are likely to sense where the barriers and defenses in different people's personalities lie. Honing your powers of observation by paying attention to what others say and their body language (see Chapter 3) is a good place to start. Chapter 13 presents a list of questions to ask yourself about others that may also be useful in gauging someone's capacity for self-examination.

Finally, do not try to help others become more self-aware unless you are willing to examine your own behavior and motivations as well. You will appear to be hypocritical if you cannot be as open about your own behavior, and as receptive to observations about yourself, as you wish others to be of theirs. Moreover, the other person's behavior has usually not arisen in a vacuum. Behavior is always influenced by or is in response to the behavior of others, and in some cases, of yourself. Peter Senge, in his book *The Fifth Discipline* (Senge 1990), notes that, "Trying to fix another person's defensive routine is almost guaranteed to backfire.... By focusing attention on the other person, the confronter has taken no responsibility for the situation.... . If we perceive a defensive routine operating, it is a good bet that we are part of it."

Before you try to help others with what you perceive to be their problems, be sure that you have given careful thought to your own role in creating, facilitating, or exacerbating those same problems. The manager's complicity in the preceding case study might have involved giving Colin subtle messages that they also were annoyed by shoddy data and that they secretly relished his acting as their proxy in browbeating other scientists. The manager might also have discovered that they had a tendency to assign projects that were clearly beyond some scientists' capabilities. In that case, Colin's annoyance and frustration might have been directed against the unwitting scientists because he could not express his anger to the manager directly. In either case, the manager's ability to reflect on and recognize their own complicity in Colin's behavior would have been a prerequisite to their conversation with him.

USING SELF-AWARENESS TO HELP GUIDE BEHAVIOR

Like the Raju case study, the one about Colin had a positive outcome, but for very different reasons. Colin had enough self-awareness to recognize his feelings and enough empathy to understand the effects of his behavior on others. Building on this understanding, he was able to change his behavior dramatically, and he was rapidly advanced in the company. Raju, on the other hand, had no such insight. Although he had to be removed from operational responsibilities, the company found a role for him in which it could access his expertise and his behavior would do minimal damage. The point is not that one outcome is inherently better than the other, but that both were achieved by understanding the impact of self-awareness on the scientist's ability to change and adapt. Without this perspective, both Raju and Colin might well have been fired or marginalized, to their own and the company's detriment.

By using self-awareness, you can gain greater control over your actions and better align what you do with the needs of the situation. The facility to become aware of your feelings and their behavioral consequences requires practice, as does the ability to decide whether the anticipated behavior is appropriate or dysfunctional. The following exercises are designed to help you acquire these skills as well as practice and improve them.

REFERENCES

Goleman D. 1995. *Emotional intelligence*. Bantam, New York.

Goleman D, Boyatzis R, McKee A. 2002. *Primal leadership: Realizing the power of emotional intelligence*. Harvard Business School Press, Boston.

Senge PM. 1990. *The fifth discipline: The art and practice of the learning organization*, p. 255. Doubleday/Currency, New York.

EXERCISES AND EXPERIMENTS

These exercises are designed to help you to develop self-awareness in three areas: thoughts, feelings, and sensations. You will become aware by paying attention and noticing, and then registering what you notice. There are no right or wrong answers or behaviors. The objective is to be descriptive, not judgmental, about what you observe. The exercises are best done in a quiet place where you will not be distracted, but see the instructions below for additional location options. Pay full attention to very specific parts of yourself.

1 Notice what you are thinking

Thoughts can take the form of words, images, or memories. They can also be about what has already happened or what will happen, and important or trivial matters.

Start by recording your thoughts after you notice them. You may find it interesting to do this under several different sets of circumstances. First, try it alone in a quiet place. You can also try it in the presence of others, perhaps during a meeting in which you can periodically take a moment to listen to and take note of your thoughts. Finally, try it during a tense or threatening situation—perhaps a difficult discussion in a meeting or a tense phone conversation. In these circumstances, you may be able to tune into yourself for brief moments without interrupting your overt participation.

Needless to say, it is probably not a great idea to take these momentary "time-outs" during a critical negotiation about your salary or a promotion. With practice, you will eventually get to the point at which you can do that, but for starters try it during situations where being momentarily disengaged will not be detrimental to you or the meeting.

What is the content of your thoughts? Are they about things, people, or events? Are they about planning for the future? Going back to past events?

2 Notice what you are feeling

Naming feelings gives us more control over them. This exercise consists of writing down your feelings as you notice them. You may wish to record your feelings in the following format: Right now I am feeling x. Yesterday I felt x when y happened. Today I feel x when y happened. For example, "Today I felt relieved when my experiment worked."

Again, practice this exercise in a variety of situations, especially ones in which you anticipate feeling uncomfortable. Ease into the process by taking momentary time-outs. At first, you may find the process distracting, but as you become more adept at identifying your feelings, you will need increasingly less time to access them. To help you find the right words to best describe your feelings, we have assembled a word list below. Of course, the list is not all-inclusive, and you may wish to add others.

Anger

aggravated	enraged	judgmental
aggressive	envious	mean
agitated	exasperated	miffed
angry	frustrated	miserable
annoyed	furious	negative
argumentative	grouchy	oppositional
burned up	grumpy	pissed off
confrontational	hostile	resentful
contemptuous	impatient	spiteful
disapproving	irritable	stubborn
disgusted	irritated	suspicious
dislike	jealous	teed off

Fear

agitated	horrified	restless
anxious	in doubt	scared
apprehensive	irritable	shy
blocked	jittery	spaced out
cautious	jumpy	tense
dissociated	mistrustful	threatened
dreading	nervous	undecided
edgy	pressured	upset
fearful	pushed	vulnerable
frightened	reluctant	worried

Happiness (and other positive feelings)

affectionate	eager	passionate
amazed	elated	peaceful
appreciated	empathic	pleasant
aroused	enjoying	pleased
assertive	enthusiastic	powerful
bold	euphoric	proud
brave	excited	relieved
calm	exhilarated	respected
capable	glad	satisfied
cared about	happy	sensual
cared for	hope	sentimental
close	hopeful	soothed
compassionate	in awe	sure
competent	interested	sympathetic
confident	joyful	tender
contented	kind	thrilled
delighted	masterful	triumphant
determined	optimistic	warm

Sadness (and other negative feelings)		
alienated	defeated	insulted
alone	defensive	isolated
apathetic	depressed	jealous
ashamed	disappointed	lazy
bad	discouraged	let down
bashful	disturbed	lonely
betrayed	envious	lost
blocked	foolish	moody
bored	humiliated	restless
burdened	hurt	sad
childish	indecisive	weak
confused	insecure	

3 Notice your sensations

Sometimes "physical" information such as bodily sensations can provide clues to emotional states that are otherwise unavailable to us. This exercise consists of recording sensations as you notice them.

Sensations are in the present, not the past or future. Sensations include warm and cool, vibrations and tingling, discomfort and comfort, lack of sensation and numbness, tension and relaxation, calm and excitation, pressure and lightness, contraction and expansion, hard and soft, increasing and decreasing energy, wet and dry, stiff and pliable, and full and empty.

Sensory experience occurs through the skin, the five senses, and the internal organs. For example, we might notice that our stomach is churning, our heart is beating quickly, our palms are sweaty, or our facial muscles feel soft and relaxed.

This exercise consists of writing down your sensations as you notice them. Like the exercises for thoughts and feelings, try this one in a variety of circumstances during your work day. Pay attention to sensations that occur in your body and write them down when you have the opportunity.

You may wish to record your sensations in the following format: Right now I am sensing x. Yesterday I sensed x when y happened. For example, "Today I had stomach cramps during my meeting with Seth."

4 Putting it all together

The thoughts, feelings, and sensations you experienced in the above situations represent your self-awareness data. With practice, it will become easier to gather this data as you learn to notice yourself and your experiences. Think of self-awareness data as the raw material required for making better decisions.

Reflect on a situation that you have experienced recently, yesterday or today. Try to reconstruct your thoughts, feelings, and sensations from that experience. Take the time to review your state of mind.

Review the thoughts, feelings, and sensations that you recorded during a situation in which you were taking time out to write them down. Do these observations give you a useful perspective on the way in which you behaved (or wanted to behave) or what you said (or wanted to say)? Do these thoughts and feelings help you to make sense of how you acted and what you said in similar situations in the past?

If you have an interaction that causes you to get "off-balance," recall your sensations, feelings, and thoughts by bringing your attention to them. This requires only seconds as you check in with yourself. You will gather important information that will help you to regain control.

5 Know your hot buttons

"Hot buttons" are sensitivities that cause you to experience strong feelings in response to something that someone says or does. A key element of self-awareness is knowing in advance what causes you to feel upset, angry, hurt, etc. The importance of this knowledge cannot be overestimated, especially if you are prone to respond in a manner that is not in your best interest.

To recognize your hot buttons:

- Recall past incidents in which someone has done or said something that elicited an extreme reaction or a strong feeling. Write down what specifically was said or done to trigger that reaction. In your own words, or from the list above, record the feelings you experienced that were associated with your reaction.

- Categorize the type of stimulus that triggered the reaction. Were you accused of lying or incompetence or were you left out of some activity, ignored, or challenged?

- Identify the trigger situations and list as many as possible that have elicited emotional reactions in the past.

- Remember the results of this exercise the next time that one of these situations arises. When you feel a hot button response coming on, do not act, but instead find some way of extricating yourself from the situation, at least momentarily. For example, excuse yourself to go to the rest room, answer an imaginary cell-phone call, or just remain quiet for a moment to give yourself time to notice what you are feeling.

- From this new vantage point, ask yourself if reacting to your feelings is the best course of action, if your reaction is appropriate to the circumstance, if you will be making a tense situation worse, and if it is best to cool off before you do or say anything. The more extreme your feeling, the more you must avoid acting on it.

Over time, you will learn to recognize hot button responses as they are happening or even to anticipate them, giving you greater control over your reaction.

6 Identifying behavioral polarities

The following table presents questions that are designed to help you think about and remember how you are likely to behave in certain uncomfortable or stressful situations. The responses are framed in the form of behavioral opposites. Use your responses in conjunction with the self-awareness exercises on page 29 to help you identify and remember how you are likely to behave in similar circumstances.

	Attack (counter-accusations, trade insults)	Retreat (refuse to take the offensive or to defend yourself)	Verbal (e.g., argue, discuss, inquire)	Nonverbal (e.g., withdraw or express feeling with facial expression or hostile emotions)	Internalize (blame one's self)	Externalize (blame others)
In a threatening situation, my first response is						
In a disagreement with a peer, my first response is						
In a disagreement with my boss, my first response is						
When I feel insulted, my first response is						
When I feel hurt, my first response is						

Gordian Knots: Solve the Toughest Problems through Negotiation

Learning negotiation, persuasion, and diplomatic skills is important for a scientist.... Obtaining these skills is a critical part of a scientist's training and is generally acquired by watching the behavior of others. Diplomacy is essential for preserving relationships that may be important for a fellow's career development, and a key step is learning how to cooperate with the very people whose help will be needed to achieve goals.

A GUIDE TO TRAINING AND MENTORING IN THE INTRAMURAL
RESEARCH PROGRAM AT NATIONAL INSTITUTES OF HEALTH*

The person who is best suited to us is not the person who shares our every taste (he or she doesn't exist), but the person who can negotiate differences in taste intelligently—the person who is good at disagreement. Rather than some notional idea of perfect complementarity, it is the capacity to tolerate differences with generosity that is the true marker of the "not overly wrong" person.

ALAIN DE BOTTON (2016)

When Carl runs workshops for scientists, his favorite exercise is to have them pair off and take part in a mock negotiation involving first authorship for a publication. He tells them to come up with a solution agreeable to both sides, and they look at him as if he is a lunatic. After all, they say, there can only be one winner. Yet 30 minutes later, after he has shown them how to "expand the pie" during a negotiation, they have become true believers.

WHY LEARN NEGOTIATION?

The quote at the beginning of this chapter by the philosopher Alain de Botton is for us illustrative of a deep truth about human interactions, whether they be in a marriage or

▸ Why learn negotiation?
▸ Why learn principled negotiation in particular?
▸ The elements of principled negotiation
▸ Not everything needs to be negotiated
▸ Negotiate in good faith
▸ Do not enter into a negotiation unless you are prepared
▸ Preparing for a negotiation
  Always negotiate with "interests," not "positions," in mind
  Maintain a collegial interaction by using "I" statements
  Learn to expand the pie
  Ask open-ended questions to get a stalled negotiation moving again
  Manage yourself: Know your hot buttons
  Body language during a negotiation
  Do not take anything personally during a negotiation
  Listen to and acknowledge the other side: Be empathic
  Defuse anger and hostility, including your own
▸ References
▸ Exercises and experiments
  1. *The right tools for each negotiation*
  2. *Identifying underlying interests*
  3. *Identifying negative listening patterns in yourself*
  4. *Correcting negative listening patterns*

in the lab. Although we may pine for the perfect mate or wish for just the right collaborator, postdoc, or Ph.D. adviser, these desires are often in response to our inability to establish a satisfying relationship with the less than perfect people we deal with on a daily basis. de Botton's point is that maybe it is less about finding the perfect partner than it is about learning how to deal with—negotiate with to be exact—the one we have now.

Scientists who are skillful negotiators and who notice and pay attention to their experiences as well as the behavior and reactions of others are better equipped to function and succeed in the complex social world of scientific research.

A surprising number of work-related interactions that scientists have daily are negotiations in one form or another. In fact, we use the term negotiation to refer to almost any interaction in which there is a difference of opinion, the interests of the participants differ, or the two parties have conflicting agendas. Discussing the design of an experiment, interpreting experimental data, planning for the next experiment, getting access to reagents and equipment, working out the specific contributions of those who are involved in the project, and, of course, every scientist's nightmare, deciding who will be an author of the resulting scientific paper or report, and in what order, are all negotiations.

If you approach these and other negotiations as power struggles or opportunities for manipulation, you may get what you want in the short term, but in the process may create such ill will and resentment that future negotiations will be far more difficult. Alternatively, if you have such anxiety about negotiating that you either avoid it or accede to others' demands at the first sign of conflict, you may find yourself taken advantage of, time after time. In both cases, the consequences of how or whether you negotiate will limit your effectiveness and productivity. Learning to be effective at negotiation can be as important for your success in science as learning to be skilled at organic chemistry, string theory, or making knockout constructs.

As you learn to be a good negotiator, you will become adept at many of the core skills that we present in this book including:

- noticing what you are thinking and feeling
- anticipating the behavioral consequences of what you are feeling
- recognizing your hot buttons
- managing anger
- identifying underlying interests
- attacking the problem, not the person

These skills are important elements of successful management in science and technology, and it is no coincidence that these same skills are the hallmark of good negotiation. Effective negotiators need to have sufficient self-awareness to choose appropriate behaviors in tense situations. They must be able to avoid responding to inflammatory comments in kind and to remain focused on the issues of the negotiation. They must learn to promote alliances, even when the other person is resistant. Good negotiators also need to accurately hear the concerns and interests of the other person to understand their viewpoint. Finally, good negotiators must observe the effects of their own behavior on others to ensure that what they are saying is heard and understood as intended. Learning to be a good negotiator is a great way to integrate and use many of the most important skills in this book. Over time, as you practice your negotiation skills and internalize them, you will find yourself becoming much more "tuned in" to yourself and others during all types of interactions, and you will manage yourself and others more effectively.

What negotiation teaches you
• Becoming a good negotiator forces you to "read" needs, interests, and beliefs of others.
• Good negotiators learn to monitor and modulate their own behavior in tense and emotion-laden situations.
• Negotiation teaches that listening can be more productive than talking.
• Effective negotiators (like effective managers) identify and focus on underlying interests rather than on rigid positions.

WHY LEARN PRINCIPLED NEGOTIATION IN PARTICULAR?

Of the many approaches to negotiation, the one we advocate most is often referred to as "principled negotiation." One of the chief architects and supporters of this form of negotiation is William Ury, who with Roger Fisher, wrote the book *Getting to Yes* (Fisher and Ury 1991) and, subsequently, *Getting Past No* (Ury 1993).

Principled negotiation focuses on trying to satisfy as many as possible of the underlying interests and needs of yourself and the person or people with whom you are negotiating. The focus on *both sides* is important. If you're only out to get what you want with no consideration for the other person, you're not engaging in principled negotiation; you're engaging in coercion or manipulation.

We should note that there are many other approaches to negotiation and it is not uncommon to find books on negotiation in airport book racks, with such titles as "Learn to get everything you want through negotiation" or "How to be a killer negotiator and give up nothing," etc. Our advice is to avoid these approaches. As a scientist, most or all of your negotiations will be with professional colleagues, employees, employers, and others with whom you interact on an ongoing basis and have long-term relationships. Whether you know it or not, these relationships are important to you and your future. The last thing you want is to use a bag of clever negotiation tricks to convince these people to come to an agreement that they may regret or that leaves them with the feeling of having been manipulated.

In fact, most of us are not very adept at being clever during a negotiation. It takes too much effort, and as a result, we almost always come across as being devious and insincere. Most people can read these signs instantly and react defensively or with suspicion and caution. This is a nonproductive way to interact with professional colleagues and should be avoided. What you learn here will be less about tactics and strategy—important aspects of some types of negotiations—and more about reaching an agreement while maintaining or strengthening the relationship with the other person.

THE ELEMENTS OF PRINCIPLED NEGOTIATION

In the following, we present those elements of principled negotiation that are most relevant to the themes of this book and to scientific or technical settings. Although the case studies and examples are ours, nearly all of the principles and their formulation are adapted from *Getting Past No* (Ury 1993). We strongly recommend that you read this book, which also contains many negotiating elements not covered here, such as how to craft an agreement that will last, how to help the other person find a solution, and how to demonstrate the consequences of not reaching an agreement in a nonthreatening way.

Principled negotiation
• Participants work to satisfy the underlying interests (not the positions) of each.
• Focusing on interests leads to creative solutions by "expanding the pie."
• A mutual expectation and desire for an ongoing relationship among the negotiators is assumed.
Underlying assumptions
• Conflict is a normal part of human interactions and need not be avoided if handled appropriately.
• Conflict can be resolved in a collegial manner.
• Negotiators are joint problem solvers, not adversaries.
• The interests of all sides deserve respect and consideration.
• Negotiators should avoid rigid or fixed positions.

NOT EVERYTHING NEEDS TO BE NEGOTIATED

Before we discuss the fundamentals of principled negotiation, let us address one question that scientists ask during every workshop Carl runs. He is invariably presented with some situation calling for a critical decision on the part of the presenter, who asks, "How do I negotiate that?" The situation may have to do with a scientist not performing their duties or an important decision that needs to be made quickly. Carl's response is that not every situation calls for negotiation.

You as a leader must use your judgment to determine what is decided by negotiation and what can or should be decided by you alone. In some cases you have the undisputed authority to make a decision, and the decision needs to be made quickly. In other instances, involvement of the other party might be inappropriate. An example might be excluding a troublesome employee from a discussion involving the type of disciplinary action to be taken against them. Similarly, a performance review is not a negotiation.

Another circumstance in which you must question whether negotiation is appropriate is when you are being subjected to disrespectful or abusive behavior. If you cannot put a stop to abuse by being direct ("I find your comments insulting and we will need to continue this discussion at another time unless we can have a more respectful conversation."), you always have the option of walking away.

Be cautious about starting a negotiation with someone who has no intention of being fair or considerate of your interests. In addition, be careful if the other party is in a position of strength at the outset of the negotiation and you are having a hard time coming up with convincing arguments to support your views. Situations such as these present a good case for postponing the negotiation until you are better prepared or until the other party shows a willingness to address your interests as well as their own. If you choose your negotiations carefully, as well as the time and circumstances under which you engage in them, you will have a greater likelihood of satisfying your interests.

NEGOTIATE IN GOOD FAITH

Sometimes you have the authority to make a unilateral decision, but you think you ought to engage in a negotiation as a way of ensuring the buy-in and involvement of the other party in the outcome. If you enter into such a negotiation, do so in good faith and with an open mind. If you plan to make the decision yourself, it's better to let people know what it is and why you're making it rather than to engage in a token negotiation with the goal of making the other person believe that they had a say in what was decided.

There may be cases in which you would sincerely like the decision to be a negotiated one but where the outcome is so important that you feel the need to make it unilaterally. Scientists, more than others, may be more prone to such an attitude, especially if they believe that, because their views are based on solid science, there is nothing to discuss.

If your intention is to make a unilateral decision, make this clear from the outset. Many people confuse trying to convince someone of their point of view with negotiating. For example, you may believe that you know how to correctly conduct an experiment, but your postdoc has a different view. You invite the postdoc into your office to "discuss" the matter, but your intention is actually to convince him or her to do it your way. Your plan is to appear open, but in the back of your mind the outcome has already been decided. In most cases the other person will almost always see what you are doing and will feel manipulated. The consequence will be a lack of trust during future negotiations.

If your mind is made up and you have no interest in hearing more on the matter, say so in a polite but clear manner, but do not dissemble. You might say, "Janice, I've decided that the experiment needs to be done on rats, not rabbits. I know that you have a different view and I respect that. But in this case, I'm going to make this decision. We can see how it goes and if I'm wrong we can decide how to proceed from there."

DO NOT ENTER INTO A NEGOTIATION UNLESS
YOU ARE PREPARED

Many times you will find yourself knee-deep in a negotiation that you had no intention of having. We often get surprised by a discussion that quickly turns into a contentious struggle. In this situation, first become aware of what is happening. If you find yourself responding instinctually to a threatening discussion, try the techniques described in Chapter 2 to gain some psychological distance on the situation. Also, use the tools in Chapter 8 to calm the other person down if they are angry or confrontational. Once you have done this, you are in a position to make a decision about whether it is in your best interest to continue with the impromptu negotiation or to postpone it until you have had time to gather your thoughts. You will most always have a better outcome if you can do the latter. If you need or wish to continue with the negotiation, the more familiar you are with the following guidelines, the more confident you will be that the other person will not take advantage.

PREPARING FOR A NEGOTIATION

Before going into a negotiation become thoroughly familiar with all of the facts and issues, even if negotiating with a good friend or colleague. Good friends can make bad agreements, if only because one of them forgets to bring up some important issue that should have been included in the discussion. What seemed like a fair and equitable agreement at the time ends up feeling like a poor agreement later. Then you are faced with either a bad agreement that you have to live with and may resent, or the prospect of reopening the negotiation later.

In *Getting Past No* (Ury 1993), Ury outlines the basic elements of preparing for a negotiation. These include the following five components.

i. Identify your and the other person's interests underlying the negotiation

Interests are the underlying needs and wants that you hope will be satisfied by the negotiation. As we will discuss in more depth later in this chapter, interests differ from positions. Most people go into a negotiation with a position—for example, "I need to use these departmental funds to expand my tissue culture facility." In this example, the interests may include ensuring that a particular set of planned experiments can be completed, increasing the throughput of your lab, or replacing an older facility that is difficult to maintain. Before you enter into a negotiation, determine the interests that you will try to satisfy. As much as possible, do the same for the other side. This will help you to be more creative when coming up with ways to satisfy those interests during the negotiation.

ii. Identify multiple ways in which interests can be satisfied

Once you have identified the interests of both parties, you are in a good position to consider the options for satisfying them. The more options you have, the greater your flexibility during the negotiation. Having a list of options in advance gives you a valuable advantage for proposing effective solutions.

iii. Identify standards that can be used to evaluate proposals

Try to find guidelines, precedents, or paradigms to help determine the fairness of a solution or whether it is consistent with similar solutions used previously. In some cases, standards can be very useful and even decisive in a negotiation. For example, standards about who gets included in authorship of a scientific paper can help resolve such thorny issues. Institutional salary guidelines or salaries from other institutions can be valuable information to have during salary negotiations. Do not overuse standards, though, and do not rely on them as your sole justification for a viewpoint. Unless the standards are universally accepted and unequivocal, you may open yourself to comments such as "Just because they did it that way does not mean that we should."

iv. Think through your alternatives to negotiation

No matter how well prepared and how flexible you are, it may not be possible for you to reach an agreement in a negotiation. In such a case, you have the option of walking away without an agreement. If you find yourself in such a situation, have a good understanding of the consequences. In scientific settings, which are often in the context of large research organizations, the most common fallback scenario will be that some higher authority will intervene and make the decision. If both parties perceive this person as neutral, this may be an acceptable solution. If you and a peer have a disagreement about whether to use rats or rabbits as an experimental model in a joint study, you might defer to the Director of Preclinical Studies. In some cases, however, you may have no acceptable alternative. If you and your supervisor cannot agree whether rodents or rabbits are the better experimental model, you may have no choice but to go along with their decision. On the other hand, if you are negotiating with your supervisor over salary or a promotion, your alternative to an agreement might be to seek a position elsewhere. Of course, you will want to think very carefully before letting the other person know that this is what you will do if you don't reach an agreement. If you already have an offer in hand, you may be able to use that as leverage. If you don't, you may wish to keep your plans to yourself.

v. Define your aspirations and limits

Define the kinds of agreements that might arise from the negotiation. Can you define an ideal outcome in which all your interests are met? Such an outcome may be unrealistic, but it is helpful to know what it would be. Next, make a similar list of outcomes that would be good but not ideal—for example, an outcome that satisfies most but not all of your important interests. Finally, how could your interests be satisfied in ways that might be just acceptable? This is your bottom line. Agreements that fail to achieve this level of satisfaction will be unacceptable and might trigger you to walk away from the negotiation and exercise whatever alternatives are available.

If possible, try to perform the same exercise for the other person as well, by putting yourself in their position. Having thought through what they might view as acceptable or unacceptable will help you to craft an agreement that is mutually satisfactory. If you feel that you do not have sufficient insight into the other person, you may not be paying sufficient attention to what they say and do during your interactions. See Chapter 9 for suggestions on improving your powers of observation.

Most simply, you can just ask the other person what they're trying to achieve or what elements of an agreement are most important to them. This approach has two benefits. First, you get information that helps you craft a fair agreement. Second, it lets the other person know that you're genuinely interested in what they want and need and not just in what you want and need.

This idea of being just as interested in the other person's outcome as you are about yours is one of the defining elements of principled negotiation. If you are sincere in your effort to understand the other person's interests, it sends a powerful message to them and can change what might have been a contentious discussion into one of openness and cooperation.

Always negotiate with "interests," not "positions," in mind

An interest is an underlying need or desire that you are trying to satisfy in a negotiation. Positions, on the other hand, are specific ways in which interests can be satisfied. Understanding this distinction is the single most important lesson in this chapter. If you learn nothing else, learn this.

Most people negotiate to get something that they have decided they need or want: a salary increase, a mass spectrometer, someone's approval of a project plan, etc. These people typically enter into a discussion with a specific objective or outcome in mind, such as "I deserve a $10,000 raise," "I need you to approve this project plan," or "We need to buy this $250,000 mass spectrometer for the proteomics group." In each case, the objective is expressed as a position. They have decided what they need, and they seek to convince the other person to go along with them. This approach feels especially natural to science and technical professionals because they typically start by analyzing a problem and then come up with a solution. All they need is to convince the other person to buy into their solution.

The problem with this approach is that staking out a position limits the other person's spectrum of responses. If the other person has a problem with your position, you have actually made it easy for them to say, "We don't have the money for a raise," "I cannot approve the plan as it stands," or "Use someone else's mass spectrometer." Suddenly you are arguing for your position and the other person is resisting.

Never go into a negotiation with positions. Instead, go in with interests in mind. To discover your interests, you need the capacity for self-reflection that we discussed in the previous chapter. Going into a negotiation with demands or rigid positions means that you have not examined your motives and the thoughts and feelings that influence them. Only by stepping back and uncovering the underlying interests that you believe your negotiating position will satisfy can you escape the rigidity of positional thinking.

Recall the lessons of Chapter 2 in which you learned to examine and identify your feelings. Knowing your feelings can be a great way to identify your interests. In many cases, you may think that your negotiating position is the result of a rational analysis of all possible alternatives, when in fact it is actually the result of choosing an outcome that "feels right" without actually understanding why.

Let us examine in detail three examples of positions and their underlying interests and ways to satisfy those interests that go beyond the initial position.

Problem 1

Positional demand by you. "I need to purchase an additional mass spectrometer for the proteomics lab to keep pace with our work requests. I know that it's not in the budget, but we're swamped."

Analysis. Why do you think that you need to buy a new mass spectrometer? Perhaps people are lined up at your door with samples to be analyzed and you continually put them off, explaining that your group is backlogged. You are under a lot of pressure. You have analyzed the workflow in the facility and can find no way to increase throughput. You decide that the only way to solve your problem is to buy another mass spectrometer, which costs $250,000.

This position feels right because it is a quick fix, and it shows your team that you can go to bat for them and come up with a solution.

What if you had analyzed your underlying interests?

Underlying interest 1. Meet the demands of proteomics facility users for protein analysis. The following alternatives will satisfy interest 1.

- Send some samples to an outside lab.
- Increase hours and/or staff in proteomics facility.
- Use alternative techniques to get needed data.
- Buy another mass spectrometer.

Underlying interest 2. Your staff is campaigning for more equipment. They feel that you have not supported them in the past and you want to show support for them now. The following alternatives will satisfy interest 2.

- The vice president tells your staff that you are campaigning for their interests and you are both working to come up with a solution.
- You find an affordable alternative to show support for your staff, such as providing free pizza and extra days off for anyone who works late helping to meet the backlog.
- Buy another mass spectrometer.

New way to start the discussion. "We are experiencing significant backlogs in sample processing times, and both our users and staff are frustrated. I have a few ideas about how to deal with the problem and wonder if we could discuss them? I'd also be interested in any ideas you may have." This approach starts with your underlying interests rather than with a rigid position and makes it clear that you are open to other ideas as well. Now the other person becomes your ally in helping you solve the problem, rather than an obstacle to your getting your way.

Problem 2

Positional demand on you. Your vice president for research and development (R&D) says, "We need to have the toxicology studies for this cardiology compound completed in half the normal time. Drop everything else and get this done."

Analysis. The company has suffered a serious setback in one of its major programs. Management is anxious and they want to put their best foot forward at the next board meeting. The vice president of R&D has just spoken with the chief executive officer, who told him to push the most promising programs so that he will have something concrete to show at the meeting. The first thing that occurs to the vice president is the new cardiology compound.

Underlying interest. The vice president is under pressure to come up with data on a product candidate by the next board meeting.

Your possible responses addressing underlying interests. Assuming that it will be impossible to complete the studies in half the time (the positional demand) you can

- offer strong interim, but not final, results
- suggest that the vice president present nearly complete data for another drug candidate instead, in the metabolic disease arena
- suggest outsourcing some of the studies to complete them on time

By responding in these ways, you are addressing the vice president's underlying interests rather than the demand to complete a specific study in half the required time.

Alternative approach for the vice president. A more effective approach for the vice president would be to approach this as a negotiation. She might say, "Tim, you know that we're under pressure because of the failure of the CD34 compound last month. I need some positive news for the board meeting next quarter and one idea is to accelerate the cardiology project. Can we do this? Do you have any other suggestions?" In this alternative approach, the vice president has used the powerful tactic of letting Tim know up front what their underlying interests are. This puts Tim in the position of being able to make suggestions that the vice president might not have thought of. The two of them could then negotiate on the most promising approaches to take. People who are honest about their underlying interests inspire trust and make it easier for others to offer help.

Problem 3

Your positional demand to division chief. "I'd like to talk to you about my salary. I think I did a great job making those knockout mice for the project, but I think that I'm being underpaid and would like to request a $10,000 per year raise."

Analysis. You have worked hard during the past 18 months on an important project to create several knockout mice. The work was successful, but you feel that your boss has used your data without any real appreciation for how hard you worked. You recently heard that a person with a position identical to yours in another division of the company has a salary that is considerably higher than yours. You go to your boss demanding a salary increase, but is this really what you want? This position feels right because you feel unappreciated and getting this raise will be a validation of your importance.

Underlying interest. What really is going on here is that you feel unappreciated by your boss and would like some recognition for all the hard work you've put in. A salary increase would be great but you understand that the lab has many fiscal constraints right now.

New way to start the discussion. You: "I was really pleased with the way those knockouts turned out and I appreciate your acknowledging me at the conference last month. Also, I'm excited about the new CRISPR project we started on. I have high hopes for it. I wonder if this would be a good time to talk about my salary. I think I'm at the low end of the scale for my position and I think a $10,000 increase would bring me in line with where I'd like to be and would be a way to recognize my contributions to our work here."

Them: "We really can't do anything about salaries now. Our budget is just too tight."

You: "I understand that. At the same time I would appreciate some form of recognition for my efforts. Can we discuss that?"

Them: "Well, what do you have in mind?"

You have managed to move the conversation from a demand for a raise to a negotiation about recognition now and possibly a raise later. You can suggest one or more of the following alternatives that will satisfy your underlying interest for recognition:

- a promotion to a higher level of responsibility with the promise of a salary review within six months

- that the supervisor helps you to increase your visibility by having you present the results of your studies at a major company review and an upcoming international meeting

- that the supervisor agrees to compare the company's salary structure with others in the industry and address inequities

- that the supervisor agrees to assign another postdoc to work under your supervision. This makes you that feel he appreciates your work and it increases your productivity.

Problem 4

The following case study relates to a conversation Carl had after one of his workshops at a large government research laboratory. It illustrates how the concept of underlying interests can be used in managing an employee performance issue.

After my workshop a senior group leader approached me and posed the following problem. He said that the scientists in his group were required to fill out timecards each week to specify which projects they were working on and for what amounts of time. He said that he had one scientist who was working on only one project and who consistently failed to fill out a timecard because it made no sense to him. The manager had called him on this several times but the behavior persisted and the manager couldn't figure out what to do about it.

I asked the manager whether he had ever indicated to the scientist that there would be consequences to this behavior. The manager said he hadn't because he really couldn't think of any meaningful consequences. He said the person was a good scientist and that he did not want to come down too hard on him because his work was generally excellent. I indicated that he couldn't have it both ways. He couldn't complain about the scientist not filling out the card unless he was willing for there to be consequences for noncompliance. In fact, the manager's not having done anything about this over the past year implicitly has given the scientist the message that it is actually okay not to fill out the cards. Moreover, the manager said that he had to "cover" for the employee at times, making it clear that he was enabling the very behavior he was complaining about.

The manager said it occurred to him to use the tactic, "I know we both agree that filling out these cards is nonsense but we just have to do it, so you have to do it even if it is nonsense," but he immediately saw that by devaluing the cards he might just exacerbate the behavior. Also, this would result in further polarization between the scientific and administrative arms of the organization, which would be undesirable and counterproductive.

I asked whether this behavior affected how the manager thought about the scientist—did he think, for example, that this person might not be as reliable or attentive to detail as a result of this and that this might affect the kinds of projects or responsibilities he would give to the scientist. The manager said that was possible. So I suggested that the manager inform the employee that his behavior would affect how the manager assessed him and his reliability and that this would affect the kinds of projects he got put on—that is, he might not get to work on the kinds of projects that interested him most, or that were important, if the manager had doubts about his attention to detail and reliability. This would appeal to the scientist's deep underlying interest of working on challenging and important projects and might provide the kind of motivation that complying with timecard guidelines could not. He needed to make it clear that he expected the employee to fill out the cards and that there would be consequences that would affect things that were meaningful to the employee (i.e., his ability to do interesting science).

I later learned that this approach proved successful.

The examples illustrate that you will have more success if you focus on satisfying underlying interests than on predetermined or rigid positions. In most cases positions are actually proxies for complex underlying interests, each of which can be satisfied in many ways. Focusing on interests gives negotiators many more ways to reach an agreement than focusing only on positions.

Positions are "take it or leave it." In contrast, there are many ways to satisfy underlying interests.

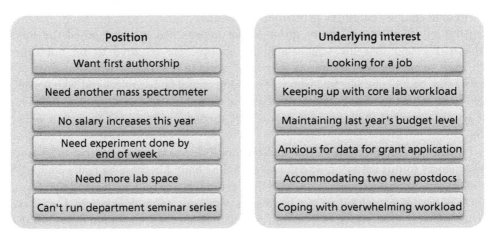

Position	Underlying interest
Want first authorship	Looking for a job
Need another mass spectrometer	Keeping up with core lab workload
No salary increases this year	Maintaining last year's budget level
Need experiment done by end of week	Anxious for data for grant application
Need more lab space	Accommodating two new postdocs
Can't run department seminar series	Coping with overwhelming workload

Once you get the other party involved in helping you to find solutions that satisfy your underlying interests, the negotiation becomes a problem-solving session and not a demand for a specific outcome. Most people respond much more positively to a request for help than to a demand that they buy into your solution.

Maintain a collegial interaction by using "I" statements

In a negotiation, or in any professional discussion, if you find yourself ready to start a sentence with the word "you," stop and think. "You left my name off this manuscript" may, in fact, be true but it starts the discussion with an accusation. The other party is placed in a defensive posture and may be tempted to respond in kind, "You really didn't contribute very much to the work." The discussion goes downhill from there.

Do not frame the problem in terms of what the other person did, but rather in terms of the impact on the project or its effect on you. "I see that my name does not appear on this manuscript. I thought it would be there. Could we talk about it?" makes the same point, with the same frankness, but addresses the issue rather than the integrity of the person.

The following represent some examples in the context of the negotiations discussed above.

Example 1

Accusatory: "You didn't allocate enough money in the budget for what we need to do in the proteomics facility. That's why we need this mass spectrometer now."

Nonaccusatory: "Our proteomics budget didn't provide sufficient funds for a new mass spectrometer. As a result we're behind in our workflow."

Example 2

Accusatory: "You do this all the time. You come to us at the last minute and make us drop everything to do the study that you think is most important. My team is really frustrated by this stop/start mentality."

Nonaccusatory: "My group functions better if they can follow through on our projects. I understand that you're under pressure, and I think that we might be able to come up with some alternative approaches to help you."

Example 3

Accusatory: "You have no appreciation for all of my work on this project. I work overtime and my only reward is more work."

Nonaccusatory: "I'm not sure that the work I've put into this project is fully appreciated. I'd like to have some concrete acknowledgment of my contributions."

Chapter 5 provides additional examples for using this technique when dealing with difficult or objectionable behavior in the lab.

Learn to expand the pie

Once you learn to negotiate based on interests, and not positions, you are on the road to opening up a broad spectrum of possibilities and solutions. Look at the interests we identified in the three sample problems earlier in the chapter. Each of these interests might not have been obvious had we not delved beneath the surface of the problem. As a consequence, we were able to extend the scope of the negotiation far beyond what it might have initially focused on. As a scientist, you may tend to pare the problem down to its essential elements: Get to the root of the difficulty, simplify as much as possible, and negotiate about the one key element. In negotiation, this is a big mistake.

Let us return to the negotiation involving first authorship of a scientific paper. In its simplest form, the negotiation seems to be a classic win/lose situation: There can only be one first author. It is true that some journals now include footnotes indicating that the first and second authors contributed equally to the work, but many (maybe even most) scientists would still opt to be the first author if they could. After all, when your paper is discussed in the "Current News" section of *Nature*, would you rather see it referred to as work by "Your Name et al. 2005" or "Someone Else's Name et al. 2005"?

■ *Case Study: First Authorship*

Alex and Pat are postdocs in Dr. Waffle's lab; Alex is one year senior to Pat, and Pat is new in the lab. For the past year, Alex has been working on a project to generate adult stem cells from rodent liver. Although he has made good progress and has established important tools and assays for the work, the project is bogged down. In an attempt to help, Dr. Waffle assigns Pat to work with Alex. Pat has experience in isolating mouse hematopoietic stem cells and introduces several powerful new approaches to Alex's experimental system.

In a very short time, Pat's modifications yield spectacular results and Dr. Waffle instructs the pair to write up a manuscript describing the work. Dr. Waffle tells Pat and Alex that they will be first and second authors, he will be third author, and it is up to them to decide which one will be first and which one second.

Pat and Alex meet several times over the course of the next month to discuss the manuscript. Each time, they have a brief discussion about order of authorship, but it quickly becomes clear that they both believe that they deserve to be first author.

Alex is planning to start looking for a job in six months and believes that first authorship on this seminal paper will be his key to a great job. Pat plans to continue working on the project in Dr. Waffle's lab for at least another year and feels that his contribution was the sole reason that the project got to where it is today. Pat feels that it is simply a matter of fairness that he be first author.

Neither will budge from his position.

If you are a scientist, the interests of the two parties may seem obvious to you: They each want to be first author. But that is not an interest; it is a position. Remember, an interest is what each expects to achieve by becoming first author. A position is simply one way of satisfying an interest. A more detailed analysis of the interests of the two parties might reveal a far more complex and useful set of opportunities.

Alex's interests (what he hopes to achieve)

- Professional recognition for the groundbreaking work
- Credit for having started and laid the groundwork for the project
- A good job in the next six months with the help of the publication

Pat's interests

- Professional recognition for groundbreaking work
- Recognition for his specific creative contributions
- Continued work on the project after Alex leaves the lab
- Continued work in this field and eventually a National Institutes of Health (NIH) grant

The first authorship issue represents the tip of an iceberg of underlying interests for each. By identifying these interests, we also identify additional topics for Alex and Pat to negotiate. Although this may seem counterproductive at first glance, expanding the pie (adding more elements to the discussion) actually gives Pat and Alex greater flexibility in coming to an agreement. If all or some of the interests listed above become part of the discussion, it no longer seems like the simple win/lose situation of who gets to be first author.

Here are some ways that Alex and Pat can use their interests in an expanded negotiation.

- Both know that a Gordon Research Conference is coming up in three months and one of them will need to present the work. Because one of Alex's interests is in having face-to-face exposure with people in his field who may have job opportunities, presenting the work at the conference might be a good way to accomplish that. In that case, Alex may be willing to agree to be second author if he can present the work at the conference.

- Alternatively, Alex may already have a job lined up and plans to start writing a grant application to the NIH for research support. In this case, being able to continue with the work that is described in the paper would be a great benefit. Thus, he might be willing to let Pat be first author in exchange for an agreement that he will have the right to take the project with him when he leaves.

- It may be that another paper is expected to arise from the collaboration. Perhaps Pat will not look for a job for another 18 months, whereas Alex is looking for one now. Pat may agree to let Alex be first author provided that Pat gets to be first on the next paper and is guaranteed to be able to keep a major part of the project as his own when Alex leaves.

The next two examples of expanding the pie show how inventing new possibilities that benefit both Alex and Pat can facilitate an agreement.

- At Alex's suggestion, Dr. Waffle promises that he can get an invitation from a prestigious journal for Alex and Pat to write a joint review article outlining progress in this field. Both will benefit greatly from the exposure, and neither could write the article on his own. For this article to be written, Alex and Pat need to be on good terms. This is a powerful motivation for them to come to an agreement. Waffle suggests that whoever is second author on the first publication should be first author on the review.

- Alex suggests that in exchange for first authorship, he will send Pat any new cell lines he derives in his new position and that they will collaborate on the work associated with these lines. Pat benefits by having access to these important cells and Alex benefits by having Pat as a collaborator.

These last examples are sometimes referred to as "options for mutual gain" because the negotiators have created new opportunities that benefit the two of them simultaneously.

The point here is that the more elements you add to the discussion, the greater the likelihood that you and the other person can find a way for your underlying interests to be met. You have "expanded the pie" by adding additional elements to the negotiation, but you were able to do this only because you were willing to look beyond the narrow confines of the authorship question. Although it is true that there can only be one first author of the original paper under discussion, the negotiation is now a much broader one including many other elements of value to Alex and Pat.

As we noted above, it is important to prepare to expand the pie before you begin the negotiation. Think through your interests and all of the possible ways that they can be satisfied and be prepared to offer new elements to the authorship discussion. Also, try to learn what the interests of the other person are as well. As noted above, showing the other person that you view their interests as being just as important as yours sends a powerful message and reduces the likelihood that they will revert to positional demands.

In addition, consider ways in which you and the other party can work jointly toward an outcome that neither of you would be able to achieve separately. The last two examples above show how adding new options like joint review papers or projects allows more opportunity to create an equitable agreement.

By discussing the authorship question in the context of satisfying interests, you turn what could have been a contentious discussion into one of mutual problem solving. By presenting your interests honestly, you become open to ideas and suggestions that may not have occurred to you.

Asking plenty of questions of the other party can help you to expand the pie by identifying the other side's interests that may not have been apparent to you at the outset. For example, you might say, "I'd like to understand why being first author on this particular paper is so important to you. Can you run through that with me?" Although you may think that you already know the answer, you might be surprised if the other person were to say, "I'm leaving research to go to law school and I want to have just one first author paper. You can have the project when I leave and you can present the results at the Gordon Conference." Once you close your gaping jaw, you'll have a very different discussion than you anticipated.

Ask open-ended questions to get a stalled negotiation moving again

There are times during a negotiation when you feel like you're stuck. Perhaps the other person is taking a hard position on some issue and you feel like you've exhausted all your strategies for getting them to focus on underlying interests. You're sitting there and you feel like you've run out of ideas on what to do or say next. A great way to get the discussion jump-started is to start asking open-ended questions. These are questions that start with phrases like "What if we…?" or "Have you considered the possibility of…?"

By asking questions like these, you are accomplishing several objectives. First, you're asking the other person what they think rather than pressing them to accept your solution. You can think of this strategy as figuratively bringing the other person to your side of the table and making them your partner in helping to solve the problem. The idea is to create an environment in which the two of you are problem solvers rather than adversaries.

This approach works especially well with scientists and technical professionals because they love being asked for advice and for their opinion. Now here's the surprising part of this strategy—the hypothetical questions you ask don't even have to make a lot of sense or need to have been thought out in advance. The idea is just to get the other person engaged in a dialogue focused on solving the problem and away from whatever fixed position they may be stuck on. The following case study shows how this works in real life.

■ Case Study: Making Friends with the Food and Drug Administration

A few years ago, my company was running a clinical trial for a new drug designed to treat a deadly form of cancer for which there was, at that time, no treatment. The trial had been running for far too long, mostly because we were having a hard time enrolling patients. In fact, it was beginning to look as though we might need to halt the trial altogether. We decided to discuss our options with the Food and Drug Administration (FDA) on a conference call.

Before the call, we thought long and hard about what options might be acceptable to the FDA. We also thought about what the FDA's underlying interests might be beyond ensuring that the trial was conducted according to the best clinical and regulatory standards. We thought it likely that in addition to its mission of ensuring good regulatory compliance, the FDA also has a strong interest in helping companies such as ours develop innovative treatments for untreatable and deadly diseases. Thus, we thought that the FDA might have an underlying interest in helping us to find a way to get this trial back on track. We just didn't know how this interest might translate into allowing us to modify the scope or elements of the trial. Because the FDA doesn't like to discuss hypotheticals, we had to send them a specific proposal that would form the basis for a discussion.

The conference call began, and after a few minutes of preliminaries, the FDA turned the call over to their statistician. In very short order, the statistician said, "We've reviewed your proposal and find it inadequate in almost every regard. There is no way we could even consider entertaining a proposal of this type. The changes you proposed in the enrollment criteria are unacceptable, as are the alterations in the statistical analysis… ." This went on for a few minutes, basically slamming everything we had proposed. We were assembled at a conference table and I could see that my team was in shock. Ellen, our statistician, who was sitting next to me, passed me a note that said, "S**t! What do we do now?" The FDA statistician was summing up and we needed to have some sort of response, so I wrote on the note pad, "Just start asking hypothetical questions. It doesn't matter what they are!"

Ellen gave me a look of panic, but as soon as the FDA statistician stopped talking, she jumped right in. After thanking the FDA for their analysis of our proposal she just started asking a series of hypothetical questions. "What if we were not to change the enrollment criteria but rather reduce the follow-up time…?" "Could there be any flexibility in altering

> the agreed upon statistical endpoint...?" And a lot more. After each of these, the FDA statistician would make a dismissive noise and then reject the idea out of hand.
>
> However, at some point, Ellen asked a question that the FDA statistician thought was at least interesting. He said, "Well, we couldn't agree to that, but we did do something similar with another company last year," and went on to outline a possible scenario that eventually turned out to be the basis for a modified agreement.

What happened here was that by asking open-ended questions, Ellen changed the dynamic of the interaction from one of confrontation to one of collaboration. By engaging the FDA statistician in a give-and-take dialogue, she changed him from an opponent into a colleague—for 15 minutes the two of them spoke as though they were the only ones on the call. A kind of professional bond was formed and eventually a possible resolution was found. Ellen didn't really have any specific idea in mind when she started asking questions—her questions weren't non-sensical but neither had they been terribly well thought out either. This may actually have worked to her advantage because it was clear that she was not trying to convince the FDA statistician as much as she was asking for guidance.

Manage yourself: Know your hot buttons

Self-awareness means not only being aware of your feelings and behavior, but also anticipating how you are likely to behave in a particular situation. Most of us have sensitivities to certain things that people do or say to us. When these sensitivities are extreme they are called hot buttons, because when "pressed" they cause us to become or act angry, defensive, or hurt. You had the opportunity to identify your hot buttons in Exercise 5 in Chapter 2.

Sometimes someone we hardly know will say something that presses a hot button. More often, it is someone who knows us well, sometimes too well as in the case of spouses who tend to know each other's hot buttons with alarming accuracy. Whatever the circumstance, if we react overtly to feelings of anger during a negotiation, we will almost certainly negatively impact an already contentious discussion. Alternatively, if we react to feelings of hurt, we may withdraw from the discussion and fail to represent our own interests adequately.

These types of reactions are what Daniel Goleman has called "emotional hijacking" (Goleman 1995, pp. 31–51), and that is exactly what it feels like. In the moment it happens, you feel as though your mind has become clouded and rational thought is impossible. You may feel overwhelmed and unable to continue effectively in the discussion. Here is an example of such a situation.

> ■ *Case Study: Hot Buttons*
>
> I once worked for someone who was familiar with all of my hot buttons and used them to great advantage for years. Whenever he wanted to win a debate, he would accuse me of being arrogant and selfish. I believed there to be some truth to these accusations and felt guilty about being this way. Moreover, I felt vulnerable and exposed, because I knew that he was

aware of these characteristics in me. As a result, every time he accused me of being arrogant, my face would become flushed, I would feel flustered, and I would lose my ability to continue with our discussion in a calm and dispassionate manner. This prevented me from thinking clearly and led to my becoming withdrawn. Needless to say, whenever this happened, I lost whatever debate we were having.

It took me a long time to learn how to anticipate this emotional hijacking. I started by practicing becoming aware of my feelings during these incidents. This, in and of itself, was a big step, because simply telling myself that I was feeling angry and humiliated gave me some perspective and control over what I was experiencing. Thinking in this way moved me from a mode in which I was reacting to my feelings to one in which I was thinking about them. Thinking about my feelings eventually enabled me to realize that I could have the feelings without reacting to them at that moment! I was able to notice what was going on inside myself, anticipate behavior that would be counterproductive or harmful to my interests, and decide to behave differently.

At long last, I was eventually able to tell my boss, "I may be a bit arrogant at times but that is not what is at issue here. Let us try to focus on the issues at hand."

Managing your hot button reactions requires you to achieve some emotional distance from yourself and your feelings. It is akin to what Ury has called "going to the balcony" (Ury 1993, pp. 31–51), meaning observing yourself and your feelings from a distance. This puts you in the role of an observer rather than a participant. If you can achieve this state of mind and maintain it during negotiation, you can represent your interests more effectively.

Body language during a negotiation

If you are concentrating exclusively on the content and outcome of a negotiation, without realizing it you may be interacting with the other person in a counterproductive manner. For example, you may speak or behave in a threatening way. A scientist in one of Carl's workshops repeatedly jabbed his finger toward the other person's chest when making a point during a negotiation on authorship. How do you think this person felt? That negotiation turned into an argument. When Carl pointed this out to the workshop participant, he admitted that he wasn't even aware that he was doing this. You can be certain that the person he was negotiating with was acutely aware of it.

Most of us are blissfully unaware of the signals we are sending with our body language. The next time you are in a meeting, perform the following experiment. Look around the table and make a quick note of how you think each person is feeling or reacting based only on what your eyes tell you. Chances are you'll see one person with their arms tightly crossed (disapproving?), another gazing out the window (bored?), and another whose leg is vibrating under the table (nervous, anxious?). Whether you are conscious of it or not, you're always looking at other people and drawing conclusions about what they're thinking or feeling based solely on what their bodies or faces are doing. And guess what? They're doing the same with you!

When you're in a negotiation and the other person's body language tells you that they're nervous or anxious, you might start feeling a bit mistrustful of them. And vice versa if it's you that's looking anxious. This is another case where periodically "going to the balcony" can be a great help.

Periodically take a mental step back from the table and look at yourself as others in the room see you. What is your body saying? Is that the message you want to convey?

Positive body language: Expressing interest or empathy	Negative body language: Expressing anger, concern, anxiety, or lack of interest
• Maintaining eye contact • Facing the other person directly, rather than at an angle • Nodding head in affirmation • Routinely saying "uh huh" as an acknowledgment • Having an empathic facial expression or smile • Leaning forward when listening or speaking • Smiling	• Consistently not making eye contact • Suddenly or transiently losing eye contact in response to something that the other person says • Crossing legs or folding arms across body • Leg vibrating • Leaning back from the other person • Doing something distracting such as picking lint from clothes, checking phone for messages, doodling • Frowning, furrowing brow, looking confused

The table above shows some common body languages and how they are typically interpreted in American culture. Be aware, however, that body language is not universal. During one of his workshops, Carl was told by a participant that in her culture looking someone, especially a more senior person, directly in the eye could be viewed as challenging that person's authority. Also, certain types of body language need to be used sparingly and with sensitivity. We have all experienced the effusive, outgoing individual who periodically touches us on the arm or shoulder to make a point. In one situation, this can feel reassuring and sincere (if the person is a good friend), whereas in another (especially in a mixed gender situation) it can feel presumptuous or threatening. When in doubt, don't touch.

Don't underestimate the impact or importance of your body language during a negotiation. It has been estimated that 65%–90% of information communicated during human interaction is nonverbal in nature (Knapp 1972). Although you may be saying one thing with your words ("I really respect your ideas and want to hear more."), if your body language is saying something else ("I have no clue what this dummy is saying to me."), guess which message will win?

> *Any words, be they ever so flawless, can have their meaning cancelled by body language—but not vice versa. There are no words capable of cancelling the meaning of transmitted body language.*
>
> ELGIN (1985)

Do not take anything personally during a negotiation

People with low self-awareness are more likely than others to say or do things during a negotiation that sound contentious or just plain insulting. Your best tactic in such a situation is to simply ignore it and move on. Dwelling on or responding to petty insults or affronts switches the topic of conversation from what you're negotiating to how you feel (insulted, angry, etc.) and will probably be a waste of time.

Most of what people say that makes you feel angry or insulted (especially in a professional situation) is simply the result of thoughtlessness. Remember, the scientists with whom you work have the same problems as you. If you have a tendency to speak before you think— to say or do things without fully considering their impact on others—you are not alone.

The same behaviors that we excuse in ourselves by saying, "Oh, I really didn't mean to say that; I just wasn't thinking," we readily attribute in others to malicious intent or a deliberate desire to affront. This is a well-studied psychological phenomenon called the fundamental attribution error (Plous 1993, pp. 180–181). That is, particular behaviors that we observe or experience in others we attribute to fundamental characteristics of their personalities, rather than, for example, careless error, ineptitude, or naïveté. If someone is strolling down the street and they bump into us, our first reaction might be that they are careless or aggressive (personality traits) rather than that they were shoved by someone else or lost their balance.

Even if you think that the other person is deliberately insulting you, ignore it anyway. Remember that your objective during a negotiation is to get your interests satisfied, not to fight. Your best strategy if someone says something insulting is to ignore the comment and continue as you were. Never trade insults.

Of course, if the insults persist and become malicious in character, you can say, "I find what you're saying insulting and suggest that we postpone this discussion until we can have a more collegial conversation." Telling the other person how you feel calls attention to the objectionable behavior in a nonaccusatory manner and is often enough to change the person's tune.

The following case study illustrates how a science consultant, Samantha, deals with a difficult client. The case illustrates how a seemingly simple interaction could have turned into a conflict if one participant had reacted differently to an insult.

▪ Case Study: Don't Take the Bait

The head of a small company that was trying to develop a treatment for Alzheimer's disease contacted me. They had tried to raise money but were unsuccessful largely because investors were skeptical about their approach. The chief executive officer (CEO) asked me to review their data because I'm an expert in this field. However, a close friend had consulted to this company a year ago, and it took the company six months to pay him for his work. He told me that he would never work for them again.

When I spoke to the CEO, I told him that I worked on a contingency fee basis and would need to be paid for the work before I started. He agreed to this and said that he would have a check for me when I arrived. After I arrived, we engaged in small talk, he led me into a conference room, and then he launched into his data presentation. I interrupted him and asked if we could get the financial formalities out of the way before we began, as we had agreed. He looked at me a bit askance, but agreed that he would go to his office and get the check.

After about ten minutes, he came back and said that his administrative assistant was out that day and that she was the only one who knew how to print out checks using their enterprise software. He suggested that we proceed, and he would send the check later.

I was taken aback by this, especially because we had an explicit agreement that I would be paid in advance. I said, "Well, I think we need to resolve this first. I'm a real stickler for details, and we did agree that I would be paid in advance. If you simply write a check by hand, that would be fine."

He looked surprised and said, "You're really not a very trusting person, are you? Aren't you being a little paranoid?"

> I felt angered and insulted by his comment, and mentally counted to three while I fidgeted with my pen. At that point, after having taken a couple of deep breaths, I said, "It's not a matter of trust, and I'm really not paranoid. But I am a bit compulsive about sticking to the terms of agreements I make. I agree that it may sound strange, but I would be distracted during our discussion and would have a hard time giving you my full attention if we didn't first get this out of the way. So if you would just humor me, we can get on with the meeting."
>
> He shrugged, went to his office, and came back with a handwritten check. The consultation went smoothly from there.

Samantha could easily have taken umbrage to the accusation that she was untrusting and paranoid. She could have said, "We had an agreement and you're not sticking to it, so maybe I should be untrusting. I've heard stories about you... ." Instead, she refused to take the comment about trust as an affront and made light of it, suggesting that although it might seem that she was mistrusting, it was just an idiosyncrasy and she should be humored. At the same time, she was agreeing with the client's observation and empathizing with his feelings, further limiting the likelihood of a confrontation.

Note that Samantha really did feel insulted by the comment, but she managed to recognize her feelings and behave in a manner that enabled a resolution. The key to avoiding taking things personally is not that you should never feel angry or hurt. In fact, it is likely that you will never be able to control how you feel in these situations. The key is to give yourself the opportunity to make decisions about how you will behave or respond, regardless of what you are feeling.

Listen to and acknowledge the other side: Be empathic

How often have you been in a meeting or discussion, and while the other person is talking, you are carefully rehearsing what you are going to say in rebuttal? If you do this, when it comes time for you to speak you may hear, "Oh, that is not at all what I said. What I said was... ."

We often spend less time listening than preparing to refute what we think we heard. This happens often with scientists, because so many of their arguments are carefully reasoned, sequential, and orderly. All of that does take some thinking. What better time to be thinking than when the other person is talking?

Adele Lynn (2002) identifies six types of what she calls "negative listening patterns." These occur when we seem to listen, but in reality we are not listening at all. Below we describe the six patterns, but we modified her list somewhat, for our purposes. Read through the list and ask yourself if you display any of these behaviors when you are supposed to be listening.

1. **Faking.** Make all of the outward signs of listening (i.e., eye contact, saying "uh huh," etc.), but do not really listen. The faker is thinking about what they are going to say next or about something else altogether.

2. **Interrupting.** Interrupt the speaker before they are able to finish and/or interrupt with questions or comments that show what you know rather than ask for relevant clarifications.

3. **Intellectualizing.** Become obsessively logical about what the speaker is saying, highlighting and focusing on minor logical flaws and not hearing the meaning or effect.

4. **Free-associating.** Pounce on something the speaker says because it reminds you of an association that you feel compelled to mention, but that has nothing to do with the content of what the speaker is saying.

5. **Gathering ammunition.** Listen only long enough to hear something with which you disagree or that you can refute, and then pounce on that.

6. **Solving the problem yourself.** Give the speaker advice about how you would solve their problem rather than listen to the full story and understand the speaker's viewpoint.

Can you identify occasions when you have exhibited one or more of these behaviors? Scientists can become obsessed with being right, so they give what they say a lot of thought to ensure that they are not making some error of logic or fact. That's all fine—provided all that thinking and preparing leaves time to listen to what others say. Listening carefully takes much more work than most people think. Remember that in a negotiation not only do you need to listen, you need to demonstrate to the other party that you hear what they are saying, and that takes even more work.

There are many ways to produce a climate conducive to an agreement, but the simplest is to behave in a way that shows the other party that you are listening to and hearing what they are saying. It also helps a great deal if you actually are listening to and hearing what they are saying.

Simple methods to convince the speaker that you are listening include repeating what he has just said, "So if I understand you correctly, you're suggesting that... ." "Ok, so your view is that... ." Other methods include making eye contact with the speaker, and nodding and saying "uh huh" every so often to assure that you are hearing, but not necessarily agreeing with, what she is saying.

You may be surprised to discover that as you act out these listening behaviors, you will actually become more attentive and responsive to what the other person is saying. That is, as any behavioral psychologist will tell you, changing your behavior can change your attitude, outlook, and feelings (Burns 1990). As you force yourself to listen to the other person and to repeat some of what they said to make sure you got it right, you actually become more empathic towards them. In most cases the other person will notice this and act reciprocally.

Finally, when listening, take special note of your own reaction to and feelings about what is being said. Pay attention to what might seem like transient thoughts or feelings that lie just beneath the surface of your consciousness. If you can capture these, you may find them extraordinarily useful. Often we intuit something about someone or what they are saying before we are fully conscious of it, and often those intuitions are far more accurate and revealing than what we consciously think.

If you do become aware of feelings of discomfort about what someone is saying, be extra careful that you do not misinterpret what they are saying. Your emotions or anxieties may lead you to jump to conclusions, make assumptions, or misunderstand what is being said. In these circumstances, it is a good idea to ask questions to clarify what the other person is saying. Repeating what you just heard and asking if you understood it correctly is a good way to do this.

Principled negotiation: What to remember

- Negotiators are problem solvers, not competitors.
- The goal is a wise outcome, not victory.
- If you feel like attacking, don't.
- If you feel like interrupting, listen.
- If you want to tell them the answers, ask questions.

Defuse anger and hostility, including your own

Do not respond to or trade insults or accusations during a negotiation. If the other person becomes angry or hostile you need to defuse or neutralize the anger before you can engage in a productive negotiation. Chapter 8 presents tools for dealing with an angry boss and these same tools can be used in a negotiation. Jump ahead to Chapter 8 for a preview of how to use the "Agree, Empathize, Ensure, and Inquire" technique. This approach helps dissipate anger and can allow you to refocus on the negotiation.

No matter how self-aware you are and no matter how attuned you are to knowing when someone has pressed one of your hot buttons you still may lose your temper on occasion. It happens to even the most experienced managers, leaders, and negotiators. When that happens you need to own your actions and work to reestablish a productive relationship. If you said something regrettable, apologize right away. If your anger led to an angry response from the other person it may not be possible to resume the discussion at the time. Call a time-out or suggest that the two of you take time to cool off and get back together the following day. After 12 or 24 hours your anger will inevitably have dissipated. During the hiatus, reflect on the situation and try to understand what it was that angered you. Often you will find that whatever it was no longer feels so infuriating. Unexpected angry outbursts frequently have more to do with an association you had with a past event or another person than with this event and this person.

REFERENCES

Burns DD. 1990. *The feeling good handbook*. Plume/Penguin, New York.

Elgin SH. 1985. *The gentle art of verbal self-defense*. Dorset House, New York.

Fisher R, Ury W. 1991. *Getting to yes: Negotiating agreement without giving in,* 2nd ed. Penguin Books, New York.

de Botton A. 2016. Why you will marry the wrong person. *New York Times,* May 28, 2016.

Goleman D. 1995. *Emotional intelligence.* Bantam, New York.

Knapp ML. 1972. *Nonverbal communication in human interaction,* 1st ed. Holt, Rinehart & Winston, New York.

Lynn AB. 2002. *The emotional intelligence activity book: 50 activities for developing EQ at work.* Amacom, HRD Press, New York.

Plous S. 1993. *The psychology of judgment and decision making.* McGraw-Hill, New York.

Ury W. 1993. *Getting past no: Negotiating your way from confrontation to cooperation.* Bantam, New York.

EXERCISES AND EXPERIMENTS

1 The right tools for each negotiation

Below is a summary of negotiation tips condensed from Ury's *Getting Past No* (Ury 1993). Make a copy and review it before your next negotiation. You will not be able to apply all of the guidelines to every negotiation, nor should you need to. For example, if you are negotiating with someone who is very aggressive, you may need to use techniques to ignore or deflect insults or innuendos. With a very passive person, you may want to use your listening skills to better understand their interests. If a situation seems to be intractable because of the limited number of options available, remember to expand the pie and identify underlying interests.

*Negotiation Guidelines**

Keep in mind the following.
- Negotiators are problem solvers, not competitors.
- The goal is a wise outcome, not victory.
- Separate "people" from "problem." Be hard on the problem and soft on the person.
- Focus on interests, not positions.
- Invent options for mutual gain.
- Insist on objective criteria.
- Yield to principle, not pressure.

Before the negotiation, prepare.
- Identify your and the other's interests. Think through options.
- Define standards to be used in reaching an agreement.
- Define what you would like to achieve.
- Define what you would be content with.
- Define what you could live with.

During the negotiation, manage your state of mind.
- When attacked, do not strike back, give in, or break off the negotiation.
- "Go to the balcony" to gain perspective.
- Buy time to think; rephrase what has been said.

Manage their state of mind. Identify with the other side.
- Find something with which to agree, and agree whenever possible.
- Paraphrase and ask for corrections to your understanding.
- Acknowledge their feelings.
- Express your views in a nonprovocative manner.
- Make "I," not "you," statements.

Use clarification to identify underlying interests.
- Do not focus on their position; focus on common goals.
- Ask open-ended questions such as "Why?," "Why not?," or "What if?"

Deflect obstacles and negative tactics.
- Focus on fairness, standards, and objective criteria as ways of evaluating an agreement.
- Ignore flat refusals or reinterpret them as aspirations.
- Ignore attacks or reinterpret attacks as assaults on the problem. Never attack back.

*Adapted from Ury (1993).

See the following table and make a list of three or four negotiations that you have had in the past year. Remember as accurately as you can how you felt and acted during each negotiation. Fill in the table, indicating which negotiation tool or concept would have been most useful to you during each negotiation. For each of the negotiations that you listed, check the tools that could have led to a better outcome.

Topic of negotiation	Negotiation technique						
	Prepare for negotiation	Focus on interests, not positions	Focus on problem, not person	Expand the pie	Manage yourself	Listen to the other side	Defuse anger

2 Identifying underlying interests

Think of a person with whom you find yourself in disagreement on a regular basis. Recall the most recent disagreement or difficult negotiation and answer the following questions (it will be most useful if you write down the answers because you will need to refer to them later in the exercise).

a. What was the disagreement about?

b. What did you hope to achieve? What were the interests or underlying needs that you were trying to satisfy? List as many as you can. Refer to the authorship negotiation case study earlier in the chapter ("First Authorship") for examples of underlying interests.

c. For each of your underlying interests, write down one or two ways that you could have added an element to the discussion, besides the primary topic, that would have enabled you to satisfy that interest.

d. What were the underlying interests that the other person was trying to satisfy? (You may need to guess if you do not know enough about this person or their interests.) Ask yourself what this person wants or needs and how these factors might have influenced the negotiation.

e. For each of the other person's underlying interests, write down one or two elements that you could have added to the discussion that would have enabled him to satisfy that interest.

f. What were you feeling during the negotiation? (Refer to the list in Chapter 2; use feeling words, not thinking words.)

g. What did the other person feel during the discussion? What did their body language convey?

h. Review the negotiation guidelines in the summary below. What did you do or say during this negotiation that violated these guidelines? What did you do or say that was in accord with them?

3 Identifying negative listening patterns in yourself

You may be a poor listener without knowing it. Over the next few days, pay close attention to what you are thinking during several discussions with people in your workplace. After each discussion, record in the following chart which of the negative listening patterns described in this chapter you found yourself using. Note any patterns. Do you use some of the patterns with specific people? Do you use some more than others?

Negative listening patterns						
Situation	Faking	Interrupting	Intellectualizing	Free-associating	Gathering ammunition	Solving the problem

4 Correcting negative listening patterns

If you exhibit any of the traits listed in the previous exercise, it is likely that you are missing or misinterpreting what is being said more often than you may think. Test this hypothesis by conducting the following experiment.

Pick three or four people with whom you may use negative listening patterns. Choose a discussion during which you will ask at least six clarifying questions in response to things those people say, even if you think you know exactly what they mean. You may find that you are mistaken more often than you think. People use words and phrases to mean different things. Moreover, they are often unclear themselves about what they mean.

- How often did you misunderstand what each person said?

- Were you more likely to misunderstand some people more than others?

- Was the quality of the discussion improved when you began to ask questions?

CHAPTER 4

Bring Them On! Interviewing, Selecting, and Hiring Scientists

INTRODUCTION

O ver the years Carl has hired many science and technical professionals for a variety of roles—from lab technician, to postdoc, to staff scientist, to Director of Research, and higher. Carl never had what you would consider an organized process for making hiring decisions. Sure, he reviewed résumés, checked references, had candidates give seminars, meet team members, etc. But in the end, he always felt that a "gut feeling" about the candidate would guide him to make the right decision. In many cases there were just a few, sometimes only two, finalists so it never felt like a process that required sophisticated algorithms. But as Carl's career progressed and moved into the private sector (biotechnology to be exact) and as the position he was hiring for became more consequential and had more responsibilities, he felt that his "gut feeling" was really being put to the test.

For example, here is a candidate with amazing qualifications, three or four pages of publications on their résumé, who gave a great seminar with terrific data, and had dynamite presentation skills. But Carl felt uneasy about the candidate for reasons he just could not put his finger on. Was he uncomfortable because the candidate was too well dressed? Did that signify a desire for an even higher-level job than the one he was applying for? Thus, would the candidate be unhappy in this job, always looking to move up the ladder too soon and making Carl's life miserable? Or was it that during the seminar Carl noted that the candidate gave what seemed to be a slightly dismissive answer to a question asked by one of the junior female scientists in the audience. This was nothing egregious, perhaps just a slight change in tone that Carl noticed. But there it was, a niggling doubt.

How to decide whether this was important or not? Did this seemingly dismissive comment mean that the candidate was resistant to hearing feedback or criticism? How to weigh these "gut" reactions to seemingly trivial observations? Would following his instinct lead him to reject a perfectly well qualified candidate who could move the company forward? Or would ignoring his instinct lead him to accept a bull-headed malcontent whose difficult personality would infect his team? Was there some way to go about interviewing and hiring that took account of both the "hard" data (résumé, experience, accomplishments) as well as the "soft" data (gut feelings, sense of "fit" with the group)? This chapter is an attempt to answer these questions.

Why you need to do it this way

The central premise of this chapter is that, left to our own devices, we often make bad hiring decisions based on our "gut feelings" about a candidate. The more experienced among us often believe that we know what a good candidate looks like because we have seen enough good candidates and enough bad candidates and we can just tell the difference. We trust our brains to guide us based on our experience. The problem with this approach is that our brain is not a neutral observer. By overreliance on our trusted "gut" feelings we fall prey to a variety of biases, cognitive distortions, and outright prejudices that influence our thinking and our decision-making process without our being aware of them.

Let us look at a few of the more common biases (for an excellent review of these and other biases see *The Psychology of Judgement and Decision Making* [Plous 1993]).

Cognitive biases	
Bias type	**Origin and/or result**
Similarity-attraction bias	We gravitate to and are attracted to people who we perceive as being similar to ourselves ("Wow, we both went to MIT!").
Confirmation bias	We place more weight on information or observations that confirm our initial (possibly ill-informed) opinion of the candidate.
Stereotype bias and implicit bias	We have predetermined and typically unconscious biases against particular characteristics (race, gender, ethnicity).
Halo effect	Some feature of the candidate (e.g., they come from a famous lab) unconsciously spills over onto and *enhances* another unrelated characteristic (therefore, they must be brilliant).
Salience	Your evaluation of a candidate is dominated by one highly visible characteristic, causing you to *ignore* other important but less visible characteristics (e.g., the candidate is highly articulate, leading you to pay *no attention* to their interpersonal skills or technical expertise). Similar to halo effect except that traits are being ignored rather than unconsciously enhanced.

These biases have been studied extensively in the psychological literature and under some circumstances have been referred to as "heuristics," mental shortcuts that we use, mostly unconsciously, to make decisions quickly. Two psychologists who contributed significantly to this field are Daniel Kahneman and Amos Tversky, and their work has been summarized in some detail in Kahneman's book *Thinking Fast and Thinking Slow* (Kahneman 2011). The "thinking fast" part of the title refers to the exigent use of mental shortcuts to make quick decisions, whereas "thinking slow" refers to more time-consuming, deliberate, and analytical thinking. In this chapter we use the term "gut feeling" as a synonym for the mental heuristics that Kahneman and Tversky have described.

In his book Kahneman describes a situation early in his career that presaged his later work on mental heuristics. Kahneman was a young man in Israel and was inducted into the Israeli armed forces. Having a background in psychology, he was assigned to the "psychology unit" of the Army, a group that was responsible for selecting which service members should be sent to officer training school. The failure rate of the selected candidates was high, and the Israeli Army knew that they were not doing a great job. Kahneman saw that the way such decisions were being made was based almost exclusively on the subjective judgments of the reviewers based on open-ended, unstructured interviews as well and on how candidates in small groups behaved when given difficult or even impossible physical tasks to perform. Kahneman discovered that there was almost no correlation between the examiners' ranking and success at being an officer. He became convinced that the examiner's mental biases were being used in place of useful data. But what data should they be using instead?

Kahneman's epiphany was that rather than having the raters rank candidates based on their subjective conclusions, he would instruct them to assign a numerical score to candidates' responses to a series of factual questions designed to reveal certain behavioral traits (sense of responsibility, punctuality, and determination, among others) that he would create. The questions probed what the candidate did or how they behaved in specific situations, and the interviewers had to assign a score to the answers. Recommendations would be made solely based on these scores. After some months Kahneman examined how the recruits selected by this new approach fared as officers and found that it worked remarkably well—far better than the old technique, which hardly worked at all. Kahneman concluded that by reducing the level of subjectivity (and influence of cognitive bias) in the selection process they had improved it.

What is most remarkable to us (and to Kahneman as well) was that the questions Kahneman devised to ask the recruits were based on little more than his hunch about what traits they should be looking for. That is, there was no advance pretesting of the validity of these questions or traits—he just made them up on the fly. That is not to say that the questions were uninformed or random; rather, they were more like informed guesses. The lesson for us is that the introduction of even moderately relevant quantitative scoring in the selection process made a huge improvement in selecting the best candidates. But hiring on the basis of objective criteria is not by itself enough to mitigate the impact of unconscious bias. In fact, believing that we are objective when in fact we harbor unconscious bias may magnify the influence of that bias on decision-making (Uhlmann and Cohen 2007). Managing the impact of bias on our behavior and decisions is no simple task. A good place to start is exploring our own bias by visiting the Project Implicit website (www.implicit.harvard.edu/implicit). For an excellent summary of the kinds of bias (especially gender bias) that influence hiring decisions in academia and how they can be mitigated, we recommend a presentation by Eve Fine of the Women in Science and Engineering Leadership Institute of the University of Wisconsin, Madison available at http://wiseli.engr.wisc.edu/docs/Present_StOlaf_2015.pdf, as well as the book *Searching for Excellence and Diversity* by Eve Fine and Jo Handelsman (2012), available as a free download at http://wiseli.engr.wisc.edu/docs/SearchBook_Wisc.pdf. Both of these contain useful guidance, references for further reading, and links to additional resources.

Years after Kahneman and Tversky's work was published in academic journals the author Michael Lewis wrote a book called *Moneyball* (Lewis 2003) about baseball. The underlying theme of *Moneyball* is that an underdog baseball team (the Oakland As) rose to ascendancy when they discarded the traditional method of selecting new players (the subjective judgment of scouts) in favor of a data-driven analytical method; this echoes Kahneman's Israeli army story that we just related. When Lewis was made aware that his baseball book was really at its heart a book about heuristics and selection biases he contacted Kahneman and ultimately wrote his own version (*The Undoing Project* [Lewis 2017]) of how Kahneman and Tversky's theories came about.

Before we get carried away with data and analytics we need to acknowledge that we are fully aware of the limitations of such approaches. Indeed, there has even been a backlash against the overuse of this approach (Lewis 2017). Moreover, as more than one scientist has said to us "I do not have the luxury of having dozens of candidates for my postdoc position—I am lucky if I have three." To this we say that you still have to make a decision and better to make an informed data-driven decision than one based on "I've just got a good feeling about this candidate."

The perils of interviewing

Perhaps one of the most cherished elements of the hiring process is the face-to-face job interview. For many of us this is the equivalent of those ill-defined exercises the Israeli army used to use—watch someone in action and make a decision. It seems that anyone who hires people with any regularity thinks that they have some unique intuitive filter that enables them to separate the wheat from the chaff, especially during a face-to-face interview.

We were fond of reading the (now-discontinued) column in the Sunday *New York Times*, "The Corner Office." There, each week over a period of roughly eight years, Adam Grant interviewed the CEO of a company ranging from huge corporations to small mom-and-pop enterprises. Each week he asked the same questions, including one on how they interviewed and made hiring decisions. Each CEO seemed to think that they had a "secret sauce"—one or more questions that they believed got to the heart of the type of person they were looking for. And, although each "sauce" was different, they all had in common the goal of obtaining some deep insight into one or another character trait of the interviewee: How creative are they? Are they self-starters or do they need to be constantly pushed? Are they organization-focused or wrapped up in their own drama?

Even if you were to follow Kahneman's advice about creating a structured selection process you still must ask questions, right? So, what are the right questions and, equally important, how do you go about asking them? That is, is there a "right" or at least a "less wrong" way to interview people?

Grant's own newspaper answered this question for us in an article by Jason Dana that caught our eye entitled "The Utter Uselessness of Job Interviews" (Jason 2017). Whereas this specific article is based on the published work of Dana and his colleagues (Dana et al. 2013), it makes reference to an entire field of psychological research devoted to how people make decisions in general and during interviews in particular. The research focuses on what the authors call "unstructured interviews"—that is interviews that follow no set path and in which the interviewer asks whatever questions seem relevant at the time or to the particular candidate. Sounds reasonable, no? No.

Think back to our discussion about biases. It turns out that if you have a conscious or unconscious bias for or against a candidate, you may end up structuring your interview questions or interpreting the candidate's answers to conform to your initial judgment, regardless of the candidate's objective qualifications. As Lou Adler says in his book *Hire with Your Head* (Adler 2007), "One of the biggest problems [in hiring] is that too much emphasis is placed on the interaction between the candidate and the interviewer and too little on the candidate's motivation and ability to do the job," and "If you like a candidate, you tend to go into chat mode, and ask easier questions, and look for information to confirm your initial impression. If you do not like someone, you put up a defense shield, ask tougher questions."

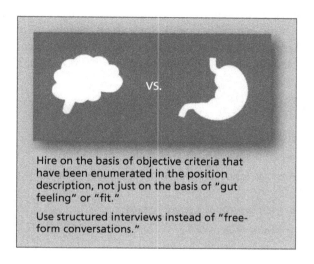

Hire on the basis of objective criteria that have been enumerated in the position description, not just on the basis of "gut feeling" or "fit."

Use structured interviews instead of "free-form conversations."

Moreover, even in the unlikely event that you have no biases, interviews can introduce "noise" that obscures relevant data. In a revealing study, Dana et al. (2013) showed that interviewers who were given accurate factual performance-based information on which to base a candidate evaluation actually made less accurate predictions of the candidate's future performance when they were allowed to interview the candidate than when they had to make the judgment on the basis of the performance data alone. That is, the interview made them poorer judges than they would have been had they just read the résumé. Dana's work is not something new but is supported by ample precedent in the literature (Wiesner and Cronshaw 1988; Meehl 1954).

So, what do we conclude? That interviews are useless? Not at all. What is useless are unstructured interviews. The solution is to use structured interviews—ones in which you create a list of questions designed to probe suitability for the position (like Kahneman did for the Israeli Army) and then ask those questions, and the very same questions, of *every* candidate (Kahneman 2011, p. 232). This eliminates on-the-fly creation of unique and possibly bias-inducing candidate-specific questions and creates a level playing field on which different candidates can be compared. We will apply this principle to every phase of the candidate selection process, from reviewing résumés to speaking with candidate references. In the remainder of this chapter we will show you how to use and follow a script during the recruitment process and how this will enable you to make more informed and less biased hiring decisions.

The people involved

If you are the principal investigator of a research lab or running a scientific group in a company, it is easy to feel that the responsibility for selecting new hires is all yours. That puts a tremendous burden on your shoulders. In fact, there are many others involved in the process and you can and should take advantage of their input.

Human Resources

The human resources (HR) group of your organization is the first place to start, and we cannot emphasize enough that you should meet with an HR representative before you do anything else. If you have a good HR group, they will be an invaluable partner in helping you through the recruiting process. Our recommendation is to find someone in HR with whom you can develop a close working relationship. They will acquaint you with all the relevant institutional, state, and federal guidelines and requirements that you will need to adhere to during the recruitment and hiring process. The table below shows the steps in the hiring process that HR can help you with.

Human Resources' role
• Accurate and thorough position description
• Job posting that accurately reflects the position description
• Application form that requests permission for records (previous employment, etc.) check
• Verification of employment history and education
• Criminal records check (if allowed)
• Offer letter
• Reference check

Your HR office can tell you which of these roles they can help you with. In some states (e.g., Massachusetts) a criminal records check is subject to strict guidelines on how and when they can be used. Obviously, the depth of the background check will reflect the level of the position being filled. Also, although some HR departments, especially in larger organizations, have a policy of doing the reference checks themselves, for reasons that will become clear, we advocate doing those yourself unless it is impossible for practical reasons.

Your team

If you are hiring a new member of a team or work group, we will show you how to elicit the views of the members of the group. There is nothing more demoralizing to team members than to be saddled with a new member who has a difficult personality and whom they have never met before and whom they had no role in selecting or evaluating. Given the degree to which we expect our science team members to work either together or in close proximity to each other, we owe them the opportunity to have their voices heard during the selection process. By interviewing the candidate and attending their seminar, your team can help you determine whether the candidate has the requisite scientific skills and whether there is a good fit with the culture of the group.

You, the leader

Ultimately, the final decision needs to be made by the person in charge—the PI or lab or group leader. As the leader, you are able to see the needs of the group from a unique perspective and one that the members themselves may not see. This is especially true of certain interpersonal characteristics that you may see as beneficial. For example, in deciding between two equally technically skilled applicants you may choose the one who seems more outgoing and social because your current group is composed of quiet introverts. That could be valid basis for making the decision if you think it will have a positive impact on the working environment and productivity of your group. However, we discuss the potential downside of trying to play social engineer below.

THE PROCESS YOU NEED TO USE

Hiring flowsheet

Create position description and advertise

Review résumés, sort and select for phone screen

Phone-screen and select candidates for in-person interviews

Face-to-face interview, seminar, meet team

Reference phone interviews

Select finalist, make offer

The figure above shows a flowsheet depicting a typical hiring process. The key to making intelligent hiring decisions is to use the same process, apply the same criteria, and ask the same questions of each candidate. To help you do this we have created Scoresheets and accompanying Scoresheet Guides for the critical stages in the hiring process.

Try adopting our approach in phases. Using the scoresheets for reviewing résumés and during phone interviews should be the easiest elements to adopt. Similarly, using the interview guide with phone references should be easy because you can have the scoresheet and guide in front of you while talking on the phone. Once you are used to the format and the questions, you can start using the face-to-face interview guide with candidates.

The following table shows the information you will collect at each phase of the hiring process as well as the names of the Scoresheets and Scoresheet Guides we have created for each phase. The Scoresheet Guides are for face-to-face and phone interviews and contain specific questions to ask and tips on what to listen for in the answers. At the end of the process the total scores from the candidate's Scoresheets get tabulated in Scoresheet 7, which gives a final total score for each candidate. This score will enable you to compare one candidate with another. The following sections will take you through the phases of this process one by one.

Source of Data	Key information gained	Scoresheet to use	Scoresheet Guide to use
Résumé review	Education, experience, skills, attention to detail.	Scoresheet 1. Résumé Scoresheet	
Candidate phone interview	Confirm above, listen between the lines.	Scoresheet 2: Candidate Phone Interview Scoresheet	Guide to Scoresheet 2. Candidate Phone Interview
Candidate face-to-face interview	Personal attributes, management skills if necessary.	Scoresheet 3: Face-to-Face Interview Scoresheet. Plus Scoresheet 6, Management Candidate Scoresheet for candidates with supervisory responsibilities.	Guide to Scoresheet 3. Candidate Face-to-Face Scoresheet and Guide to Scoresheet 6. Management Candidate Scoresheet.
Seminar	Technical, communication, presentation skills.	Scoresheet 3: Face-to-Face Interview Scoresheet	
Meet the team	Compatibility, knowledge.	Scoresheet 4: Team Member's Scoresheet	
Reference phone interview	Confirm above, social, interpersonal, work habits.	Scoresheet 5. Reference Phone Interview Scoresheet	Guide to Scoresheet 5. Reference Phone Interview Scoresheet

Creating the position description

The foundation for an informed hiring decision is the position description. The position description accurately describes the job itself as well as the kind of person you are looking for in terms of scientific and technical skills and personal characteristics. This description forms the basis of everything that follows in the selection process. If the description is accurate and you adhere to it, you have a good chance of making an informed decision. If it is vague or applied differently to each candidate, you are more likely to make biased judgments.

The basics of a good position description include the following:

Position description

- Position title
- Reports to (although this may seem obvious, some positions have dual reporting structures, i.e., responsible to different people for different activities.)
- Direct reports (who if anyone reports to the individual)
- Position description
  - Specific responsibilities. Technical, group, organization.
  - Do not forget to include "... and other duties as assigned" and special requirements (see below).

- Skills/background needed
  - Education
  - Technical (specific skills and/or years of experience)
  - Managerial experience (if needed).
- Personal characteristics (more on this below), for example,
  - Team player
  - Highly organized, manage multiple priorities
  - Self-starter, highly motivated
- Performance expectations
  - Technical/scientific deliverables (what do you expect the individual to accomplish, create or produce?)
  - Managerial/organizational goals or deliverables
- Salary/compensation

In addition to the position description, you may want to have an addendum or private version of the description that expands on some of the important skills you are looking for. This might apply to the "personal characteristics" category, which lists key traits the new hire needs to have. In the position description outline above we have listed just a few of them. In the face-to-face interview you will be probing for more traits that do not typically show up in an advertisement or position description for which you will want to screen (see Scoresheet 3, Face-to-Face Interview). Having an addendum helps you to select the characteristics that are most important to you, so you can focus on those during the interview, and, more importantly, so that you are focusing on the same traits with each candidate.

From the position description comes the position advertisement or posting. Although the advertisement must, for practical reasons, be highly abbreviated, it must nonetheless accurately reflect the position description and can often refer to a more detailed position description via reference to a web link. Your HR department may have a standardized position description format for your organization. It is best to consult with them first.

Reviewing résumés

Once applications with résumés start coming in, the first stage of selection starts. This should feel familiar to scientists because it is basically data collection. Just as we can be biased during a face-to-face interview, we can be biased in reading a résumé. We may have an opinion of the candidate's graduate school, thesis adviser, or choice of font. To make this first phase of the selection process as objective as possible, our recommendation is to use Scoresheet 1, the Résumé Scoresheet (all forms shown in this chapter are available for download in editable Word or Excel formats at www.sciencema.com/downloads) to create a numerical score for each résumé. If you are screening many dozens of résumés you could first go through a triage process of removing those that are either clearly unqualified or less qualified than the ones you will score. If possible, wait until you have all or nearly all of the résumés and then score them together rather than as they arrive. This minimizes the chance that you will change your evaluation criteria from résumé to résumé (Basu and Savani 2017) and has also been shown to reduce the impact of implicit gender bias (Bohnet et al. 2015). If you prefer not to wait, make sure you are using the same scoring criteria for every résumé. One way to do this is to review your previously scored résumés before you score new ones.

Scoresheet 1: Résumé Scoresheet

Candidate name		Score (0–4)
	Parameter	**Score (0–4)**
1	Technical skills List specific skills if you need to rate them separately. Skill 1 Skill 2, etc.	
2	Appropriate and relevant educational background Appropriate and relevant work background	
3	Managerial or supervisory skills (if needed)	
4	Language and attention to detail in résumé and cover letter	
5	Quality of previous scientific work and or publications	
	Notes	
Total		(0–20)

Note: in this and the following scoresheets, dark gray shading in the "Score" column indicates that this item is for notes or comments only and does not get a score.

In Scoresheet 1, we start with a list of the key skills, technical and other, the candidate needs to have. Because the document is editable, you can add your own list of skills. It is important that the skills, educational background needed, and other categories listed in this scoresheet be the same as those that appear in the position description. Note that although you can list several key skills in line 1 of this scoresheet, you will only be assigning one overall score (0–4) to the skills section. Thus, you may wish to indicate in advance on your "private" version of the position description which skills are most important to you so that when you assign a score to this entry on the scoresheet you will using the same criteria for each candidate.

Most of the entries on this and the following forms call for a judgment on your part, which is fine and unavoidable. The key to using this data-driven approach is not to eliminate judgment or subjectivity but rather to provide a framework that facilitates making judgments based only on what you are seeing in the résumé (the data) or, later, in the interview, and to use the same criteria in making these judgments for every candidate. If you do that you will end up with a stack of résumés that you can rank by their score and use to determine which candidates should advance to the next step.

The phone interview

Once you have whittled your résumés down to a manageable number (for some people that will be three, for others it might be five or six), you are ready to set up phone interviews with the candidates. Phone interviews are great time savers because they help you identify weak candidates without going to the trouble of a face-to-face interview. You do not need to spend a lot of time on the phone with a candidate to get the data you need. Twenty or thirty minutes might do it. The topics listed below refer to the numerical entries in Scoresheet 2, The Candidate Phone Interview scoresheet (shown below).

1. Technical skills and background

Use the phone interview to verify your assessment of the candidate's skill set and technical qualifications (item 1 in Scoresheet 2) as well as to get a general impression of their suitability. As with reviewing résumés, it is important that your assessment of the candidate's qualifications relate back to the qualifications that you enumerated in the position description. The phone interview will then become your second source of data for skills verification (the résumé being the first). Our goal is to have multiple sources of data for as many of the characteristics or requirements that you will be screening for.

2. Communications skills

You will be talking to the candidate on the phone, which is always more challenging than face-to-face when judging communication skill. Nonetheless, you will get a good preliminary idea of their speaking skills. Be careful not to be biased by candidates who have less than stellar phone skills. Some candidates who do not have English as their native language may not come across well on the phone. Unless you conclude that this will impact their job performance (communication skills are a key part of any job but more important in specific jobs), try not to let that bias you.

3. Managerial and supervisory skills

If these skills are part of your requirements (and they have been spelled out in the position description), you should ask a few questions to make sure that the level of management experience they have had is what you are looking for. It is easy to embellish such experience in a résumé, but a few thoughtful questions may be all it takes to help you rank candidates. ("Tell me the specifics about your supervisory experience.") We discuss managerial skills in depth later in this chapter.

4. Their preparation

Item 4 in the scoresheet relates to the questions they have for you and whether they did their homework by researching your lab, group, or organization. We believe that candidates who are serious about a position ask good questions during phone interviews as well as during face-to-face interviews.

5. Red flags

Red flags are anything that came up in the conversation that gave you pause—a reticence to explain some facet of their background, a comment that seemed inappropriate (e.g., criticizing current or former supervisors). This does not get a score (the score box is dark gray), but space is available to record comments for future reference.

6. Position-specific questions

Item 6 gives you space to record anything you want to remember about their having the requisite position-specific skills in which you are particularly interested. Make sure you verify their understanding and acceptance of any position-specific requirements that the job may have—for example, "This position requires you to fly by helicopter to the Yukon to help us collect sedimentary shale samples. Will that be a problem for you?" Although we do not score these responses, you do need to take note of them as they can easily disqualify a candidate.

Scoresheet 2: Candidate Phone Interview Scoresheet

Candidate name		
Item	**Parameter**	**Score (0–5)**

1	Technical skills and background
	  List specific skills if you need to rate them separately.
	  Skill 1
	  Skill 2, etc.
	  Appropriate and relevant background	

2	Communication (articulate and clear phone skills)	
3	Managerial or supervisory skills (if relevant)	

4	Their preparation
	  Did their homework about us?
	  Quality of questions they asked	

5	Red flags?	
6	Position-specific questions (schedule, physical capabilities, travel, etc.)	
Total		(0–20)

Guide to Scoresheet 2. Candidate Phone Interview Guide

Candidate name		Contributes to score of item # in the Phone Interview Scoresheet
	Question	
1	"Why are you interested in this position?"	1,4
2	Tell them about the position (include position-specific qualifications like technical skills, travel, lifting, etc.).	
3	"Tell me about yourself. Talk me through your background. Why are you a good fit for this position?" (including technical skills)	1,2,3,4
4	"Talk me through the work transitions you have made and why you made each one." (especially important for candidates with multiple short-term positions on their résumé)	5
5	"Tell me about your management or supervisory experience." (if relevant)	3
6	"What do you know about us?"	4
7	"What questions do you have for me?"	4
8	"As you read the position description and heard me describe the position, what aspects of the role do you think will be the most challenging?"	4,5

In the Guide to Scoresheet 2 we have created a series of questions that can walk you through the interview process. The answer to each question in the Guide contributes to the score you assign to one or more of the entries in the Scoresheet. The right-hand-most column in the Guide shows which questions in the Scoresheet are addressed by each question in the Guide. Because it is easy to have the Scoresheet and Guide in front of you during a phone interview, you can take notes on the Guide and then record your scores in the Scoresheet immediately after the phone call.

No-nos. Questions not to ask

Now that we are starting to talk to candidates on the phone and, in the next phase, face-to-face, there are certain topics that you need to be careful to either avoid or treat with care. The following list shows some questions you should never ask.

Questions not to ask the candidate
• Are you married?
• Do you have children?
• What's your health like?
• How old are you?
• Can you work weekends?
• What do you do for child care?
• What is your political affiliation?
• Do you plan to get pregnant?

The reasons for not asking these questions should be obvious—it is unfair and, in some cases, illegal to make hiring decisions on the basis of gender, physical disability, age, race, religion, or marital status. Your hiring decision must be based on an objective assessment of the ability of the individual to perform the job, and these assessments must be applied in a uniform manner to all candidates. So, for example, you could legitimately say, "This position requires that you be in the lab a full 40-hour work week, sometimes more, and may require you to work late sometimes or come in on a weekend," or "The nature of your position requires you to be here to set things up before others arrive—so you will need to be here by 8:00 AM every day." You must make these statements to all applicants for the same position. Before you include job responsibilities that require long or unusual hours, consider that such requirements may selectively disadvantage applicants, especially women, who have family or other personal responsibilities.[1] If your requirements are legitimate, ask yourself what kind of accommodations (flexible hours, availability of backup support when unexpected family circumstances arise) you could provide to make the position accessible to the broadest array of applicants. Similarly, for requirements relating to physical capability, you could ask, "This position requires you to be able to lift or move heavy objects like CO_2 gas tanks. Are you able to fulfill this requirement?" if this is legitimately part of the job. However, be aware that individuals with

[1]National Academies of Sciences, Engineering, and Medicine. 2018. *Sexual harassment of women: Climate, culture, and consequences in academic sciences, engineering, and medicine.* National Academies Press, Washington, DC.

physical handicaps are protected against workplace discrimination by the Americans with Disabilities Act which requires that "reasonable accommodation" be made to enable individuals with physical or other disabilities to perform a job. As we noted on page 68, we strongly advise that you consult with someone in your organization's HR office for further guidance on this matter.

Employers sometimes ask candidates for their past salary or salary history. In some states and cities this has been made illegal (check with your HR department to see if this applies to your institution) because it can selectively disadvantage certain groups of employees (women and minorities in particular) who are historically underpaid. (See the U.S. Department of Labor Report "Women's Earnings and the Wage Gap."[2]) If you want to make sure the candidate's expectations match your proposed salary range, you can ask instead what their salary expectations are.

The face-to-face interview

The face-to-face interview is the time when most hiring managers or Principal Investigators think that they finally get to put their special insights into the human psyche to work in choosing the best candidate. And as we discussed in the introduction to this chapter most of them are wrong in this conclusion. As stressed by Adler (2007), you need to reframe your thinking so that you are using the interview *to gather data, not to make a decision.* Because being face-to-face with someone summons up our biases, stereotypes, and preconceived notions, it is dangerous to form a conclusion in that moment. Discipline yourself to withhold judgment and decision-making until you have all the data. We have created a form-based interview script that we believe enables the interviewer to evaluate all candidates in the same way rather than changing interview styles from one candidate to another.

Because the face-to-face interview is typically coupled with a visit by the candidate to the lab or organization, we have put together a typical schedule for a lab/group visit below.

Typical candidate visit schedule

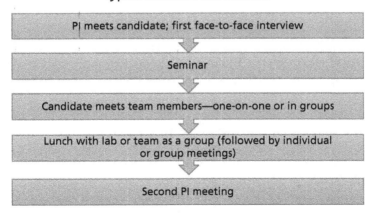

The schedule starts with the PI or hiring manager orienting the candidate at the beginning of the day, followed by a seminar given by the candidate, one-on-one or small-group

[2] Available at https://www.dol.gov/wb/resources/Womens_Earnings_and_the_Wage_Gap_17.pdf.

meetings and lunch with team members, and ends with a follow-up meeting with the PI or hiring manager. The first meeting with the PI is the time when the face-to-face interview takes place (below). The seminar and meetings with team members enable you and your team to see the candidate in action and for the team members to gather the data they need to make their evaluation. The final meeting with the PI allows you to ask the candidate follow-up questions and, more importantly, to field any questions the candidate may have developed over the course of the day. The quality, depth, and thoughtfulness of these latter questions are an invaluable gauge of how much attention the candidate paid while meeting your team and how engaged they were. Below is Scoresheet 3. The Face-to-Face Interview Scoresheet.

Scoresheet 3: Face-to-Face Interview Scoresheet

Candidate name		
Item	**Parameter**	**Score (0–10)**

1	Background and skills verification
	   A. Technical skills and know-how
	   B. Relevant background and experience
	   C. Speaking and language skills during interview	

2	Their preparation
	   A. Interest in and knowledge about lab or project
	   B. Quality of their questions	

3	Seminar
	   A. Clarity and organization
	   B. Speaking skills
	   C. Thinking on feet
	   D. Responsiveness to questions
	   E. Relevance/novelty of topic	

4	Personal attributes
	   A. Feedback
	   B. Manages emotions
	   C. Team mindset
	   D. Problem solving
	   E. Conflict management
	   F. Independent thinking/autonomy	

5	Managerial skills	
6	"Fit" with the group	
7	Strongest characteristic	
8	Red flags	
9	Position-specific questions verification	
Total		(0–60)

Conducting the face-to-face interview

Before we begin, let us address a concern that you may have regarding using this form-based interview process. You may wonder, "How am I supposed to remember all this material while I am sitting with a candidate in my office? Is not looking at this interview form going to feel awkward?" The answer is that you do not need to remember the questions and that you can refer to the interview guide and scoresheet during the interview. Yes, that sounds awkward, but it is OK.

One of the most common reasons that we fail to get the information we need during an interview is that we act as though the interview is a casual conversation. Like those CEOs in Adam Grant's newspaper column, we go into an interview thinking that we will just have a casual conversation and at the end our instincts will take over and determine our decision. We already debunked that myth earlier in the chapter. The alternative is to treat the interview for what it is, an exercise in data collection. That means it is OK and even necessary to refer to the interview guide and take notes. If you feel awkward about it (and you probably will) you should be up-front with the candidate, "I just want you to know that I am going to be referring to some papers in front of me during this conversation so I can make sure I learn as much as possible from our time together." During the interview make it a point to look at the candidate as much as possible, especially when they are talking. Nod and show that you are interested and listening. Once they have answered a question you can say something like, "Great, give me a second to jot down some notes so I can refer back to what you said." And then move on to the next topic.

When we refer to having the interview scoresheet in front of you during the interview we mean this literally—a physical piece of paper. Do not use a computer, laptop, or iPad for this purpose. We all know how alienating it feels to be separated from a person sitting across from us by the barrier of a screen.

Finally, after using the Face-to-Face Interview Guide a couple of times you will almost certainly start to remember the questions by heart and have no need to read from it.

Scoresheet 3 has six entries that get numerical scores and we have weighted each of these entries greater than on the previous forms. Each entry gets a maximum score of 10 points on Scoresheet 3 compared to four or five points on Scoresheets 1 and 2. This reflects the greater import we attach to the face-to-face interview. The impact of the weighting will show up in the total score for each candidate that we will calculate at the end by adding all the scores from each of the scoresheets. At the end of this chapter we will discuss how you can adjust the weights of different questions to reflect your own priorities.

Scoresheet 3 is used in conjunction with the "Guide to Scoresheet 3" (reproduced and discussed below). The guide walks you through the interview process, gives specific questions you can ask for each entry in the scoresheet, suggests the kinds of answers you should listen for, and provides space for notes and scoring. The scores for each entry in the guide get recorded on Scoresheet 3.

Note that although there are multiple elements to some of the categories ("Preparation," "Seminar," "Personal Attributes," etc.), each category gets only one numerical score. It is up to you how you create this score. Some people like to score each element in each category and then after the interview compute the average combined score for the category. Below we go through each numerical entry on Scoresheet 3.

1. Background and skills verification

During the face-to-face interview you will get a better sense of whether the candidate has the qualifications you think they had from your résumé review and phone interview. A brief but open-ended conversation about their and your science should suffice. You will get further data on this from their seminar and from debriefing your team. This same item also includes their communication skills during the interview, which you will also be able to assess during their seminar and from your team. Our goal is to have overlapping or redundant inputs for most of the parameters you score.

2. Their preparation

This entry records your evaluation of how well they prepared for the interview. How much do they know about your lab, department or organization? Have they read your publications? Do they ask thoughtful questions? Their score on this item is a measure of how interested they are in this position. It is also a measure of their due diligence—how much they think ahead, plan for important events (a job interview should be high up on the scale of importance), and take an interest in the people they are going to be meeting.

3. Seminar

This is a straightforward assessment of their seminar: How well was it organized and delivered, how responsive were they to questions, and how adept were they at "thinking on their feet" when responding to questions. You will also be getting feedback from your team on this.

With the possible exception of administrative personnel and technical assistants, most of your candidates will give a seminar highlighting their past work. Judging a candidate by their seminar probably comes as second nature to you and is possibly the easiest part of your selection process. We have broken down the seminar evaluation into five subtopics.

A. Clarity and organization

We have all heard seminars that we eagerly anticipate because the topic is compelling and then walk away disappointed because the data, while interesting, was presented in a way that seemed deliberately designed to confuse. Although muddled organization is not a death sentence and can in most cases be remedied by coaching and practice, it may also be a sign of muddled thinking, especially if hints of this show up in other parts of the interview.

B. Speaking skills

By this we literally mean speaking—are their words understandable and is their speaking voice loud and clear? Also, are they looking at the audience to see if they are being followed or if there are questions? Very junior scientists will often not be great speakers, and these are skills that can be learned by coaching and practice. But if lack of these skills shows up in other parts of the interview, it might signal a communication issue.

C. Ability to think on their feet

How did the candidate respond to questions during or after the seminar? If a question was asked during the seminar were they able to pivot briefly to answer it and then get back on track? If a challenging question was asked were they able to work through the issue on their feet or did they draw a blank? If you are looking for an agile thinker, these will be important questions. Remember, though, many deep thinkers need time and space to process before coming up with a brilliant answer and will not shine under this kind of pressure.

D. Responsiveness to questions

Did the candidate actually answer the questions being asked or did they deflect or resist an alternate viewpoint? Be alert for candidates who seem dismissive of certain questions or who rush to answer without fully understanding what is being asked.

E. Relevance/novelty of topic

We see this parameter contributing bonus points rather than being a deal-maker or -breaker. For postdocs and graduate students, the novelty of their project is often dependent on their adviser's agenda. Relevance of the seminar topic should not be overemphasized unless you are looking for a very specific knowledge and experience base in the candidate.

4. Personal attributes

Item 4 is the heart of the face-to-face interview—what we refer to as personal attributes—so we have given this its own section.

Let us start from the end. What are the main reasons that most new hires do not work out? If you guessed that it is because of a lack of or deficiency in technical skills, you would be wrong. In our experience when people do not work out in the world of science it is because of an issue with one or more personal traits. These traits include the ability to hear and use useful feedback, the ability to manage their emotions, and the ability to productively deal with conflict. Others have reached this same conclusion for new hires in general, not just in science and technology (Murphy 2012).

So, if your interview and hiring process is based exclusively or even in large part on technical skills, chances are that you are going to be making poorer hiring decisions than you otherwise could. Because judging a candidate's technical skills seems rather straightforward (listen to their seminar, ask questions, look at their publications, and ask for their references), it makes sense that most of us would focus on that. Assessing their personal attributes is another story.

The Guide to Scoresheet 3 (below) gives you specific questions to ask the candidate that will allow you assess the personal characteristics we are interested in. The basis of the approach is what has been called "situational'" or "behavior-based" interviewing, which we expand upon below. The guide also gives suggestions on what kinds of answers to listen for. In the following, we review each of these attributes, why they are important, and how you can assess them.

We have chosen six personal attributes that we believe are important attributes of successful scientists working either independently or as part of a team. You may decide that you want to screen for only two or three of these personal attributes, or you may have others that you want to add or substitute for ours. That is fine as long as you use the same attributes for every candidate for a given position.

Remember that this entire category of six attributes gets one score between 0 and 10. You can score each attribute separately in the guide during or after the interview and then average the scores to get one combined score that gets entered on the scoresheet. Alternatively, you can listen to the responses to the questions for each attribute and then assign a group score at the end reflecting your best judgment.

Something to keep in mind when asking these questions during the face-to-face interview is that the candidate may not have a ready answer for every question. Some of the questions in the guide ask the candidate to think of a situation or circumstance from the past. It may be that the candidate has not been in such a situation or that their mind has gone blank (interviews can be stressful!). If that happens just move on to the next question. They may think of an answer later or you may decide to skip that question entirely.

A. Ability to hear and use feedback

Let us start with the first entry on the list—"Feedback." How a candidate hears and responds to feedback is one of the most important traits of success in almost any field, and in science in particular. Especially with younger hires (e.g., junior postdocs), you are going to (or should) be spending a fair amount of time guiding them on their performance and direction. The utility of your guidance and feedback will be determined by their ability to hear, accept, and act on your input. We have all met people who when given guidance of any type immediately adopt a defensive posture and try to justify why their way is best or attempt to absolve themselves of any responsibility for doing what they did. Sometimes this is a natural reaction to inept feedback ("That was a dumb way to set up the experiment. Do it this way instead." Who would not get defensive at that?) But other times the feedback is delivered in a thoughtful and considerate manner ("I'd like to give you some feedback on that last experiment. Let us look at how you set it up.... .") and you are met with a defensive attitude ("Well, it worked OK last time. That's the way we have always done it.").

How would you determine how a candidate hears and responds to feedback? Would you ask them, "Tell me how you respond to feedback?" I doubt it. (You: "Are you open to feedback?" Them: "Sure!" What have you learned? Nothing). Instead of asking a candidate how they respond to feedback, ask them instead what they did in situations where they received feedback in the past. How about this: "Can you think of a time when someone gave you feedback on something you did that was uncomfortable to hear? Tell me about that. What was the situation? What did you do?" And then listen for evidence that they can hear feedback that may be uncomfortable, they can manage the feelings associated with that, and that they can use the feedback to improve their performance without getting defensive.

It is important to note that we are asking what they did in the past, not what they would do in the future. We believe that past behavior and performance (assuming the candidate is reasonably honest in their answers) is a better predictor of future behavior than a candidate's assertion of what they *might* do in the future.

The approach we have just outlined, situational interviewing, is the one we will use for each of the six personal attributes we have listed on the scoresheet. The Guide to Scoresheet 3 walks you through the interview and has tips on what to listen for when you ask each of the questions. Have this guide in front of you during the interview. After using it a few times you will likely remember it.

For the record, we have been asked, "Can't these questions be gamed? What if the candidate read this book and they frame their answers to make it look like they are the ideal candidate?" To this we answer, "If they read this book and they went to the trouble of learning how they ought to respond to feedback, manage conflict, and all the other attributes we discuss, hire them right away."

Note that in many of the Personal Attributes questions below, we have included two categories of information gathering. "Past" refers to questions that you pose about past behavior or performance during the face-to-face interview process. "Present" refers to information you gather by watching how the candidate reacts to events during their day visiting your lab. So, in the section of how the candidate responds to and uses feedback, under "Past" you ask how they responded or reacted to some situation in the past when they got feedback that was hard for them to hear. For the "Present" you would be watching in real time how the candidate reacts or responds to feedback or questions during their seminar or during the interview process. In the former case you are getting information from a story they are telling you, whereas in the latter you are getting information by watching their behavior.

Guide to Scoresheet 3: Face-to-Face Interview Guide

Candidate name	
Checklist item 1	Background and skills verification Briefly review the organization, the responsibilities, the department, and the requirements for the position. Ask them again why they are a good fit. Listen for appropriate background and technical skills. A and B. Review their answers to phone interview questions and probe or clarify areas where you have open questions (background and technical skills). Discuss the technical aspects of the position and pose questions whose answers require the technical knowledge that you need. Again, make sure you are asking every candidate the same questions. ("We're having a problem in the lab with creating stable cell lines with the *XYZ* gene knocked out; is that something you could help us with?"). C. How well do they communicate and relate during the interview? Score item 1 on checklist.

Checklist item 2	Their preparation
	  A. Interest in and knowledge about lab or project
	    PI Interview "Tell me about your interest in our lab/group/ department."
	  B. Quality of their questions
	    PI interview. "What questions do you have for me?'
	Score item 2 on checklist

Checklist item 3	Seminar During or after their seminar score the elements listed in the Scoresheet and enter either an average of those elements or one global score in line 3.
Checklist item 4	Personal attributes

	A. Ability to hear and use feedback
	
	  1. Past: "Tell me about a time when you were told that what you were doing was not correct or needed to be done differently? What was the situation? How did you handle it?"
	    Listen for openness to feedback, desire for feedback or seeking feedback, lack of defensiveness.
	
	  2. Present: Watch how the candidate responds to questions and critiques during their seminar or interview. Are they open to hearing critiques? ("Yes, that's definitely an alternative explanation. Thanks for the suggestion.") Or are they defensive ("No, we thought of that. It is not likely. Let me keep going... .").
	
	Score for A

	B. Ability to manage their emotions
	
	  1. Past: 'Tell me about a time when you got angry or upset about something in the lab. What happened, and what did you do?"
	    Listen for: Ability to manage feelings of anger, frustration, anxiety, and loss of control without "acting out," losing temper, blaming or "shutting down" (withdrawing). Listen for ability to self-reflect ("I thought a lot about why that happened and learned a lesson.").
	
	  2. Present: Watch for candidate's reactions to interview schedule changes, delays, unexpected occurrences, etc.
	
	Score for B

(*Continued*)

Candidate name	
	C. Team mindset 1. Past: "Tell me about a time when you worked as part of a team on a project. Was any part of that difficult for you? Explain that to me." And "Do you prefer to work as part of a team or on your own project? Why is that?" Listen for whether the candidate expresses a preference for working independently or as part of a team. Your evaluation will depend on the role for this person. 2. Present: Listen for the candidate's use of "I" versus "we" when describing group or team projects. Score for C
	D. Problem solving—accessing help 1. Past: "Tell me about a time when you were stuck on a difficult problem." Let them tell a story. "How did you resolve it?" "Tell me about a time when you made a mistake. What happened?" "Have you ever felt that you were at a dead end in a project? How did you handle that?" Listen for whether they asked for or sought out help or just persisted in trying to solve it themselves. 2. Present: "Do you know anything about analyzing hemi-haplotype-associated retrotransposon patterning? What would you do if you needed to do that in a project?" (Substitute an actual technical challenge relevant to your work for this nonsense one.) Listen for whether they ask for help or say that they would figure it out themselves. Score for D
	E. Conflict management 1. Past: "Can you think of a time when you had a serious disagreement or conflict with someone in the lab? Tell me about it. How did you handle it? Whose fault was it?" "Have you ever disagreed with a decision that your supervisor or organization made? Tell me about that time and how you managed your disagreement." Listen for ability to negotiate solutions, ability to see the other person's perspective, ability to manage a conflict in order to get work done. Be alert to a tendency to blame others, attribute fault, bear grudges, become passive in the face of conflict and avoiding conflictual situations that need to be addressed. If their answer to "Whose fault was it?" is to quickly assign blame to others without taking any responsibility themselves, this could be a red flag. Listen for an ability to accept some responsibility for their own role in conflicts. 2. Present: Be alert to any minor conflicts that may arise, especially during the seminar question period. Watch how the candidate fields challenging questions or criticisms. Score for E
	F. Independent thinking/autonomy 1. Past: "What do you when you get to a point in a project where you need to make a decision about direction or strategy? Can you think of an example?" "Can you think of a project where you came up with the focus, strategy, or approach?" Listen for an ability to arrive at independent conclusions and a comfort level with taking a risk through independent thinking.

(Continued)

Candidate name	
	2. Present: At your second face-to-face candidate meeting at the end of the day, "Now that you have had a chance to meet with the lab members and hear what we're working on, do you have any ideas for us?" Score for F
Checklist item 5	Managerial Not every position will have significant managerial responsibilities. For those that do, there is a separate section later in this chapter. If the position does not require significant managerial or supervisory skills but you still want to get an idea of how this candidate thinks about managing others you can ask a couple of open-ended questions such as the following. 1. Past: "How would you describe your management philosophy or approach?" This is a softball question. Once they answer it, then ask "Can you give me an example of a management challenge that you faced with an employee and how you handled that?" "Tell me about a time when you had to discipline someone or when you had an employee who was not performing to your expectations. What did you do? How did it turn out?" Listen for a comfort level with giving people performance feedback, not ignoring the problem. 2. Present: "What if we had a person who was a management challenge—they had a hard time meeting deadlines and did not communicate well. How would you handle them?"
Checklist item 6	"Fit" with the group. Several dimensions. Technical fit; personal/interpersonal fit —can you imagine working with this person? Do you foresee challenges?
Checklist item 7	Strongest characteristic. What stands out as the greatest strength of this applicant? No score, just comments.
Checklist item 8	Red flags. Did anything the candidate do or say stand out as concerning, inappropriate or worrisome? Gaps in their technical knowledge? Insensitive comments? Ignoring certain people in the group? No score, just comments.
Checklist item 9	Position-specific questions verification. Are there specific skills (technical, managerial, interpersonal, etc.) that the candidate needs that were not already covered? No score, just comments.

B. Ability to manage emotions

Working in a lab with others or as part of a group can be anything from exhilarating to infuriating. Science is frustrating enough by itself. Adding other people into the mix is sometimes too much for some people to take. Because frustrations with work and with others are inevitable, the ability to manage these frustrations while maintaining collegial working relationships is a key factor in being successful at what you do. Some scientists feel that it is their moral duty to express their frustration with others in the service of being "honest." More often than not being "honest" translates into unloading unfiltered emotional vitriol onto an unsuspecting recipient. We advise scientists that their job is to get their work done, not to give others the benefit of knowing what they happen to think of them as human beings in the bluntest possible way. As we show in Chapter 6, you can and should critique the work without critiquing the personality of the person who did the work. Question B in the Interview

Guide shows one way in which you can assess the candidate's ability to manage strong emotions.

C. Team mindset

We imagine that many group leaders have the fantasy that their groups will work as a harmonious team, selflessly helping one another, troubleshooting for one another, sharing equipment and reagents, and going on group outings on weekends. The reality is often that people are working at their benches with earbuds in their ears to drown out the din of the lab, may hoard scarce supplies, and deliberately come in at odd hours to have the luxury of not having to interact with or be bothered by others. As we note in Chapter 11, many labs are run as collections of individuals working on their own projects and interacting with others in the lab only to coordinate use of shared resources (space, equipment, lab assistants). But even in these labs there are often certain complex projects that benefit from or require two or more lab members to collaborate. A scientist's ability to shift gears and think in a collaborative way will often be the determining factor in the success of such projects. If your lab relies heavily on such collaboration, pay attention to the questions we have formulated for this attribute.

D. Problem solving and accessing help

We have all seen the symptoms of the scientist who for some reason or another refuses to seek help with their stuck project. They are working extra hours, they are muttering to themselves at the lab bench, being uncharacteristically quiet at lab meeting, and in a foul mood for days or weeks on end. It can be admirable when people show determination and focus and try to figure things out for themselves, and we often encourage scientists to do so to foster independent thinking. But there comes a time when good judgment dictates that you need to ask for help. For us, the ability to ask for help is one of the most valued traits of a productive scientist. They understand that sometimes it is more important to seek help to get the project going again than to wait (possibly forever) to figure it out themselves. Of course, there are external factors that come into play here as well. If you run a lab in which asking for help is treated as a sign of weakness or is met with ridicule by you or others in the lab, you should not be surprised if people are reluctant to admit that they are stuck. The questions for this attribute help you get at this quality.

E. Conflict management

Conflict arises when two or more people have different goals, needs, or desires that interfere or seem to interfere with each other. Conflict per se is neither good nor bad, but rather an inevitable part of human existence. What is important about conflict is not that it exists but how you manage its inevitable occurrence. Those of us who are conflict avoidant will ignore uncomfortable disagreements, leaving them to fester. Those of us who impulsively charge headlong and vociferously into an argument every time a conflict arises may wonder why our working relationships and friendships seem to be deteriorating. Those who are adept at managing their own behaviors in difficult situations are able to navigate conflicts using the tools of negotiation (Chapter 3) and get their work done. These questions will help you identify those candidates.

F. Independent thinking/autonomy

In our view the best and most productive lab directors or supervisors work hard at providing opportunities for their team members to be creative, independent, and eventually autonomous. Of course, this is a two-way street because the scientists themselves must be capable and desirous of thinking independently. We have seen many excellent scientists who are superb at the bench and highly productive but who require frequent guidance and direction on what to do next. If you run your lab as a top-down enterprise where you decide what everyone is going to do, such a person may be perfect for you. If you would rather have someone who is responsive to direction and guidance and also thinks independently and comes up with ideas and directions that would never have occurred to you, pay attention to how they respond to these questions in the Interview Guide.

5. Managerial skills

If the thought of selecting and hiring the people who will have the power to make or break your research project keeps you up at night, then the prospect of selecting and hiring the people who will supervise some of them may send you to the sleep clinic. Hiring someone with poor managerial skills can have a deep and long-lasting negative effect on your work because they impact others in your team or the team as a whole. Conversely, hiring a savvy manager can lift everyone's sprits and productivity.

    In this section we focus on hiring scientists who will have responsibility for managing one or a small group of other scientists and/or technical staff. If you are hiring someone with broader management responsibilities, including managing those who are themselves managers, setting strategic direction, and playing a leadership role in the organization, use the section later in this chapter entitled "Hiring Managers and Faculty" and use Scoresheet 6 (Management Candidate Interview Scoresheet) in that section to generate the score for this entry. Also, read Chapters 12 and 13, which focus specifically on identifying and selecting leadership candidates.

    If you are hiring a postdoc who will have supervisory responsibilities, you will want to know something about their supervisory experience. Very often a candidate for a postdoctoral or staff scientist position will not have had any managerial or supervisory experience. In those cases, you will rely on the Personal Characteristics score as a guide to how they might manage others. Look for candidates with an ability to manage their own emotions, who deal productively with conflict, and who are eager to hear and learn from feedback. For candidates who have had some managerial or supervisory experience, have them describe how they dealt with those they managed—specifically, how they set direction and goals and how they delivered feedback and guidance. Refer to the section below "Hiring managers and faculty" for more specific guidance and examples of questions to ask. Assign a score between 0 and 10 to the "Managerial" section of the scoresheet for candidates who will need this skill.

6. "Fit" with the group

Here is where you can use that intuition that you have been dying to exercise. There are many dimensions in which "fit" can be measured. You may need to bring in someone with a particular skill set that no one else in the group has. You may be looking for a senior postdoc who can serve as a resource and adviser to more junior people. You may be looking for a good "social fit"—a candidate outgoing and social enough to fit into the tight-knit group of extroverts in your lab. Conversely, you may have a lab of introverts and think that adding an extrovert will bring some social interaction in the lab. Frankly, we are a bit leery of this type of social engineering via hiring. Although you may be focusing on one prominent characteristic of the candidate you noticed during the interview, their day-to-day behavior and character may be very different. Besides, transplanting that bubbly extrovert into your impenetrable clan of introverts may depress the extrovert rather than transforming the introverts.

7. Strongest characteristics

This and the next two entries do not get scores but are there to make sure you think carefully about each candidate and record your observations and conclusions right after the interview. If you wait too long your memory will be diminished by time, distraction and intervening matters.

8. Red flags

Was there anything that you heard or saw during the interview or visit that gave you pause or concern? Here we really do recommend that you pay attention to your instincts and "gut feelings." We often meet with someone and come away with a feeling that they are just not right. Perhaps it was that they seemed distracted during the interview (looking at their phone!) or that they made a disrespectful comment about a colleague. Red flags could also have cropped up in any of the checklist entries above—a disjointed seminar presentation, lack of familiarity with a field they claimed to be an expert in, or some personal attribute that seems too glaring to ignore (say, a history of falling out with colleagues). None of these by themselves might be reason to rule out the candidate, but in conjunction with the rest of the data you collect they may be sufficient to tip the scale if they end up in a tie with another candidate. Although this section may seem to fly in the face of our admonishment not to go with gut feelings because they are so prone to being influenced by unconscious bias, we see no great harm in including these red flags as notes. We suggest that before you add red flags ask yourself explicitly what the basis of your flag is. Insist that you explain it to yourself and do not settle for "I've just got a feeling." Just because a candidate's looks remind you of your terribly annoying brother-in-law is not a good reason to flag them. This section does not get a numerical score.

9. Position-specific skills verification

Here is the chance for you to verify one more time that they have the requisite skills, capabilities, and willingness to make that helicopter trip to the Yukon to collect samples with you. Again, no score here, just make sure you cover what the position needs.

Getting input from your team

Bringing a new person into your group is a bit like adopting a new member in your family. The people in your group will be working with and often in close proximity to the new hire. Whether and how you involve your team in making the selection of a new member can go a long way to determining how readily, if at all, a new member is accepted and integrated into the group. In the following, we recommend asking each group member (or a relevant subset who will be working closely with the new hire, if it is a large group) to do individual evaluations of each candidate that can be averaged into a combined score for comparison between candidates. We recommend this route instead of having the group evaluate a candidate together because the latter process can be unduly influenced by one lab member with a strong opinion or loud voice.

Below is Scoresheet 4, Team Member's Scoresheet, which should be completed for each candidate by each relevant lab member. Importantly, before you ask team members to score candidates in this way you must meet with them as a group to go over the scoring process and to agree on how you will score each entry. Discuss what a score of 0, 5, and 10 means for each of the four major categories in the Scoresheet.

The highest score a candidate can achieve on this scoresheet is 40 points (compared to 60 points on Scoresheet 3). Once you have the candidate scoresheets from your team members, total the scores and compute the average. The average team member score for the candidate gets recorded on line 5 of Scoresheet 7, the Candidate Summary Scoresheet.

Scoresheet 4: Team Member's Scoresheet

Candidate name		
	Parameter	**Score (0–10)**
1	Skills Technical skills and know-how Relevant background and experience Speaking and language skills	
2	Preparation Interest in lab and/or project Quality of their questions Knowledge about the lab/group/department, or organization	
3	Seminar A. Clarity and organization B. Speaking skills C. Thinking on feet D. Responsiveness to questions E. Relevance/novelty of topic	
4	"Fit" with the group	
5	**Strongest characteristic**	
6	Red flags	
Total		(0–40)

Second PI meeting

If you refer to the schedule of the candidate visit (p. 76), you will note that there is a second meeting with the PI at the end of the day. There are several things you want to accomplish during this time. The first is to see whether the candidate has any questions, comments, or observations about what they have heard during the day. Although the bulk of the day will have been spent with you and your team questioning and listening to the candidate, the candidate will also have heard a lot about your group and the work it is doing. Asking the candidate what they learned and whether they have any questions or ideas for you and your team is a great way to determine whether they were paying attention and whether they are thinking on their feet about your lab and its work. Second, this is your opportunity to sell the candidate on you and your lab. As we noted earlier, the initial script-based interview may feel awkward to you and the candidate, despite your assurance that this is simply your way of gathering the data you need to make the best hiring decision. The second interview is the time to turn on the charm and show the candidate what a warm and empathic leader you are and that you are thrilled that they are interested in joining your group (assuming, of course, that is the message you want to send). Coming at the end of the visit, this second meeting with you is the one they will remember.

Although the focus of this chapter is on choosing the best candidate, the candidate is also making a choice. If you are just starting a lab in an academic setting or are at an institution that is not in the top 10 best places in the world to do science, you may feel that you need to convince applicants that taking this job will be in their best interest. Of highest importance will be the nature and quality of the science you do. This will hopefully speak for itself, but there is nothing wrong with talking up the most exciting aspects of your science. Although being in a lab doing exciting science is important, an applicant's choice of lab is not only about science. They will also be thinking about how working in your group will prepare them more generally for their career or next position. Here are a few things you can offer job applicants that might make your job stand out from others and that echo the latest recommendations for best practices in graduate and postgraduate training from the National Academies of Sciences, Engineering, and Medicine (Leshner and Scherer 2018).

- Tell applicants that you take your job as mentor seriously and that they will get the kind of personalized attention they might not get in a larger lab.

- Tell applicants that you will do your utmost to prepare them for whatever career they choose —regardless of whether it's in an academic, industry, or other setting.

- Tell applicants that everyone in your group creates an Individual Development Plan (see Chapter 6) that the two of you will use to set their career and learning objectives and to track their progress toward them.

- Show applicants the adviser/advisee compact (see Chapter 6) you use that documents your commitment to them as well as theirs to you.

- Explain to candidates that you provide consistent feedback on their performance as well as routine performance reviews (see Chapter 6) so that they will never be in doubt about how they can improve their science and performance.

- Tell applicants that part of your career preparation process will be to give them opportunities to learn how to write scientific papers and grants as well as how to critique them if that is relevant to the career path they will choose.

If you do all of these things for your group members and let applicants know that, you will find that you can attract higher-quality candidates than you might otherwise.

Phone reference checks

Checking a candidate's references is one of the most important sources of data you have. To our surprise, we hear not infrequently of candidates being hired without anyone speaking with their previous employers. Make this mistake at your peril. Besides interviewing the candidate, speaking with their references is one of the jobs that the PI or hiring manager should never outsource to HR (the exceptions being if you are working with a professional recruiter or a highly skilled HR staff member, who will invariably be much better at interrogating references than you are).

Fortunately, you do not have to reinvent the wheel when speaking with references. You will be looking for information on the same parameters as those in the face-to-face candidate interview except that the questions will be framed differently. Consequently the Reference Phone Interview Scoresheet (Scoresheet 5) is nearly identical to Scoresheet 3 for the face-to-face interview. The difference is that you will be asking the reference about the candidate rather than asking the candidate about themselves.

Scoresheet 5. Reference Phone Interview Scoresheet

Candidate name		
Reference name		
	Parameter	Score (0–10)
1	Confirm their position and employment	
2	Confirm position responsibilities and or project goals	

3	Technical skills
	  Skill 1
	  Skill 2
	  Skill 3	

4	Other skills
	  A. Presentation skills
	  B. Writing and language skills
	  C. Work habits (organization, planning, etc.)
	  D. Creativity and idea generation	

5	Personal attributes
	  A. Handle feedback
	  B. Manage emotions
	  C. Team mindset
	  D. Problem solving
	  E. Conflict management
	  F. Independent thinking/autonomy	

(Continued)

Candidate name		
Reference name		
	Parameter	Score (0–10)
6	Managerial skills	
7	Overall performance	
8	Hire again?	
9	Greatest strength?	
10	Areas for improvement?	
11	Is there anything you would want to know if you were hiring this person?	
Total		(0–60)

To guide you through the reference phone interview process, we have prepared a Guide to Scoresheet 5 that lists questions to ask for each Scoresheet entry. It is much easier during the phone reference check than during a face-to-face interview to have these forms in front of you and to record your observations, scores, and notes.

Guide to Scoresheet 5: Reference Phone Interview Guide

Candidate name	
Reference name	
1	Confirm their position and employment. No score, just comments.
2	Confirm position responsibilities and or project goals. No score, just comments.
3	Technical skills "How would you rate X on their lab skills (specify specific skills if appropriate)?"
4	Other skills "How would you rate X on their presentation skills? On their writing and language skills?" "How would you characterize X's works habits? Is X organized in the lab and/or in X's lab notebook? Is X good at planning and scheduling so their work gets done in a timely manner?" "How would you rate X's ability to be creative and/or to generate ideas?"
5	Personal attributes
	A. Ability to hear and use feedback "Can you think of times when you gave X some feedback or critiqued their performance or work? How did they respond to that? Were they able to take responsibility for errors or mistakes?"
	B. Ability to manage emotions "Can you think of a time when X got angry or had any kind of emotional reaction? Tell me about it." If answer is "No," then ask, "How does X react when criticized?" and "How does X react when under stress? Can you give me an example?"

(Continued)

Candidate name	
Reference name	
	C. Team mindset "Did X work as part of a team? Can you describe X's relationship with other team members? Did conflicts or disagreements arise at any point? How did X manage these? Can you give an example?"
	D. Problem solving "Can you think of a time when X was stuck on a tough problem? What did X do? How did X handle it?"
	E. Conflict management "Was X ever involved in a conflict or disagreement with anyone in the lab or group? Can you tell me about how X handled that," "How does X react when someone disagrees with them?"
	F. Independence "Can you give me an example of X's independent thinking?" "Do you consider X a self-starter? Can you give an example?"
6	Managerial skills If the candidate will have little or no supervisory responsibilities skip this section. For those who will, the amount of time you spend on this entry will depend on what fraction of their job will involve managing others. "Did X have supervisory experience?" "How would you characterize X's effectiveness as a manager?" "Can you think of a specific example that illustrates X's ability as a manager or supervisor?" "How would you rate X as a manager or supervisor relative to others who have worked for you?" For candidates expected to take on significant managerial responsibilities you will also want to ask questions exploring some or all of the managerial parameters that are listed in the Management Candidate Scoresheet (Scoresheet 6) and Interview Guide, in the following section.
7	Overall Performance "On a scale of 1–10, how would you rate this candidate's overall performance relative to others who have worked for you in similar capacities?"
8	Hire again? "Knowing this candidate as you do now, if you had the opportunity to hire them again for another position, would you? Rate the likelihood on a scale of 1–10."
9	Greatest strength? "What would you say is this candidate's greatest strength or asset?" No score, just record their comments.
10	Areas for improvement? "Can you think of any areas where this candidate could improve, either scientifically or in other areas such as collegiality for example?" No score, just record their comments.
11	"Is there anything you would want to know if you were hiring this person?" This gives the reference an opportunity to mention anything they may think worth noting and that you have not already asked about. No score, just record their comments.

Reference checking by phone is a more complex and demanding process than it may seem. We know professional recruiters who spend their careers perfecting their skills at eliciting accurate information from references who are at times reticent to convey any but the most anodyne comments on the phone. Sometimes what the reference does not say is more important than what they do say. Listen for telltale comments like the ones listed below.

Listen for "weak" or "telltale" phrases when interviewing references
• "They worked hard." versus "They accomplished a lot."
• "They follow instructions very well." versus "They are a creative and independent thinker."
• "They mostly got along with people." versus "They were able to work productively with everyone on the team."
• Does "They can be quiet at times," mean "They are deeply introverted"?
• Does "They are not shy about expressing his opinions," mean "They are a loud and domineering person"?

Because you will likely be speaking with more than one reference for each candidate use the average of the reference interview scores in the candidate summary scoresheet (Scoresheet 7, below).

Hiring managers and faculty

Select and score candidates using the scoresheets and guides presented above. In addition, you will use a separate set of criteria, shown below in Scoresheet 6, The Management Candidate Scoresheet, which will help you identify those with the best managerial skills. As with the Face-to-Face Interview Scoresheet, the Management Candidate Scoresheet is accompanied by a Guide (Guide to Scoresheet 6. Management Candidate Interview Guide), which walks you through the interview process, shows you how to frame questions that can reveal the candidate's approach to managing others, and gives examples of desirable and less desirable answers to your questions.

We have not included entries for some skills that relate to academic faculty recruiting, most specifically teaching. Evaluating this skill can involve having a candidate conduct a mock-class on an assigned topic. You can customize the scoring process to include such skills, using criteria that are specific to your institution.

The importance you assign to the management skills in the selection process will depend in part on how much of the new hire's time will be spent managing and how many people they are managing. If it's 50% or more of their time and more than one person being managed, you may wish to assign more points to the management section of the scoresheet than we have (see "Customization" at the end of this chapter). For those who will have a lighter management load, the management section of the scoresheet can serve as a tiebreaker to distinguish two candidates who are otherwise equally well qualified.

Your scoring of the following elements can take place on the Guide during the interview or right after the interview based on your notes. Once you have a score for each of the 10 entries on Scoresheet 6, add them up and take an average (divide by 10). That number gets entered into line 5 of Scoresheet 3.

The elements that we are looking for in potential managers are described below.

A. Coaching and mentoring

Good managers are good at coaching and mentoring. As we will see in Chapter 6, giving feedback and doing performance reviews are key skills that enable managers to ensure that scientists are performing up to expectations. Being a good mentor means showing genuine interest in your employee's career path, learning, and professional development and giving them guidance to help them succeed.

B. Negotiation and conflict management

Conflict and managing it well using negotiation skills is so important that we have included it here, even though it is also covered in the previous interview guide for scientists. Here we have framed the questions in the interview guide in the context of management issues—as in mediating disagreements between two direct reports or with a supervisor.

C. Managing performance

Managing performance is one of the most important roles of a manager. Knowing how to set clear goals (see Chapter 6), give useful feedback, and keep people focused are the hallmarks of excellent managers. Asking a few key questions can reveal whether a candidate has the requisite skills or not.

D. Managing up

Managers must navigate the choppy waters between what they can reasonably expect from those they manage and the (sometimes) demanding expectations of those who manage them. The skills required are those of negotiation. Good managers know and support the priorities of their own managers and are adept at balancing those priorities with what can actually be delivered.

E. Managing to a deadline

This skill is critical, especially in the private sector where time is money. Many scientists newly arrived from academia have a hard time adjusting to the importance of timelines and deadlines. Managerial candidates need to be able to articulate how they hold people accountable and that they know how to deliver feedback and guidance to their reports to make sure work is getting done according to timelines.

F. Hiring

Although we do not expect new management candidates to have memorized this chapter, it will be a distinct benefit to you if they have some sense of how to identify and select new hires for their team. It should come as no surprise that our interview guide suggests you look for evidence that they understand the importance of selecting candidates with certain desirable personal traits, even if they cannot enumerate a long list of such traits.

G. Managing uncertainty

Even junior-level managers find themselves in situations where they need to keep their group motivated and working hard in the face of an uncertain outcome or future. In the private sector, fear that a beloved project will be suddenly killed by senior management for no apparent reason is nearly universal. In academia, the rejection of a manuscript or grant application can engender uncertainly and worse in a project team. The ability to keep the team focused and motivated under such circumstances is one of the hallmarks of good managers and leaders and is discussed at greater length in Chapter 5.

H. Managing silos

Silos are domains within an organization housing specific departments or functions (see Chapter 11 for further discussion). Unless managed adeptly, such domains engender clannish thinking, isolationism, scapegoating, and more. Just being aware of this phenomenon and the challenge it can pose to cross-department or intergroup communication is a huge positive in a management candidate. Knowing how to counteract it is an even higher-level benefit.

I. Failure and disappointment

We spend so much time trying to find people who will get things done in the lab that we sometimes forget that for the majority of time nothing gets accomplished. Managers who can keep scientists motivated during times of uncertainty, failure, and dead ends are worth their weight in gold. Even the recognition that scientists need someone in a position of authority to convey optimism, hope, and a path forward in such situations is a strong endorsement for a new management candidate.

J. Disciplining and letting go

Most new managers probably will not have much experience in this category of skills because they are called upon rarely. On the other hand, finding that someone has a good intuitive sense of the progression of steps that might lead to the creation of a performance improvement program (see Chapter 6) and eventual separation of an employee from the organization might push them up a notch on your hiring list.

Scoresheet 6: Management Candidate Interview Scoresheet

Use at least five (or use them all) but do the same with every candidate.

Candidate name		
	Parameter	**Score (0–10)**
A	Coaching and mentoring	
B	Negotiation and conflict management	
C	Managing performance	
D	Managing up	
E	Managing to a deadline	
F	Hiring	
G	Managing uncertainty	
H	Managing silos	
I	Failure and disappointment	
J	Disciplining and letting go	
Total	Divide this score by 10 and enter result into Scoresheet 3 line 5.	(0–100)

Guide to Scoresheet 6: Management Candidate Interview Guide

Candidate name	
1.	Coaching and mentoring "Have you ever had an employee who was not performing up to your requirements? How did you handle that? How did it turn out?" "Have you ever had an employee who was a very high performer? How did you acknowledge that?" Listen for (a "+" means a desirable answer, whereas a "−" signifies a less desirable answer): +: An ability to provide useful performance feedback that helps an employee improve. An awareness that employees need and want feedback. An ability to recognize and encourage an employee's work and to help them advance if appropriate. −: An inability or lack of awareness of how to provide constructive guidance to poor performers. Not being aware of the need to recognize high performers.
2.	Negotiation and mediation "In your management role have you ever had a situation where you had a disagreement with one of your reports? Tell me about that and how it turned out." "Have you had a time when two or more people reporting to you had a disagreement? How did you handle that?" Listen for: +: Recognized the conflict and dealt with it productively. Intervened successfully and helped mediate a solution. Open mind about whose "fault" it was. −: Ignored it, thinking or hoping it would go away by itself; let junior people "work it out" so they can learn.
3.	Managing performance and holding people accountable "How do you go about letting people know what's expected of them?" "Have you had a case where someone was not performing up to your expectations? How did you manage that?" Listen for: +: A clear idea of how to set goals and how to manage to expectations. Frequent meetings with feedback and guidance. For poor performers, seeking the root cause, giving clear advice and feedback about performance before it becomes a big problem. −: "People just know what I think of their work," "We cover that in lab meetings," "They get feedback from other lab members," "I like to give people ultimatums to shake them up." "Around here, it's sink or swim."
4.	Managing up "Have you had occasion to disagree with a decision or request made by someone senior to you? What did you do?" Listen for: +: A willingness to engage authority in collegial discussion about differing views. An ability to "choose your battles." A recognition that sometimes people in authority can call the shots. −: A tendency to denigrate or badmouth authority figures whose decisions they disagree with.
5.	Managing to a deadline "Have you had a situation where you needed to get something done by a deadline and were concerned that those responsible for the work could not or would not be able to make that happen? What did you do?" "Have you had to manage someone who failed to deliver on time on multiple occasions? What did you do?" Listen for:

(Continued)

Candidate name	
	+: An ability to convey priorities to employees and to sense which ones need help in developing strategies to meet deadlines and prioritize work. An ability to coach under-performing employees.
6.	**Hiring** "Have you hired people? Tell me how you go about that and what you look for." "Have you ever made what turned out to be a bad hiring decision? Tell me about that." Listen for: +: Good balance between focusing on technical skills and personal traits. Consulted with team. Spoke with references. −: Exclusive or overemphasis on technical skills at the expense of personal traits. Made the decision solo.
7.	**Managing uncertainty** "Have you been in a situation where you were unable to give your team the information or guidance they wanted or needed because of uncertainty within the organization? How did you manage that?" "Have you ever felt that as a manager you lacked crucial information about a project or program in your organization? How did you manage that?" Listen for: +: Ability to be able to keep team motivated in spite of uncertainty. Absence of resentment or "acting out" against senior management. Absence of blaming.
8.	**Managing silos** "In your previous positions, were there silos or organizational domains separating different parts of the organization? Did this affect your ability to get work done? How did you manage that?" Listen for: +: Ability to build bridges and form alliances with other parts of the organization. −: Any sign of disdain or stereotype focused on parts of the organization.
9.	**Dealing with failure and disappointment** "Have you had a project or program fail? How did your team or group take it? What if anything did you do or say to help them?" Listen for: +: Willingness to confront failure openly and to learn from it; appreciation that employees need closure and support. −: Resistance, fear, or inability to address a failure. Lack of appreciation for how this affects employees.
10.	**Letting people go** "Have you ever had to let someone go? Tell me about that. How did it come about?" Listen for: +: The ability to pick up on poor performance early before it becomes a problem. Willingness to get HR involved in problem cases rather than waiting for performance failure. −: Be alert for a tendency to wait too long before either coaching a poor performer or letting go of someone who is unlikely to improve.

Committee interviewing

When hiring candidates for managerial, supervisory, or faculty positions it is common for a search committee to be responsible for interviewing and making the final selection. If candidates are having one-on-one interviews with each committee member, it is a simple matter to ask each member to fill out their own Scoresheets (Scoresheets 3 and 6) for each candidate. Because of the demands of time and expediency, candidate interviews are sometimes done as a group interview with the entire committee. In these cases, we suggest that each committee member fill out their own Scoresheet for each candidate before having the committee as a whole discuss the candidates. Requiring scoring before group discussion minimizes "groupthink" decision biases and can limit the outsize influence of certain committee members who may be overly influential or vocal.

Whether committee members will be interviewing candidates one-on-one, or as a group, it may be useful to assign specific parts of Scoresheets 3 and 6 to specific members. This is especially useful when interviews take place one-on-one because it eliminates the awkward experience of a candidate being asked, "Tell me about a time when you had a conflict with a colleague…" by five separate interviewers. Dividing the "Personal characteristics" questions on Scoresheet 3 and the "Parameters" on Scoresheet 6 among the interviewers solves the problem.

For committee-based decisions the Scoresheets can be used to rank candidates by simply adding all the members' scores for each phase of the screening process or averaging them.

Candidate summary and scoring

Once you have all the data in Scoresheets 1–6, enter the total score from each one into the appropriate place in Scoresheet 7, the Candidate Summary Scoresheet. The result will be a total score for each candidate with a maximum score of 200. There is nothing magical about 200—it is just the sum of the maximum scores from all of the scoresheets. If you use a different scoring system (e.g., you may decide to assign a greater numerical score to specific personal attributes or technical skills that you consider especially important for a position; see below for more on this), you will end up with a different maximum score. Use the same scoring system for every candidate for a given position. The scoresheet also has space for you to record the candidate's notable strengths and any red flags that you or your team might have noted. Scoresheet 7 is the document that you will use to compare candidates and make a final selection.

Scoresheet 7: Candidate Summary Scoresheet

Candidate name				
Scoresheet			Range	Score
1	Résumé		0–20	
2	Phone interview		0–20	
3, 4	Face-to-face interview		0–60	
5	Team members interview (average of all members)		0–40	
6	Reference phone interview (average of all references)		0–60	
Grand Total			0–200	
Strengths				
Red flags				

Customization

We encourage you to customize the process for your own use. You may wish to assign different weights or numerical values than the ones we assigned to some of the interview questions or categories. We have made it easy to do this by providing access to editable Word and Excel versions of each of the forms at www.sciencema.com/downloads.

Here we discuss two possible customizations. The first involves increasing the maximum possible score of the "Personal Attributes" sections in the Face-to-Face Interview Scoresheet and in the Reference Phone Interview Scoresheet (Scoresheets 3 and 5, respectively). In the scoring system shown in this chapter, the Personal Attributes contribute a maximum of 10 points out of 60 (17%) to Scoresheets 3 and 5, and a maximum of 20 points out of a possible 200 (10%) in the Candidate Summary Scoresheet (Scoresheet 7). If this seems like too low a weighting in light of the importance of these attributes to employee performance, you can assign a higher maximum score to this category. For example, assigning a maximum possible score of 50 points to the Personal Attributes in Scoresheets 3 and 5 gives the Personal Attributes a 50% contribution to those scoresheets (50 possible points out of a new maximum possible score of 100) and a 36% contribution in the Candidate Summary Scoresheet (100 possible points out of a new maximum possible score of 280).

A similar approach can be taken with the weighting of Managerial Skills in Scoresheets 3 and 5. In this chapter that entry has a maximum possible score of 10 points, but you can easily assign a higher maximum score to this category to increase the importance of this characteristics just as we did with "Personal Attributes" above.

If you find that you have two or more candidates with overall scores that are very similar or too close to distinguish you can use some of the specific scores as tiebreakers. For example, if the overall score between two candidates is nearly identical, you could choose the one with the highest "Personal Attributes" or "Managerial" score. "Red Flag" notations can also be used as tiebreakers if one candidate has one or more Red Flags while another has none.

SUMMARY

This chapter introduces an approach to hiring scientists and staff that may seem awkward and involved at first, but we are confident that after you have used it you will quickly appreciate its merits. If this approach helps improve your hiring decisions by even a modest degree, the minimal time invested in learning it will, over time, be more than repaid.

Try adopting our approach in phases. Using the scoresheets for reviewing résumés and during phone interviews should be the easiest elements to adopt. Similarly, using the interview guide with phone references should be easy because you can have the scoresheet and guide in front of you while talking on the phone. Once you are used to the format and the questions you can start using the face-to-face interview guide with candidates.

Before you start using this approach, decide what you mean by a score of 0 and what you mean by a score of 5 or 10 as well as the numbers in between. This will help you score candidates consistently. By itself, scoring may not be sufficient to ensure a level playing field and may even magnify bias in some cases (Wennerås and Wold 1997). The key is to follow all of the steps we present, from acknowledging that you may harbor implicit bias to preparing written criteria for candidate evaluation in the position description, and finally treating all candidates the same during the screening and interviewing process. Using the same questions, criteria, and scoring for every candidate can help to reduce bias and subjectivity. The advantage of this system is that it enables you to obtain the same information from every candidate and to evaluate each candidate in the same way. By minimizing differences in evaluation between candidates and using predetermined and consistent assessment criteria (remember Kahneman and the Oakland A's at the beginning of the chapter), you will make the hiring process more reliable.

Once you start using this process you will find that it gives you a greater sense of control over what is often thought of as a high-stakes guessing game. Hiring with your head and basing your decisions on data will change all that.

REFERENCES

Basu S, Savani K. 2017. Choosing one at a time? Presenting options simultaneously helps people make more optimal decisions than presenting options sequentially. *Org Behav Human Decision Processes* **139:** 76–91.

Bohnet I, Van Geen A, Bazerman M. 2015. When performance trumps gender bias: Joint vs. separate evaluation. *Management Sci* **62:** 1225–1234.

Fine E, Handelsman J. 2012. *Searching for excellence and diversity. A guide for search committees at the University of Wisconsin–Madison*, 2nd ed. WISELI–University of Wisconsin–Madison, Madison.

Leshner A, Scherer L, eds. 2018. *Graduate STEM education for the 21st century*. National Academies Press, Washington, DC.

Lewis M. 2003. *Moneyball: The art of winning an unfair game*. W.W. Norton, New York.

Lewis M. 2017. *The undoing project*. W.W. Norton, New York.

Meehl PE. 1954. *Clinical versus statistical prediction: A theoretical analysis and review of the evidence*. University of Minnesota Press, Minneapolis, MN.

Murphy M. 2012. *Hiring for attitude*. McGraw-Hill, New York.

Plous S. 1993. *The psychology of judgement and decision making*. McGraw-Hill, New York.

Uhlmann EL, Cohen GL. 2007. "I think it, therefore it's true": Effects of self-perceived objectivity on hiring discrimination. *Org Behav Human Decision Processes* **104:** 207–223.

Wennerås C, Wold A. 1997. Nepotism and sexism in peer-review. *Nature* **387:** 341–343.

Wiesner WH, Cronshaw SF. 1988. A meta-analytic investigation of the impact of interview format and degree of structure on the validity of the employment interview. *J Occupational Psychol* **61:** 275–290.

CHAPTER 5

A Herd of Cats: Managing Scientists

Whether you are a molecular biologist or a particle physicist, chances are that you do much of your work in teams. If you manage a team of scientists or technical professionals, you may find that most of your time is spent mediating disputes, ironing out misunderstandings, and placating bruised egos. As we have seen, there is a good reason for this: The people you are managing are focused on technical and quantitative aspects of their jobs, at the expense of the interpersonal and social aspects.

But there may be another reason that you spend a lot of time on these matters: You may be laboring under the same interpersonal deficits as those you manage. This chapter shows you the kinds of problems to expect when managing groups of technically oriented professionals, as well as how to deal with them should they arise. You will also learn about your possible blind spots that can make your job as a scientist/manager harder than it has to be. Our approach to helping you to become a better manager of scientists starts with helping you to be a better observer of yourself and others.

A GROWING AWARENESS OF TEAMS IN SCIENCE

During the past 50 years, science has increasingly been done by groups of scientists with complementary or overlapping skills. This is especially true in the private sector in companies of all sizes. Although the era of the scientist as individual practitioner is by no means past, it is clear that more and more of the dollars spent on scientific research (both in the public and private sectors) are spent on projects involving groups or teams of scientists.

There is increasing awareness that scientific endeavors are becoming more interactive and social. An NIH conference on "Catalyzing Team Science"[1] speaks to the importance of

[1]Conference BECON 2003. Symposium on catalyzing team science. June 23–24, 2003, Natcher Conference Center, National Institutes of Health, Bethesda, Maryland (http://videocast.nih.gov/Summary.asp?File=9924).

teamwork in the life sciences. More recently, The National Academy of Sciences published "Enhancing the Effectiveness of Team Science" (Cooke and Hilton, eds., 2015) which provides guidance on assembling science teams and reviews organizational structures that support them. Increasingly, but slowly, some graduate programs are attempting to impart communication and other "meta" scientific skills to their students.

To say that science has become a social occupation is not to say that scientists themselves have become social creatures or even that they should if they are not already. Increasing the size of groups adds scientific skills, expertise, and sometimes simply more hands to do the work. In practice, however, the accessibility of members' skills and information to the group depends on how well the member scientists relate to one another. In the worst cases, information and expertise are shared selectively or not at all, data are hoarded like a scarce currency, and team members lie in wait for the most public opportunities to demonstrate superior knowledge. More often than not, science managers and leaders fail to recognize or deal with such behaviors, much to the detriment of the group or organization.

Anyone who has managed science and technical professionals working on complex projects requiring collaboration and interaction knows how difficult this can be. Let us examine some of the challenges and some of the ways of meeting them.

SCIENTISTS MANAGING SCIENTISTS: CHALLENGES AND OPPORTUNITIES

If you are a technical professional, it would probably not come as a big surprise to learn that some people in your organization think that you are hard to manage or are a poor manager. In the private sector, scientists are often sent to management training seminars. These typically teach participants to set goals and objectives, give feedback, do evaluations, and manage projects. These are all important skills and worth learning. However, your success at applying these skills is not determined by how well you know them or even how long you use them. It is determined by how well you understand yourself, and how well you relate to and respond to the people to whom you need to apply them. If you are oblivious to your own motivations and feelings, you probably do not pay attention to or understand the motivations and feelings of those you manage. If you interpret silence as agreement, repeated absences as laziness, and failure to follow instructions as forgetfulness, you cannot be an effective manager.

Some of the studies we cited in Chapter 1 suggest why science professionals make such misattributions: They may not notice interpersonal conflicts, discern underlying motives, needs, and expectations, or listen carefully. We would also add that they are probably not very self-aware.

The good news is that as a scientist, you are the best possible choice for managing other scientists, either as a team leader or as an executive. In the following, we introduce skills and concepts that will improve your ability to manage scientists. Case studies illustrate how improving self- and interpersonal awareness can help in real world situations.

Ignoring problems and conflict in a team

The interpersonal difficulties of scientists often stem from an aversion to admitting that a problem exists or an inability to notice it. If you are unsure what to do about a problem, or are uncomfortable thinking about it, you will likely avoid it. But if you are a team or group leader, ignoring problems in your group can have a detrimental effect on team morale and productivity.

Some common problems with which scientific team leaders have difficulty dealing include:

- Employee performance or attendance problems. Your technician consistently has difficulty with crucial experiments, keeps undecipherable notes, and routinely misses one or more days of work each week. You compensate by double-checking their protocols, having them e-mail their raw data so that you can analyze it, and rescheduling experiments because they are absent. What you do not do is address the problem with the technician.

- One member of your group consistently complains about the lab, other people, and every piece of equipment that they use. They also find fault with every lab policy and voice these complaints to anyone who will listen. People in your group tell you that this negative attitude is getting on everyone's nerves. Your response is that the complaints are harmless and should be ignored.

- A new member of your lab has personal hygiene habits that others find offensive and distracting. They bathe infrequently, which makes working near them unpleasant. When a lab member complains about this to you, you are sympathetic, but find it impossible to imagine how to broach the topic. The person becomes a pariah in the lab and wonders why no one talks to them. Their work, and yours, suffers because no one gives them the advice or assistance that they need.

- One of your technicians uses their cell phone while doing experiments. They speak in a loud voice that you can hear in your office across the hall. Although no one mentions this to you, you suspect that others in the lab are having difficulty concentrating. You decide to ignore it until someone complains.

- One of your most capable scientists is a bully who routinely manipulates others into relinquishing equipment time, technician help, and supplies. They behave as though their work is more important than others'. You do nothing because they are productive and you figure that the group members need to work it out among themselves.

- A female employee in your group has a hard time getting her point across in meetings because she is soft-spoken and gets routinely interrupted by several loud males. You observe this but decide to do nothing, figuring that she needs to learn how to roll up her sleeves and jump into the fray if she is going to succeed.

- A female employee complains to you that a male coworker is viewing pornographic images on his laptop in the conference room. You are at a loss about how to broach the subject to him, so you do nothing.

These situations, to which we return shortly, are all examples of conflict avoidance, one of the most common and damaging mistakes made by managers of science teams. Recall that lack of awareness of conflict among team members was one of the personality characteristics that Gemmill and Wilemon (1994) found to be more pronounced in scientists than in nonscientists. If you do not pay much attention to other people, miss subtle cues in their behavior or manner, and if you ignore, dismiss, or trivialize what makes you uncomfortable, you will not be aware of conflicts simmering all around you.

If the conflict erupts into a full-fledged war, as happens in the case study below ("Ignoring Conflict"), it may be impossible to ignore. In this case, you may take another path to avoid conflict by simply doing nothing about it. We have heard many scientists in responsible positions assert that the reason that they do not intervene in whatever conflict is under discussion is that they think the warring parties ought to "work it out themselves." But if you have gotten this far in the book, you may suspect that this is code for, "I don't have a clue to how to help resolve this problem." This is the main reason we avoid conflict: We do not know how to deal with it. Moreover, many feel that conflict is to be avoided at all costs—it should never occur in the first place.

This, of course, is all wrong. First, conflict is inevitable and can even be a useful and necessary mechanism for bringing out differing views. Second, working through a conflict does not have to involve angry confrontation, insult, or accusation. You can learn to work through conflict in a collegial and productive manner. Many of the negotiation techniques that we introduced in Chapter 3 are also good tools for resolving conflict. If you suspect that you ignore or avoid important issues because you are conflict-averse, try one or two of the tools in Chapter 3. Take your time, especially if you have spent your life ignoring conflict.

The following case focuses on a team leader who is unaware of a simmering conflict in his lab, with unfortunate consequences for the project and the group's productivity.

■ *Case Study: Ignoring Conflict*

Ralph was a senior environmental policy analyst in the Environmental Protection Agency (EPA) and he ran a group specializing in industrial groundwater contamination. He was preparing to submit a lengthy and technically detailed report on behalf of the agency regarding a groundwater contamination suit being reviewed by a state court.

While working on the report, he noticed some discomfort on the part of two of his three junior associates, Richard and Teresa, about what exactly should be included in the report, but he did not ask them about it. He finished a close-to-final draft and gave it to all three junior associates for review.

Richard and Teresa objected to including the third associate, Tony, as an author of the report. Ralph explained that Tony had contributed some of the analysis cited and should be named. Richard and Teresa said that they did not trust Tony's results and demanded that Ralph omit Tony's name from the report. Ralph was stunned. He had no reason to believe that Tony's analysis was suspect, but he did notice animosity among the three associates.

Ralph responded that Tony had been part of the team all along and was deeply involved in its planning, analysis, and strategy. He concluded that the project had been a team effort and that Tony's name should remain. Richard and Teresa refused to have their names included on the report unless Tony's name was removed. Ralph was anxious to resolve the situation before the end of the day, which was the latest that the report could be overnighted to reach the court in time. He pushed the discussions, but Richard and Teresa grew more and more agitated. Ralph called their behavior outrageous and likened it to blackmail. At that point, Richard stormed out of the office and did not return until the following day, after the deadline had passed. Because the internal rules of the organization required that all contributors sign off on the document, the EPA lost its chance to present the work to the court and Ralph started disciplinary action against Richard.

There are many ways of looking at this situation. One is to take an approach that might come naturally to technically minded people, that of seeking the truth. What were the facts? Did Tony's name actually belong on the report? How much did he contribute? Was his work, in fact, unreliable? Is there some set of rules by which these facts could have been ascertained and weighted? These are all relevant questions, but we suspect that answering them would not have solved Ralph's problem. The real problem was that Ralph had not recognized and addressed the interpersonal conflict that blew up in his face. If Ralph suspected some animosity between Tony and the others, he needed to address it. Ralph explained why he did not address the problem earlier:

"I noticed some discomfort among these people, but I had never had any trouble with Tony, and I couldn't see why Richard and Teresa objected to what he was doing. Richard and Teresa never said anything critical of Tony's work in our group meetings, and frankly, I thought that there was some personal reason that they didn't like him."

The result was that the report was filed too late to have an impact on the court case, Richard was disciplined and eventually left the agency, and the entire group, especially Ralph, was viewed as incapable of meeting deadlines. Everyone lost.

What would have happened if Ralph was aware of his tendency to avoid and ignore conflict? He might have made a special effort to be on the alert for signs of animosity in his group, if only because he knew that he often missed such cues. In this case, he might have noticed that there was a problem, tried to determine what was bothering Richard and Teresa, and done something about it. Perhaps they had misinterpreted something Tony had said or done. Perhaps Ralph would have discovered that Tony's work really was shoddy. What if Richard and Teresa had been able to separate their personal animosity toward Tony from their professional concerns and discussed these concerns with Ralph? What if Richard had been able to control his temper and had enumerated his objections in a way that Ralph could understand?

If even one of the participants had had more insight, self-awareness, or ability to handle conflict, the outcome would have been different for everyone. Even a single individual with good self-awareness and interpersonal skills can have a profound impact on how a group functions. It is easy to focus on Ralph as the one responsible for the debacle. If Ralph had been more attuned to the interactions of his team, and if he had confronted them with observations that team interactions were deteriorating, this incident might never have happened. However, Ralph is not solely to blame. Each of the others had responsibility for their own role and inability to deal with the conflict.

If you are not a government employee, before you start feeling smug about the ineptitude of narrow-minded civil servants, try this experiment. Reread the above case study and change the participants from a team of EPA analysts to a team of engineers in dispute over the design of a Mars rover, a group of middle managers developing a plan for a new business unit, or a group from any technical discipline in which you work entrusted with reaching a goal. Our guess is that you will find a lot that feels familiar in the example regardless of the type of work you do.

The right words: Focus on the problem, not the person

Finding the right words to say in a difficult or uncomfortable situation can be challenging, so in the following we offer some specific suggestions. Notice in the examples below that we are following the advice of Chapter 3 and focusing the discussion on the problem and its consequences, rather than on the person. We will be developing the approach we introduce below in more detail in Chapter 6 where we provide a more general framework for giving performance-related feedback. The following suggestions are framed in the context of a conversation with an employee or colleague about their behavior or performance.

1. Focus on starting in a way that doesn't assign or assume blame. You could say, "There is a work-related matter I'd like to discuss with you. Is this a good time?" or "We seem to be having a problem." Starting with "we" immediately frames the problem as one you want to help solve.

2. State the problem, being as concrete as possible. For example, "You have been late for work six times in the last three weeks and that has affected our experiments" or "There have been eight mistakes this week in dosing the rats in the inflammation study." Be specific; not "a lot of mistakes" but "eight mistakes." By focusing your comments on what was done wrong rather than on what's wrong with the person ("You made eight dosing errors" rather than "You have become careless"), you make it easier for the other person to take what you say as constructive criticism.

3. Ask for their view or their comments on what you said.

4. Make sure that you both understand what the other person is saying. Ask if anything you said is unclear and repeat what you heard to confirm that you understand what they said ("I want to make sure that I understand your point of view. Here is what I am hearing…"). Ask the other person their version of the problem. Ask for clarification frequently ("What do you mean by…?"). Spend as much time as needed explaining or asking questions until you are sure that you are both discussing the same problem. If you skip this step and move to solutions too quickly, you may each be solving different problems.

5. Often this process itself leads to solutions. At this point, you are not necessarily pushing for a solution, but one may present itself. Defining the problem illuminates things that may not have been visible before.

6. Generate solutions to the problem. Start by asking them what ideas or solutions they have. If they do not suggest anything, offer suggestions yourself about what they or you can do to improve the situation. Do you need to make the experimental protocol clearer? Are your expectations for when you want people to be in the lab clear to everyone? Is there some new circumstance preventing the employee from meeting their commitments to this job? Keep the focus on solving the problem and improving the situation. If it looks as though the problem is theirs and not yours (i.e., unclear instructions or expectations), then your objective is to help the other person change their behavior. As potential solutions arise in your discussion, write them down if necessary.

7. Continue to exercise your self-awareness and communication skills during the process.

Let us apply these guidelines to five of the examples cited at the beginning of the preceding section.

Example 1. Every week is a bad week

Your technician, Jim, consistently has difficulty with crucial experiments, keeps indecipherable notes, or routinely misses one or more days of work each week. You compensate by double-checking his protocols, having him e-mail his raw data so that you can analyze it, and rescheduling experiments because he is absent. What you do not do is address the problem with the technician.

Start by making a list of several specific instances of the behavior in question. Frame the list in terms of how it has affected the work, the project, or the lab. A hypothetical conversation follows.

You: "Jim, I'm having this conversation with you because your work isn't as good as it could be. Last week, you forgot to add buffer to half of the sample tubes, and the week before you left the samples incubating for twice as long as needed. As you know, we had to repeat those experiments, and they're very costly. I know that you can do better. Is there anything affecting your work that I should know about? Are my instructions clear when we discuss the experiments?"

Jim: "Your instructions are fine. It was just a bad week. I had too much to do."

You: "Tell me what you mean."

Jim: "Well, in addition to these experiments, I had to make buffer and culture medium for Ali."

You: "OK, let's run through what you had to do last week and see if we are giving you too much to do."

You review the schedule and it doesn't look too bad.

You: "Jim, what you do is very important to the lab. So, what I'd like to do is to find a way to help you do your job more effectively. Let's look at your work schedule for the next two weeks and see if it looks like too much work." (You both review the work schedule.)

You: "OK, I think we agree that your schedule for the next two weeks looks manageable. Do you agree?"

Jim: "I guess so."

You: "Good. Now I'd like us to set a goal for no experimental errors during the next two weeks. Can you agree to that?"

Jim: "I'll definitely try."

You: "Great. Also, if you start to feel like you have too much to do, I want you to come to me and tell me. OK?"

During the next month, Jim's performance does not improve. In fact, it deteriorates. You do not ignore the problem. You call him into your office and have the following conversation with him.

You: "Jim, I'd like to review your performance during the past month with you. Specifically, we need to talk again about your absentee rate and your attention to detail in experiments 34 and 35. These are the same issues that we've been working on for the past month. Are you having any problems that might be impacting your work?"

Jim: "Not really."

You: "In that case, I need to tell you that if your work doesn't improve during the next two weeks, we'll have to discuss whether this is the right job for you. Now, let's go over those experiments and see if we can figure out what happened."

Example 2. The complainer-in-chief

One member of your group, Melanie, consistently complains about the lab, other people, and every piece of equipment that she uses. She also finds fault with every lab policy and voices these complaints to anyone who will listen. People in your group complain that her negative attitude is getting on everyone's nerves. Your response is that her complaints are harmless and should be ignored.

You: "Melanie, do you have some concerns about the way the lab is running that you'd like to discuss with me now?"

Melanie: "I don't have any concerns. What do you mean?"

You: "I know that you've expressed dissatisfaction with the way people are assigned to maintain equipment, as well as the allocation of travel funds. You may have some legitimate concerns, but I can't deal with them unless you talk directly to me. I'd like your commitment that the next time you find something that can be improved in the lab, you'll come directly to me. I promise that I will listen carefully to what you have to say. Can you agree to do that?"

Example 3. An olfactory challenge

Alan, a new member of your lab, has personal hygiene habits that others find uncomfortable and distracting. He bathes infrequently, which makes working near him unpleasant. When a lab member complains about this to you, you are sympathetic, but find it impossible to imagine how you can broach the topic. The person becomes a pariah in the lab and wonders why no one talks to them. His work, and yours, suffers because no one gives him the advice or assistance that he needs.

You: "Alan, this is an awkward conversation for me to have with you, but I think that you'll be thankful that we spoke when we're done. Everyone has different personal care habits, and for the most part, these are their own business. But once in a while those habits interfere with other people unintentionally. In your case, I've noticed that you don't seem to wear a deodorant. Although that is your personal business, and there may even be a health reason,

I need to mention this because, frankly, your 'scent,' if I can call it that, really distracts me and, I suspect, others as well. Is this something of which you are aware?"

Alan: "No one ever mentioned that to me. I don't wear deodorant because it seems unnatural. People should smell like people, not perfume counters."

You: "I respect your view. But in this case, we also need to consider that you work in close proximity to others who probably find it hard to share the lab with you. I don't think that this is healthy for the lab and it's not great for your relationships with the lab members. Do you have any thoughts about what we could do?"

Alan: "Well, I guess I could shower every day before I come to work."

You: "Terrific. Please try that and let's see how it works out."

Example 4. Reach out and touch someone, but quietly

One of your technicians, Natasha, talks on her cell phone using a Bluetooth headset while doing experiments. She speaks in a loud voice that you can hear all the way into your office across the hall. Although no one mentions this to you, you suspect that others in the lab are finding it hard to concentrate. But you decide to ignore it until someone complains.

You: "Natasha, I don't know if you're aware of this, but I can hear you on your cell phone all the way into my office. I suspect that others in the lab may be bothered by this, although no one has said anything to me about it. In the past, I have asked people in the lab not to play music that others can hear and to keep their personal phone calls to a minimum. Is there some particular reason that you need to talk on the phone while you work? Are you dealing with any problems?"

Natasha: "No, there's no problem. I just think that it saves time to talk and work at the same time. That way, I don't have to take time away from my work."

You: "Well, I wish everyone were as concerned about maximizing their time at the bench! Nonetheless, although your talking this way may enhance your productivity, I fear that it will decrease others' productivity, including mine. I don't like to make rigid rules for the lab, but I'm going to ask you to restrict your cell phone conversations to the lunchroom. I'll make sure everyone on the team knows about this new rule, and in a few weeks we can discuss how you are doing."

Example 5. Too much alpha noise

Sandrine, a female employee in your group, has a hard time getting her point across in meetings because she is soft-spoken and considered in her speech and routinely gets interrupted by several loud males. You observe this, but decide to do nothing, figuring that she needs to learn how to roll up her sleeves and jump into the fray if she is going to succeed.

Here, you have the option of either telling the interrupters to pipe down or helping Sandrine to become more insistent on getting airtime. You may decide on a bit of both.

You: "Sandrine, I've noticed that you often have a hard time getting a word in at team meetings. I wonder why you don't stand up to Fred when he interrupts you."

Sandrine: "I just don't like to argue with him. He keeps talking louder and louder. I prefer to keep quiet until he calms down."

You: "If you're willing, I'd like to help you learn to assert yourself a bit more in those situations. It's important for both the lab and you that you get the opportunity to express your views in our meetings. You know a lot about what we are working on and the other team members need your input. Are you willing to try?"

Sandrine: "Sure, I'll try anything that may help."

You: "The next time that you are interrupted, try saying, 'Excuse me, I haven't finished.' If Fred or others interrupt again, say, 'You'll get a chance to respond as soon as I finish' or 'I'd like to hear what you have to say, Fred, as soon as I'm finished.' You might say, 'If you keep interrupting, it's just going to take me longer to get to my point.' If you do this consistently, Fred will get the point. Be patient and don't attack or insult him; keep focused on saying what you have to say. Are you comfortable with these suggestions?"

Sandrine: "Yes, they sound great. Thanks."

You can also send clear messages to your group that suggest that although lively discussion is important and stimulating, rude interruptions are inappropriate. In the long term, your own behavior provides a model. If you interrupt while others are speaking, or shout over them when they are talking, your group will likely feel free to do the same.

Example 6. X, but not the chromosome

A female employee complains to you that Juan, a male coworker, is viewing pornographic images on his laptop in the conference room. You are at a loss about how to broach the subject to him, so you do nothing.

You know that you cannot ignore this problem, but no matter how hard you think about it, you cannot find a way to approach Juan about this accusation. You decide to seek the advice of Harriet, a friend in the human resources department. During a meeting, Harriet explains the institution's policy on the use of computers at work and you draft a more specific policy about personal use of computers in the lab. After you announce the new policy to your lab, you hear no further complaints about Juan.

The key in this case was in seeking help from a friend in human resources. If you do not have a friend in human resources, make one. One of the most important tasks of a human resources professional is to help managers find solutions to employee-related problems.

The preceding tools and concepts should help you to recognize and deal with some of the issues that science teams face. We have emphasized how self-awareness can help you to sense difficulties that both you and your team may be experiencing. We also introduced the notion that both team leaders and members must be comfortable dealing with conflict, which is inevitable in teams. When conflict goes unrecognized and unaddressed, the best-case result is lost opportunity, and the worst case is project derailment or failure.

Refer to Exercise 3 at the end of this chapter for more practice in finding the right words to use in difficult situations.

Technical turf wars

The following hypothetical case study involves a conflict that was both overt and destructive to the team and its progress. Despite being readily apparent, the conflict was handled poorly by all involved. In the sections that follow, we use this case to illustrate and address other challenges in managing teams of scientists.

■ *Case Study: Technical Turf Wars*

One of my jobs in the semiconductor industry thrust me into the middle of a multimillion dollar collaboration between my small research company, Monotech, and a high-profile electronics company that I will call BigTech. On my first day, I was sent to observe one of the weekly project review meetings attended by teams from both companies. I felt a sense of eager anticipation. The meeting was in one of BigTech's elegant conference rooms with a commanding view of an urban river. As I settled into a seat that had more levers, buttons, and adjustments than I had ever seen on a chair, I had a great view of sailboats taking advantage of the last good days of fall. My enjoyment evaporated quickly: Almost as soon as the meeting started, everyone on my company's team, including Andrew, the project manager, started shouting. It was a scientific free-for-all and the invective was almost unbelievably intense and hostile. I actually felt nauseous. I kept thinking that I was watching a train wreck in progress. In my role as vice president of Monotech, I was ultimately responsible for this impending disaster—and I had just started work that day. What had I gotten myself into? How had this project deteriorated to this point?

Andrew was an experienced individual with a PhD in electrical engineering who had been specifically recruited to manage the collaborative project with BigTech. On paper, his background and experience suggested that he was well suited to managing this project. Moreover, the project was actually very promising and could result in a very profitable product. But the project was moving very slowly because of technical problems.

The agreement between Monotech and BigTech required that our senior project scientists meet once every week to review progress. These meetings (and the one that I had seen that first day was the rule, not the exception) were like rugby matches with players from both sides having a scientific brawl in the mud. Having worked in science for almost 25 years, I was used to a certain amount of posturing by aggressive scientists. What I was not prepared for was that much of the mudslinging and invective was between members of my own team! As if this were not disconcerting enough, even Andrew himself lost his cool more than once and routinely turned red with annoyance at his colleagues.

Back at Monotech, Andrew complained about his team to anyone who would listen, especially about the fact that members of his team rarely listened to him—even though he was the manager. It did not take me long to see a multitude of causes for this disastrous situation.

First, the scientists felt that they, not Andrew, were the experts in this particular area; Andrew had no scientific credibility with them. On the occasions when Andrew had the presumption to present Monotech's data to the joint project team, the scientists became furious.

Second, Monotech's internal project team meetings invariably degenerated into accusations of incompetence or worse. During these meetings, Andrew himself became furious and lost his capacity to continue the discussion. More than one meeting ended with a crimson-faced Andrew ready to explode and a smug group of scientists congratulating each other for so effectively pushing his buttons.

Neither Andrew nor the scientists had the skills to resolve the conflict within which they found themselves. I suspected that the scientists were using Andrew to vent their frustrations with a technically difficult project. For his part, Andrew could not see that by losing his temper, he was just exacerbating the situation. Monotech was also responsible: They had made Andrew project manager because he had a history of managing big projects in other companies, but what they failed to predict was that his limited scientific background in the specific area of the project would be a red flag to the headstrong scientists, who had already burned out two previous managers. Andrew's limited self-awareness, self-control, and people skills simply fueled the flames.

To make matters worse, Monotech badly needed this project and the associated funding that BigTech provided. Everyone knew this, and they also knew that the project was not going terribly well. Yet Monotech's senior managers never openly discussed with the scientists the consequences of the project being canceled. As a result, there was considerable anxiety, uncertainty, and wild speculation among the scientific staff. The outcome was a disaster with multiple causative factors. Not long after I joined the company, BigTech canceled the project.

This case study illustrates a series of errors in managing science teams—some of commission and some of omission. Andrew is not the only one who contributed to the problem; each of the participants as well as the senior managers of the company had a role. Some of the problems in this case include the following:

- Management failed to recognize that this group needed a project manager with scientific credibility who would be immune to the hostilities of the group.

- Andrew lacked self-awareness and self-control. This made it impossible for him to weather challenges to his authority and keep the group focused on task.

- The team had a propensity for channeling their frustration into hostility toward one another. Management never intervened to stop this.

- Andrew usurped the scientists' data during project presentations, cheating them of the opportunity to get credit for their own accomplishments.

- Management failed to be open with the scientists about the status of the collaboration, and what might happen if it were to be terminated.

How can disasters such as the Monotech case be avoided? In the following sections, we review the causes and consequences of the team-based problems that we identified in the Monotech case. Building on the information we presented in Chapter 1 regarding psychological characteristics of scientists, as well as our suggestions for improving self-awareness in Chapter 2, we show how you can minimize the chances of these and other common problems from derailing your team.

Providing individual recognition in a team-based project

In the Monotech case, Andrew's failure to champion his team members' contributions and his usurping their results when he presented their data were two of many errors that contributed to the chaos of that team.

Explicitly acknowledging the individual contributions of team members provides essential recognition, even when team members do not own a specific element of the project. Going out of your way to notice and mention the roles of each of the team members can be the single most important contribution you can make to the team's performance.

According to Katzenbach and Smith (2015), "A team is a small number of people with complementary skills who are committed to a common purpose, performance goals and approach for which they hold themselves mutually accountable." In addition, we often think that team members benefit from mutual support and are in some way bonded or connected to each other, if only through shared objectives.

However, we believe that several elements and implicit assumptions in this view of teams are at odds with the reality of teams in the scientific workplace. First is the tension between the team being the primary work unit in many organizations and the scientist's need for individual recognition. Second, and closely related, is the dichotomy between the need to sublimate individual needs and interests to the team's objectives and the need to work on some part of the project that can be identified as one's own.

Scientists are typically individual contributors, not team players. Recall that in Chapter 1, we identified autonomy and independence as two characteristics prevalent in scientists. As one pharmaceutical company scientist interviewed by McAuley et al. (2000) explained it, "The scientific side is a collection of one-man bands who amazingly get things done." In our experience, the principal challenge in managing teams of scientists lies in the genuine need of creative scientists to obtain individual recognition for their contribution to the work of the team.

Indeed, if you do not provide opportunities by which the achievements of individual team members can be identified and recognized, the members may attempt to create the opportunities themselves. They may try to design a subproject of their own or initiate an experimental approach that has not been discussed. Although such activities can be valuable to the project, they may also subvert or distract the team from its task. If scientists go off on tangents (no matter how creative or inventive) as a way of showcasing their independence and resourcefulness, they are probably not focusing on their primary responsibilities. The team leader must encourage creativity, independence, and inventiveness while balancing these with the need to maintain focus on assigned tasks.

As noted by Katzenbach and Smith (2015), "Real teams always find ways for each individual to contribute and thereby gain distinction." If the authors' use of the word "real" was meant to convey the notion that most teams find it hard to accommodate this ideal, we are in full agreement. This is probably the single most important concept for the leader of a scientific team to grasp.

If the team grows larger and members' tasks begin to overlap or become shared, it becomes increasingly difficult for individual members to own unique domains of responsibility. Moreover, as teams and projects become more complex, more work may get done in the background. Individual contributions to the project may become secondary or tertiary in nature: A technician figures out why the cells growing in the lab are dying and fixes the problem; a colleague helps a team member get urgently needed results on a mass spectrometer. These are important contributions that take time and creativity on the part of the participants, and it is a serious mistake not to acknowledge these to the individuals and the group.

Balancing task ownership with team participation

Your role as a team leader is to ensure that the owners of specific tasks are recognized for their role in the project and are acknowledged and rewarded for their contributions, no matter how small. Leaders who recognize this create teams in which each member has a clearly defined and preferably nonoverlapping function. This makes ownership and identification of individual contributions transparent. Moreover, it enables assignment of responsibility and credit.

A team of technical professionals needs to be no more than a group of individuals who bring disparate skills and input to a project. The only one who truly owns the team's objectives is the project manager or team leader. The members of the team each own tasks for which they will be held individually accountable and for which they will seek and expect individual recognition.

Although each team member may operate as an individual contributor, the group can work together to integrate information, test ideas, and, in the most creative groups, catalyze new ideas. If you are a team leader and your group has a high level of cohesion, sharing, and collaboration, continue doing whatever you are doing. If your team does not have these characteristics, but the work of the team is getting done, there may be no need for concern.

Rather than trying to create a forced camaraderie, focus instead on modeling respect and tolerance among team members. Team leaders who belittle or devalue others or do not acknowledge their contributions, but who go out for a beer once a week with the lab to promote cohesion, are only fooling themselves and shortchanging their team members. Do not fall into the trap of believing that just because you—the team leader—see the group as "all for one and one for all" that everyone else does, too. Although teams can and do accomplish more than individuals, and members can and do derive satisfaction from group achievement, do not forget that scientists need ownership of and responsibility for well-defined accomplishments.

Creating more effective teams, one member at a time

The team concept is perhaps one of the most overworked themes in management literature. Management pundits espouse the view that the secret to success lies in inducing team members to work selflessly toward a common goal. Some of the approaches used to foster such attitudes work well in the corporate world. But we believe that most of what passes as common wisdom about managing teams cannot be applied in the science workplace.

From what we have said above, creating effective teams in science may be more a matter of meeting the needs of the individual members than of creating a shared vision or an all-for-one and one-for-all attitude that so many team-building courses and exercises promote. These needs can be met if leaders provide clear ownership of tasks and recognition of accomplishments. Beyond this, the best way to create an effective scientific team is to provide its members with the skills to interact productively in a group setting. These include the ability to manage themselves in contentious and stressful situations and to recognize and resolve conflict in a productive manner. These are precisely the skills that the Monotech scientists lacked and their leaders failed to either model or provide.

Christopher Avery (2001), in his book *Teamwork Is an Individual Skill,* suggests that the most important learning for successful teamwork takes place at the individual level, and not at the level of the group as a whole. Because science teams, especially in academia, frequently lack cohesion and a sense of shared ownership of goals, optimal team performance only occurs when each member takes responsibility for his own part of the project as well as his individual behavior and performance. It is the job of the manager or leader to encourage and support such behaviors.

Helping your team to see the big picture

As a manager or leader, one of your most important roles is to help your team understand the overall mission and goals of your team or organization. This can take several forms.

Interpreter of internal events

Scientists may be myopically focused on their specific project but have both a need and desire to relate what they are doing to the big picture. It does not take much for a leader to routinely update the team on how their project is related to and impacts the goals of the lab division or the organization as a whole.

New team leaders often mistakenly assume that because they themselves see the big picture and the relationship of their project to the organization as a whole, everyone else on the team does as well. Conversely, they may assume that because scientists are so focused on science they do not care about the larger objectives. Both are false assumptions that result in team members feeling disconnected from the lab's, company's, or organization's goals, and not understanding how their tasks relate to those goals. Leaders of science teams in the private sector need to pay special attention to helping the team relate their project to the company's overall mission and goals.

A team leader must serve as a buffer between decisions made by senior management and their team members. As discussed in Chapter 11, science professionals become very committed to and identified with their projects. Having a project terminated for lack of progress or—far worse—for business reasons that have nothing to do with how well the project is progressing can be a frustrating and disillusioning experience for young science professionals. This is what happened in the Monotech case. Although the scientists feared that the project might be terminated, they were unprepared when it happened and had little understanding of the reasons behind the termination. It is the leader's responsibility to ensure that team members understand the reasons for such decisions. Andrew could have played an important role in this process by serving as a conduit between senior management and his team members.

Often, team leaders who themselves are bitter or disillusioned about terminated projects will share these feelings with team members under the mistaken impression that this creates a bond with the team or that honesty is the best policy. In fact, when the team leader behaves in this way, the team's sense of alienation will grow and it will be more difficult for them to commit to future projects.

Interpreter of external events

In your role as buffer for the team, you serve as interpreter of external events and the business or funding environment in which the organization functions. This helps the team to manage and cope with the uncertainty and ambiguities that can arise from sources outside of the organization. Team members may feel uncertainty about the future of the organization: Will the company be acquired by a competitor? Or about external events: Will the Air Force buy the company's new guidance system? Or, will changes in the NIH budget impact funding prospects? And about the industry: What does it mean that the biotechnology stock index is down 30% for the year? The team leader can have a pivotal role in providing factual information and helping the team come to grips with the uncertainties that they engender.

One of the biggest challenges that teams of technical professionals face is instability in their field of work or organization. This has been an especially vexing problem in the life sciences and the volatile biotechnology sector but is common also in start-ups in general and in many areas of endeavor. Unstable organizations may negatively impact commitment and encourage an "every man for himself" mind-set. The team leader can mitigate the impact of such a climate by helping team members to understand and discuss the underlying issues and how the organization can and will respond to them.

Helping your team to manage uncertainty

Good leaders address discomfort, distress, and ambiguity in the group and in its members. This means that leaders must be open to signs of distress in their team and actively help the team to manage ambiguities and uncertainties. Just as important, leaders need to recognize these feelings in themselves.

If you are feeling anxious and uncertain about your project or company, chances are that your team members feel the same way. By becoming adept at sensing and identifying these feeling and emotions in yourself, you can use them as sentinels to alert you to incipient problems that need attention.

In his book *Connect,* Edward Hallowell (1999, p. 123) suggests, "Think of these emotions as noises in your car engine. Investigate them." Treat your own feelings and reactions as data that alert you to situations requiring attention. Self-awareness is just as important for managing others as it is for managing yourself.

When you do sense concern about the future, of either the project or organization, go out of your way to help team members see the big picture. Hallowell points out that because creative professionals often keep to themselves and may be reluctant to share their thoughts and feelings, concerns and fears may become magnified and exaggerated in their minds. He advises, "Get the facts. Very often stress and worry emerge from the imagination, not from reality, particularly for creative people working alone" (Hallowell 1999, p. 123). We would add that even when fears have a basis in reality, we often exaggerate the consequences. An effective leader promotes communication and provides a reality check for the team. Glen (2003) adds, "If you don't take explicit control of the meaning of situations, they will either remain ambiguous or be defined by others. In the absence of clarity, rumor and innuendo often take over. A group of smart geeks can develop some wild theories about what's going on... ."

As a manager of scientists, you must give your group a realistic picture of its future. This is especially important for young professionals new to the private sector whose only working experience has been in academia, where basic research projects can go on for years. Although young professionals in the for-profit sector may claim that they understand that industry norms differ, they may still react in a bitter or disillusioned manner when unexpected changes in projects, leadership, or organization occur.

Helping your team with uncertainty can have important benefits. A senior executive in a major Midwestern company told Carl that he went out of his way to post a list of projects waiting to be tackled as soon as the current projects were completed. He explained, "My people used to get very nervous when it looked like we were going to deep-six a project. Many of them really thought that if we killed the project, they would lose their job because there would be nothing for them to work on. This had the disastrous consequence that they would perform all kinds of technical acrobatics to keep a demonstrably bad project alive for as long as they could. Posting this list made a tremendous difference because everyone in the organization could see that we had a large pool of exciting projects waiting in the wings. Now, nonproductive projects get killed much earlier."

Of course, in some cases, the leader cannot provide unambiguous answers to the questions that trouble team members the most. In the Monotech case, Andrew need not, nor could he, have given the team solutions for their problems or answers to their questions. But by openly discussing the possibility of project cancellation as well as its consequences, he could have helped them to manage their uncertainty. He could have said, "I'm feeling a bit uncomfortable about the way BigTech executives are talking about this program, and wonder if the rest of you have picked up on that" or "I've noticed that some of you seem to be anxious about what happens if this project is canceled. Is that true?" The process of discussing these feelings with the group can create a sense of psychological relief by showing them that they have common concerns. If this results in even a small decrease in anxiety and preoccupation with dire outcomes, the work of the team will improve.

Ronald Heifetz, in his insightful book *Leadership without Easy Answers* (Heifetz 1994, p. 110 ff.), argues that one of the most important attributes of a leader is the management of uncertainty. Leaders do this by creating what he calls a "holding environment." Heifetz provides a cogent analysis of this skill in the context of world politics (Heifetz 1994, p. 110 ff.). Paul Glen also noted the importance of this skill for managing technical professionals in his book *Leading Geeks* (Glen 2003).

When projects do get terminated, scientists will have a natural and appropriate sense of disappointment. The disappointment may become bitterness if the scientist has been working overtime and is psychologically committed to the project. When this happens (and it will), managers can help team members to effectively deal with their disappointment. Do not underestimate the value of holding formal project wrap-ups in which employee contributions are openly recognized and lauded or having one-on-one meetings with project members to let them know that you are aware of their hard work and contributions. In an article entitled "Celebrating failure in a tough drug industry,"[2] *The Wall Street Journal* highlighted a company in Cambridge, Massachusetts (Ironwood Pharmaceuticals) that made it a formal ritual to celebrate the hard work that went into projects that they chose to abandon. AstraZeneca, another company mentioned in the same article, holds a "Science Oscar" ceremony celebrating "scientists who pursue promising lines of work no matter what the outcome." It is easy to neglect these small acts of recognition in the day-to-day rush of a research lab and even easier during the turmoil of a reorganization, merger, or change in management, but they can make the difference between the success and failure of a company adapting to the change. See Chapter 12 for more on these topics.

Modeling behavior

Leaders who wish their teams and team members to behave in certain ways will always have more success if they themselves behave in those ways. This is called modeling the behavior. The leader's behavior sets the norms for the team. For example, when members observe the team leader behaving in a civil manner in a tense situation, they see this as the norm and they learn to behave that way themselves. Simply telling team members that you expect them to behave in a civil manner is useless because it offers no clue for how to go about it. What you expect may seem obvious to you when you say, "I'd like to see more respect for other people's views in our meetings," but others may be genuinely baffled until you model the behavior. However, if what you say is contradicted by your own behavior, its effect is nullified. We saw in the Monotech case that Andrew exhibited many of the same destructive behaviors as the group members. It therefore would not have been effective for Andrew to lecture his team on how he wanted them to behave in team meetings.

As a leader, if you find that you are repeatedly having to exhort your staff to be collaborative or "team players," chances are that you or others in leadership positions are not modeling those behaviors yourselves or are behaving in a contradictory manner. This type of behavior modeling contributes to the culture of the group or lab, which we discuss further below.

Delegation

At one of the Cold Spring Harbor Workshops on Leadership in Bioscience we had a panel discussion to which we invited six scientists who had been running research labs for three to seven years. The topic of the panel was "What were some important lessons you learned as a new principal investigator?" The emerging leaders attending the workshop were eager to hear what those who had been running labs could share with them. A comment that one of the panelists shared was memorable. They said (we are paraphrasing), "One of the most powerful moments for me as a new lab head came when I needed to leave the lab for four or five days to attend a scientific meeting. As I settled in on the plane I suddenly realized that for the first time ever in my career, data was being collected in my absence. Up until then, I was involved with all data collection. Now that I was running the lab, others were doing it." The point that this individual was making was that delegating some or all of the science to others was both liberating and a bit scary at the same time.

[2]https://www.wsj.com/articles/celebrating-failure-in-a-tough-drug-industry-1488568710.

For new PIs, delegation can be one of the hardest skills to learn. As a new PI you got to where you are by virtue of your scientific expertise. Giving up some control of the science to others may seem like a risk not worth taking. But if you do not take it you will limit how much you can accomplish. PIs who do not delegate may soon find themselves overwhelmed with work, decisions, and tasks that could just as well be done by others. Some years ago, Carl was speaking with the CEO of a local hospital who was legendary in his ability to inspire loyalty and achievement in his staff. His secret? "I make sure that everyone in the hospital from the Chief of Neurosurgery to the maintenance and kitchen staff has some area in which they can make a decision on their own" (we are paraphrasing). By giving every employee some authority, this CEO created motivated employees who were invested in their work. If you find yourself wondering why people in your group aren't coming up with their own ideas, perhaps they do not feel sufficient ownership of their projects. Delegation may be a way to change that.

Although delegation can be powerful, it needs to be exercised with appropriate caution and oversight. If you are going to delegate, remember the following guidelines.

Delegation Guidelines

- The delegated task or responsibility must align with the individual's capabilities. If you are not sure of their capabilities, delegate incrementally more complex items until you find out. Delegating tasks or decisions that may seem trivial to you can be felt as empowering to an employee or scientist because it conveys trust and confers responsibility.

- If you are not sure of the individual's capacity to do what you are delegating, monitor their performance and decisions. Give them feedback and guidance when needed. Eventually they will learn, and you will get more comfortable with their judgment.

- Delegating does not mean that decisions get made and you never hear about them. Make it clear to the person you are delegating to what you want to know and with what frequency. Asking for updates doesn't have to feel like you are micromanaging them if it is done to keep you up to date rather than as a mechanism for second-guessing or criticizing them.

- When you do delegate, make sure that you can live with the result. Authorizing someone to make a decision and then criticizing what they decide because it is not what you would have done is setting a trap for the employee. You are asking them to guess what you would have done rather than to decide what they think should be done. If the decision is that important to you, make it yourself.

- Don't delegate just to get something off your desk. If that is your main motivation, chances are you will be disappointed with the outcome. Use delegation as a management and training tool. Depending on the level of experience of the delegate, you may need to review a decision or a work product at some point. Let the delegate know what level of involvement you will have once you delegate something to them.

If you find it difficult to delegate, here are some ideas to get you started.

- If you need to be away from your lab or group, delegate a lab or group member to "be in charge" while you are away. This could mean as little as having the person be responsible for communicating with you if issues that you need to know about arise or to make certain decisions or approvals while you are away. Examples might include signing purchase orders (up to a specified dollar amount) or dealing with equipment or infrastructure matters. Be sure to let the group know that you are doing this before you leave.

- Periodically assign group members to lead your weekly lab meeting, whether you are away or not. This not only empowers and motivates those who run the meeting but also demonstrates that you expect your group to communicate with each other about their science and help each other think about and troubleshoot their work.

- Delegate members of your group to lead a journal club discussion or to be responsible for updating the team on a new technique or methodology during a lab meeting.

- Delegate a team member to be responsible for acting as the host for a scientist or new-hire candidate who is visiting the lab. Do the same with a seminar speaker invited to address your group or department.

- Delegate someone who has been to a scientific meeting or conference to update the rest of the team on what they learned there. If there were multiple attendees from your group assign each one a part of the meeting to review.

- Delegate responsibilities associated with drafting manuscripts for publication, parts of grant applications, and manuscript reviews for scientific publications (ensuring that they maintain confidentiality).

Just because you have delegated a decision does not mean that you forever relinquish your right to comment, advise, or intervene. As PI, the buck stops with you. If you fundamentally disagree with a decision or course of action you may need to step in. The key is how and how often you do that. Very often, at the time a delegated decision is made it is the right decision but as more data accumulates it becomes apparent that a different course of action is needed. Supporting the initial decision, while counseling toward a new course of action, is both appropriate and necessary in your role as PI.

If you have not done much delegation, we predict that you will be surprised by how much members of your group will appreciate your show of trust and how receiving this trust makes those in need of greater self-confidence more motivated.

Lab culture

As a leader you have a responsibility to create an environment that fosters your and/or your organization's objectives and values. Depending on your organization, these objectives and values could include scientific productivity in all its various forms (discoveries, publications, grants), as well as science-related elements such as patents, products, education, and training. In a sense, this entire book is about how to create such an environment. How you manage and lead both directly and indirectly creates the environment for accomplishing your objectives. The environment that you create or foster can be characterized by a shared set of beliefs, practices, values, and goals often referred to as the culture of your lab or organization. In this sense we think of culture not so much as a passive attribute that characterizes a group but rather as one of the tools at your disposal that enable you to accomplish your objectives.

On the scale of importance of the tasks you must do as a new leader, it would be understandable if culture weren't at the top of your list. After all, you need to get that big experiment done to complete that big paper to support that big grant you're submitting soon. Before you know it, five years have passed, and your lab's culture has been defined by the de facto sum of all the urgent ad hoc decisions, interactions, and behaviors that have happened in between. Upon reflection you discover that you've created a culture of mistrust, expediency, and internal competitiveness. It's not pretty, you never saw it coming, and it's going to be hard (but not impossible) to undo. If you're thinking it would have been better to do it right from the start, we agree.

The value in thinking about what kind of culture you want in your lab is that culture influences and, in some cases, determines the attitudes and behaviors of your team. We believe that a culture that fosters collegiality, promotes psychological safety, and encourages respect for others facilitates creativity and productivity. From this perspective you need to be as thoughtful and intentional in thinking about the culture you're going to foster as you are about what your scientific focus will be.

In our view culture has three overlapping and mutually interdependent elements, shown in the figure below. The preceding chapters of this book each relate to one or more of these elements in one way or another, so without saying so explicitly we have been talking about culture all along.

Science. Your behaviors, attitudes, and expectations relative to how science is done in your lab or organization set the norms for your group. How many times must a complex experiment or measurement be replicated before you will publish it? Do you allow a "screwy" result to be ignored because it doesn't fit with the other five results? What actions do you take when you or someone in your group learns of someone else's confidential data that could affect your work? It is the leader's behavior in situations like these that sets the norms for the group and that helps to create the scientific culture. If you have trainees in your lab (postdocs, graduate students, etc.), they will likely take this culture with them and adopt it themselves. If you wish to instill scientific integrity and rigor in your group, then you must display it consistently, even when the temptation to loosen standards in the service of getting that important paper published seems overwhelming. Involving trainees in your own internal dialogue in reaching such decisions is a powerful teaching tool.

Three domains of lab culture

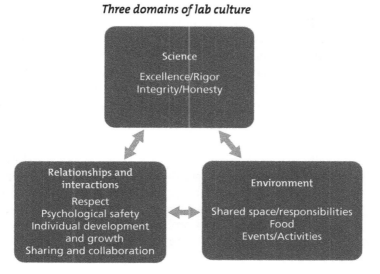

Relationships and interactions. A lab culture that fosters productive and collegial interactions is perhaps one of the most sought-after attributes by scientific group managers and leaders. They imagine a world in which their postdocs or scientists are eagerly sharing their results, reagents, and equipment with one another, critiquing each other's results in a helpful manner, and respecting viewpoints with which they may not agree.

Respect for others in the group is one element of this cultural domain. The first step in fostering a culture of respect is for you, the leader, to show respect to your lab members, administrative and technical staff, and others outside the lab. If you have come this far in this book you know that exhorting others to be respectful if you are not, or are so inconsistently, creates cynicism. The second step is to reinforce respectful behavior by giving positive feedback to those who exhibit it in challenging situations. The third step is for you to coach and advise others to be respectful when you see them behaving otherwise. That is not to say that you are trying to create a culture in which everyone is artificially nice all the time. It is a given that challenging others and honestly critiquing their science is essential to scientific progress. The key is to challenge ideas and proposals and not the integrity, motives or intelligence of the person espousing them (more on this in Chapter 7 on Meetings).

We discuss the importance of psychological safety in Chapter 7 in the context of meetings but it's not only important in meetings. Showing your team members through your own behavior that it is alright to ask naïve questions or for them to question your opinions or advice sends a powerful signal regarding how you expect others to behave.

Another facet of lab culture relates to the value placed on learning, development, and growth (scientific and intellectual) of group members. Lab leaders who demonstrate through their words and actions that they take seriously their responsibility to mentor and train their team members create loyalty and commitment. Concrete actions include one-on-one meetings, frequent informal feedback, performance reviews, and annual review of individual development plans. By showing in these ways that you care about your team members and about their future and not just about how they can help you get your next grant funded, you promote loyalty and a willingness to go the extra mile. In the language of negotiation from Chapter 3, you're helping them satisfy their underlying interest (growth, advancement), while satisfying your own (hardworking lab members and productivity), creating the vaunted win/win scenario.

The final element of how lab culture is embodied in relationships and interactions are your norms for sharing and collaboration within your team and with others outside the team. Again, you are the most important exemplar of the behaviors you want to foster. Are you engaged in one or more collaborations, or do you send the message that others are not to be trusted or just not worth the trouble? How do you talk about your collaborators with your team? With enthusiasm and trust or with hesitation and dread? Similarly, do you encourage others to share materials, reagents, or equipment in the lab, but then turn down a request from a new faculty member to borrow a spare PCR machine?

As a mentor, you model behaviors that demonstrate your willingness to help your team in challenging situations through coaching and guidance without actually doing the work for them. In this way you are acting as a role model for how your team interacts with each other. In "*What makes for a great team? Individual vs. team approaches to intelligence analysis*" (Hackman and O'Connor 2004), J.R. Hackman, a prolific contributor to the science of team dynamics (Hackman 2002), shows that the degree to which team members help each other through peer-to-peer coaching correlated with high team performance more highly than any other parameter they measured. Your attitudes and behaviors regarding sharing and collaboration show your team members how you expect them to interact with each other.

In aggregate, the culture you create and foster around relationships and interactions can be one of the most powerful tools you have to promote a productive lab and one in which people work with enthusiasm.

Environment. Depending on your point of view, the design of new open concept work spaces, including lab spaces, is either the greatest piece of social engineering ever conceived, or the bane of the existence of anyone trying to concentrate. We had a colleague who had recently moved into a very large open lab space consisting of dozens of lab benches arrayed in parallel in a large open room. She reported that because of the cacophony of noises—conversations, vortex mixers, small centrifuges, and more—she and most of the others in the lab had taken to working with earbuds in their ears listening to music to block out the noise. The point, of course, is that the open space concept had been designed to foster communication.

How you design and manage your space speaks volumes about your values and respect for your people. We have always felt that insisting on a clean, well-maintained lab shows respect for your working environment and, by extension, for the people within it. As a leader you promote this respect by assigning responsibilities to your lab or team members for such things as equipment maintenance, supply ordering, creation and distribution of protocols, and SOPs. In labs where such infrastructure is handled by technical or support staff you can and should find other ways to involve your team members in the operation and life of the lab. Examples include having lab members host a weekly journal club, introduce and chaperone an invited speaker, take a turn at facilitating or being responsible for refreshments at your weekly lab meeting, or even being the responsible person in the lab while you are away for a week. All these responsibilities foster a sense of shared ownership in the lab as a whole and in its projects.

When you give people such responsibility, make sure you're willing to live with the consequences. Asking a postdoc to introduce a speaker and then getting annoyed publicly when they don't say exactly what you would have said will make others dread taking on similar responsibilities. It's better to give them private feedback on how they did after the fact and to frame it as helpful advice rather than as a criticism.

Finding ways to bring people together for either scientific or nonscientific activities is a common tool for fostering a communal spirit. In any such event the presence of food is a universal and indisputably powerful motivator. In the lab where Carl did his postdoc, the PI had imported the British teatime ritual. Every day at 3:00 PM tea and cookies were served in the lab's conference room (true to his scientific roots, our leader had imported from England a teapot that met the British Standards Institution specifications along with an SOP for making tea!). Not everyone showed up every day, but most did on most days and the occasion created a sense of warmth and friendship and made the day a bit more pleasant. Although none of us postdocs ever verbalized it, I believe we all sensed that our mentor sponsored these daily get-togethers not because he was a tea fanatic but for the community-enhancing value of the gathering.

Finally, creating extra-scientific events and get-togethers is a tool that many lab leaders use to promote a community spirit or simply as a way to recognize or thank lab members. If you enjoy doing such things then by all means do so, but don't feel that they are obligatory. We've seen productive and collegial labs that never get together outside the lab and dysfunctional groups whose members resent obligatory get-togethers with people they spend enough time with during the workday.

They all work together. The three domains have bidirectional arrows connecting each other because an imbalance or dysfunction in one can negate the others.

If you have spent millions creating an open plan facility with spaces for people to congregate, interact, and have casual conversations, but the culture is one of high levels of competitiveness, lack of sharing, little collaboration, and siloed thinking, your sky-high architect fees will have been wasted. Architecture cannot compensate for a toxic culture.

If you espouse respect and collegiality in the lab, but have outdated equipment that is poorly maintained, and fail to set standards for lab cleanliness and inventory, people may get the message that while you expect them to respect each other, you don't respect them.

Finally, if you insist on scientific rigor but routinely ignore glaring safety deficiencies in the lab or complain about onerous regulatory requirements relating to biohazards, radioactivity, and animal use, people may wonder whether your support for scientific standards are just so many words.

REFERENCES

Avery CM. 2001. *Teamwork is an individual skill: Getting your work done when sharing responsibility.* Berrett-Koehler Publishers, San Francisco.

Cooke NJ, Hilton ML, eds. 2015. *Enhancing the effectiveness of team science.* National Academies Press, Washington, DC.

Gemmill G, Wilemon D. 1994. The hidden side of leadership in technical team management. *Res Technol Management* 37: 25–32.

Glen P. 2003. *Leading geeks: How to manage and lead people who deliver technology.* Jossey-Bass, San Francisco.

Hackman JR. 2002. *Leading teams: Setting the stage for great performances.* Harvard Business School Publishing Press, Cambridge, MA.

Hackman JR, O'Connor M. 2004. What makes for a great analytical team? Individual vs. Team Approaches to Intelligence Analysis. FAS Reports, https://fas.org/irp/dni/isb/analytic.pdf.

Hallowell EM. 1999. *Connect: 12 vital ties that open your heart, lengthen your life, and deepen your soul.* Pantheon, New York.

Heifetz RA. 1994. *Leadership without easy answers,* pp. 104 ff. Belknap Press/Harvard University Press, Cambridge, MA.

Katzenbach JR, Smith DK. 2015. *The wisdom of teams: Creating the high-performance organization.* Harvard Business School Press, Boston.

McAuley J, Duberley J, Cohen L. 2000. The meaning professionals give to management and strategy. *Hum Relat* 53: 87–116.

EXERCISES AND EXPERIMENTS

1 Recognizing conflict among others

If you are aware of simmering conflicts within your group right now, list them and ask yourself whether they are impeding progress or will in the future. Pick one of the conflicts and write down three ways in which you might acknowledge or help resolve it with one or more of the people involved. Enact the conversation in your mind and notice your reactions and feelings. By remembering your responses to these imagined conversations, you will be in a better position to notice when you are avoiding the conflict in the future.

2 Recognizing conflict between yourself and others

If you are conflict-averse, you may be choosing to ignore a difficult working relationship with someone in your group.

- Can you identify one of these? Determine whether it is in your own best interest, and that of your group, to continue ignoring this situation.

- List three things that you might say to acknowledge or begin discussing the problem with the other person. As above, note and remember your reactions and use these as signals to alert you to avoidance reactions in the future.

3 The right words: Attack the problem, not the person

The following is a hypothetical conflict that requires attacking the problem, not the person. For the exercise, we suggest that you start by attacking the person, so that you can get some idea of how this would sound. Then attack the problem as described in this chapter.

> You are explaining a complicated concept to your technician, Samuel, so that he can better understand the project. He does not seem to be listening. He keeps glancing at his iPhone. You have noticed that since he has been hired, he does not seem to pay attention to what you say and goes ahead and does what he thinks is best.

What would you say to address this conflict that attacks the person? What would you say that attacks the problem? If you wish, write down what you would say or do in each case. Suggested answers follow below.

- Ineffective responses. Use avoidance; do not say or do anything. Do not speak to the technician; let him do what he wants.

- Attacking the person. "You're not listening to what I'm saying. In fact, you never seem to listen to what I have to say. I feel like I'm talking to the wall. If you don't shape up, you will be replaced."

- Attacking the problem. "We don't seem to be communicating well. I notice that when I talk to you, you don't really pay attention. Is that true? [He may tell you that he is listening.] It would help if you gave me some indication that you are listening to me. Is there anything I can do to be clearer? [He tells you that he is better at seeing things than listening to words.]

Let's go to the white board and I'll describe it to you there. Then we can go back to the lab and I'll show you."

Write down a problem that you have with someone in your team or group.

- What are one or more things that you could say to attack the person?

- Write down your likely responses or reactions. Create a dialogue that allows you to experience this hypothetical conversation. Using self-awareness, try to capture your feelings during this dialogue. You may find that you were angry, and that your attack was a reaction to that feeling. Practice noticing when you feel anger or frustration as a result of someone's behavior. Once you notice your anger, you can consciously redirect it to the problem.

- What are one or more things that you could say to focus on the problem?

- Write down your likely response. Create a dialogue that allows you to imagine this hypothetical conversation.

Were you able to find a way that clearly communicated the problem using the second approach? If not, reread the examples in this chapter to find other ways of presenting or explaining the problem.

CHAPTER **6**

How Am I Doing? Setting Goals, Giving Feedback, and Doing Performance Reviews

I n Chapter 5 we introduced some of the challenges you face as a scientist managing other scientists as well as several approaches you can take to navigate these challenges especially in a team setting. Here we will take a more detailed approach to many of the issues raised in Chapter 5 and will focus specifically on how to manage individual scientists in ways that promote and enhance their productivity and learning. We'll start by reviewing data from two studies that bear on the productivity and state of mind of postdocs but that we believe are applicable to all scientific and technical personnel. Then we'll introduce a series of tools that you will use to set expectations and scientific goals for your team members and to ensure that progress toward these goals stays on track. Finally, we'll show you how to integrate Individual Development Plans (IDPs) into this scheme and how they can be used as both a motivational and productivity enhancing tool.

▶ Depressed, me?

▶ An amazing survey you have probably never heard of

▶ Laying the groundwork for performance management

▶ Dreaming the impossible dream: Setting goals for scientists
SMART goals
Match the goals to the person

▶ "Can I give you some feedback?"
Now or later? public or private?

▶ Performance reviews
Technical and support personnel need reviews too

▶ "Take my advice..."

▶ Individual development plans

▶ You tried all that, now what? Employee performance issues

▶ Performance management summary: An organized way to provide structured oversight for your team

▶ References

▶ Appendices
1. Compact between postdoctoral trainee and mentor
2. Combined postdoc individual development plan and performance review
3. SMART goals

DEPRESSED, ME?

In 2010 The Massachusetts Institute of Technology (MIT) performed a survey among its postdoctoral scientists.[1] The "MIT Postdoctoral Life Survey" was designed to capture information about an impressively large number of facets of postdoctoral life at MIT. The responses to one question of the report stood out to us. The question was "Have you received a written or

[1]http://web.mit.edu/ir/surveys/pdf/Postdoctoral_Life_at_MIT_Report_June_2011.pdf.

oral performance evaluation while a postdoc?" The responses, taken from that report, are shown in the figure below (the title of that figure is ours). Of the responding postdocs (there were 844), 54% said that they had never had either a written or verbal performance evaluation, 31% said that they had received informal feedback, but it is unclear exactly what that means, and only 15% claimed to have ever received what they recognized as an actual performance review (written or verbal). A similar survey from the University of Chicago, done in 2011, focused on postdocs in the biological sciences. That survey, asking the same question as the MIT survey revealed very similar statistics. 16% answered "Yes" to the question, whereas 84% answered "No."

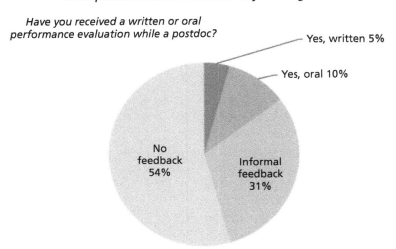

Most postdocs have no idea how they're doing

Have you received a written or oral performance evaluation while a postdoc?

Yes, written 5%

Yes, oral 10%

No feedback 54%

Informal feedback 31%

MIT Post Doctoral Life survey, 2010, *n* = 844

We have no way of knowing how representative these data are, but our personal experiences suggest that they are reflective of the majority of academic research institutions both in the United States and internationally. It is understandable why this would be so.

First, from the perspective of the principal investigator or lab head, running a research group, writing grant applications (and their seemingly interminable resubmissions), writing manuscripts for publication (and *their* interminable resubmissions), teaching courses, and worrying about promotion and tenure are enough to keep any academic researcher working overtime. Given these pressures, it is not hard to understand why many lab heads fail to find time to provide thoughtful performance feedback and career guidance for their staff. It is much easier to do what you know how to do well—research—than what you have never been trained for—managing people.

Second, from the perspective of academic institutions themselves, it is very hard to motivate faculty and research staff to undertake tasks that they do not see as directly contributing to their scientific productivity (in this chapter we will argue that this assumption is entirely wrong). Periodic performance reviews, career guidance, and other mentoring activities can be seen as onerous administrative burdens forced upon busy researchers by "those people in human resources." As we will see in Chapter 10 (The Slings and Arrows of Academe), except in cases of frank misconduct or egregious behavior, academic administrations, and human resources departments especially, have limited leverage over faculty and how they run their research enterprises. Although we know that there are many caring mentors who willingly provide thoughtful and useful guidance to their mentees, the above data show that there are also many who do not. Unless these mentors see that it is in their own best interest to conduct performance reviews and provide thoughtful guidance to their team, there is no reason to expect that their behavior will change. This chapter aims to show that it is very much in the interest of the team leader, mentor, or manager to provide thoughtful reviews, timely feedback, and well-thought-out goals to their team members.

In some segments of the scientific community attitudes are already changing. Starting in 2009, the National Science Foundation required that new grant applications containing funding for postdocs contain a mentoring plan:

> In no more than one page, the mentoring plan must describe the mentoring that will be provided to all postdoctoral researchers supported by the project, regardless of whether they reside at the submitting organization, any subrecipient organization, or at any organization participating in a simultaneously submitted collaborative project … . Examples of mentoring activities include, but are not limited to: career counseling; training in preparation of grant proposals, publications, and presentations; guidance on ways to improve teaching and mentoring skills; guidance on how to effectively collaborate with researchers from diverse backgrounds and disciplinary areas; and training in responsible professional practices.[2]

Some academic institutions are also beginning to follow this lead. Specifically, MIT (presumably having taken note of its own 2010 survey results) now requires its faculty to prepare individual development plans for their postdocs. The MIT website for postdoctoral researcher's states:

> Postdocs and their supervisors should plan for an introductory meeting to set overall goals and construct a mentoring plan, which must be provided with the initial appointment. The annual renewal of an appointment must include a progress report on the career development plan including future goals that is co-signed by both the postdoc and advisor and reflects an in-person meeting.[3]

Before we get too excited about career development plans, which we will cover in detail at the end of this chapter, let us note that they were not specifically what the surveys cited above asked about. The survey focused on performance reviews. Although it is true that updating a career development plan could and should include a performance review, it is also possible that it will not. This is especially true if the career development plan is being done by an adviser simply to "check off the box" and get it over with. So, to get back to our earlier question, if you are a PI or team leader, why should you spend a lot of time on this? Let us tell you.

Not to pick on MIT, but because they have been so generous with their survey data, let us take another look at it. In another part of the survey they asked a question about the postdocs' state of mind.

---

[2] https://www.nsf.gov/pubs/policydocs/pappguide/nsf16001/gpg_2.jsp#IIC2j.

[3] https://postdocs.mit.edu/career-development/mentoring-and-advising.

So, is it any surprise that postdocs are stressed out?

During the current academic year, how often, if ever, have you...

| | | Degree type (self-reported) | | |
		Master's	Doctoral	Overall
Felt so depressed it was difficult to function	Rarely or never	71.7%	61.3%	64.8%
	Occasionally	20.7%	26.2%	24.4%
	Often	5.1%	8.5%	7.4%
	Very often	2.5%	3.9%	3.4%
	N	80	1616	2422

MIT Post Doctoral Life survey, 2010

39%!

When asked "During the current academic year, have you felt so depressed it was difficult to function?," 3.9% of postdocs answered "very often," 8.5% answered "often," and 26.2% answered "occasionally." So that is about 39% of postdocs who at least occasionally felt too depressed to function. Think about that. Not just depressed—we all feel a bit depressed on occasion—but too depressed to function.

It is quite likely that these data are representative of other academic research institutions as well and a recent publication confirms that this is the case for graduate students. In a survey of 2279 graduate students in multiple scientific disciplines, Evans et al. (2018) used clinically validated measures for scoring anxiety and depression. They found that 41% of graduate students scored as having moderate to severe anxiety (compared to 6% of the general population) and 39% had moderate to severe depression (compared to 6% of the general population).

There are many facets of pre- and postdoctoral life that are stress-inducing, including career uncertainty, publication pressure, experiments not working, and feelings of isolation (especially among pre- and postdocs from distant countries and other cultures). As if all this were not enough, if you are a pre- or postdoc and you have no idea how you are doing or, perhaps more importantly, how your adviser thinks you are doing (our discussion above suggests that might be as much as 85% of all postdocs), then it is understandable that you would feel depressed on occasion. In the Evans et al. survey, the majority of graduate students who reported anxiety or depression disagreed with the statements that their adviser is an asset to their career and that they felt valued by their adviser or mentor.

So, let us spell it out. If you are not giving your postdocs (or grad students or technicians) feedback and helping them know how they are doing and how to improve their performance, you are shooting yourself in the foot. Unhappy, confused, and uncertain postdocs and graduate students are unproductive postdocs and graduate students. It is that simple. "More proof!" you demand? Read on.

AN AMAZING SURVEY YOU HAVE PROBABLY NEVER HEARD OF

Take a trip back in time with us. The year is 2006. George W. Bush was president, the last entry of the Human Genome Project was published in *Nature*, and (drumroll) Geoff Davis (visiting scholar at Sigma Xi in 2006 and now a quantitative analyst at Google) completed a survey of postdocs for the Sigma Xi Scientific Research Society.[4,5] This was not just any survey it was a SURVEY. Geoff contacted 22,400 postdocs at 47 U.S. institutions (40% of all U.S. postdocs at the time). Amazingly, 38% (8512) responded. For the nonscientists among our readers, take note, big numbers (and these are big numbers) can mean robust data.

The title of the survey was "The productive postdoc: Do working conditions affect outcomes?" What outcomes were they interested in? The very same ones that every principal investigator, lab head, or manager is or should be interested in: scientific productivity and job satisfaction among others.

Broadly speaking, the strategy of the survey was to see whether correlations existed between certain facets of the postdoctoral experience and both subjective outcome measures (job satisfaction, quality of adviser relationship) and objective ones (publications submitted, grants received, absence of conflict with adviser, etc.).

First, let us look at those working conditions that showed no correlation with productivity. Sadly (for those seeking higher pay), neither salary nor benefits showed any correlation with productivity (as measured by papers submitted per year per postdoc). Similarly, postdocs with independent funding were no more productive or satisfied with their experience than where those funded by the principal investigator's grants.

Structured Oversight

	High structure	Low structure
% satisfied	80	60
Advisor grade (0 = F, 4 = A)	3.4	2.7
% reporting conflicts	9	21
Papers submitted/year	1.4	1.0

$N = 8{,}360$; all differences significant at $p < 0.001$

What does make a difference? Here's where the survey gets interesting. Davis found that postdocs whose training offered a higher than average "structured oversight" were at a significant advantage (see figure above). Structured oversight includes *formal performance reviews* for postdocs, the existence of IDPs, clear lab or institutional polices guiding research conduct, and written letters of appointment. Labs that scored high by this measure had more of these activities or structures than those that scored low.

The table above shows that postdocs in labs that scored "high" in these structured oversight parameters had 40% more publications per year per postdoc than those from labs that scored "low." That should be music to every PI's ears. Moreover, postdocs in "high structure" labs reported fewer conflicts with the PI and in the lab and gave their advisers higher ratings than those in low structure labs.

Other facets of the lab environment also had significant correlations with productivity. Davis measured what he called "transferable skills," which included such things as training in grant writing, project/lab management, exposure to nonacademic career options, training in negotiation and conflict resolution, English language, etc. What kind of difference did such training make? Look at the table below.

[4]A copy of the summary PowerPoint deck can be found at http://slideplayer.com/slide/6195082/.

[5]"Doctors Without Orders. Highlights of the Sigma Xi Postdoc Survey," Special Supplement to *American Scientist*, May–June, 2005.

Transferable Skills Training

	High training	Low training
% satisfied	83	56
Advisor grade (0 = F, 4 = A)	3.4	2.7
% reporting conflicts	10	17
Papers submitted/year	1.3	1.1

$N = 8,360$; all differences significant at $p < 0.001$

Those receiving "high" levels of transferable skills training rated their advisers higher, reported fewer conflicts with their adviser and in the lab, and—hold on to your hats—submitted 30% more publications per year than those whose training scored "low" in transferable skills training. These are impressive data.

Good scientist that he is, Davis addresses the burning question about these results that is in everyone's mind—namely, does correlation equal causation? Are structured oversight, feedback, and transferable skills training responsible for the greater productivity outcomes, or are structured oversight, training, and productivity all themselves related to or caused by some other unmeasured parameters that are responsible for the outcome? You probably already know that the answer is, "We're not sure, and although we have lots of reasons to believe in a causal relationship, we cannot prove it." Davis suggests two possible mechanisms for causation. One is that labs that have better training and oversight tend to attract intrinsically more productive postdocs. Another is that the structure and training themselves promote productivity. Both are plausible. We cannot know whether these individual elements directly impact performance or productivity or rather are simply proxies, reflective of well-organized institutions and labs that provide a supportive environment to which postdocs respond by being more productive and collegial. Either way, however, if by instituting such practices you can make your lab or organization more supportive, it will be to your and the trainee's benefit.

For us, the bottom line is this: If there is even a chance that providing routine performance feedback and instituting training and career development plans for postdocs enhance their productivity, why wouldn't you spend some of your time doing this? Do you know of anything else that has even the possibility of boosting your postdocs' productivity, let alone by 40%? We're willing to bet that you do not. For us it is kind of like Pascal's argument for believing in God. He argued that even if God's existence can't be proven, it is relatively painless to believe in God and if it is true, the payoff (eternal salvation) is huge. In the remainder of this chapter we will show you how to give your postdocs feedback, performance reviews, and advice that they can use, how to include individual career plans in this process, and, coincidentally, how to help yourself along the way.

LAYING THE GROUNDWORK FOR PERFORMANCE MANAGEMENT

The remainder of this chapter will deal with what is referred to as performance management in the Human Resources (HR) world. Performance management refers to the approaches you take, the tools you use, and the systems you put in place to make your science team as productive as possible. The sections above suggest that structured oversight, performance reviews, and feedback are all important elements. Let us review how they come into play in your lab.

Before you put a scientist or trainee to work in your lab, lay a solid foundation by communicating how your lab operates and what the two of you can expect of each other. The first step is to orient them to your lab and its rules and practices. Some labs have a routine introductory program when new hires are introduced to standard operating procedures (SOPs), equipment use rules, and mandated training (biosafety, radioisotopes), whereas in others the lab head or a senior postdoc is responsible for educating the new hire in the way you do things. In any case, do not ignore this.

The second step is to review with them and jointly sign an advisor/advisee compact that defines what you expect of them and what they should expect of you. We have included an example of such a compact in Appendix 1. This document defines high-level expectations for both mentor and trainee in matters of interactions, behaviors, performance, and ethical matters. We believe that signing such a document has both practical and symbolic importance. By signing it, you signal to the trainee that you take your job as mentor seriously and that they can expect certain kinds of support and guidance from you. Conversely, by signing it, the mentee commits to taking their job and role seriously and to work and perform according to certain reasonable expectations.

Once you have taken these two steps you are ready for the next steps—setting research goals with them and putting them to work.

DREAMING THE IMPOSSIBLE DREAM: SETTING GOALS FOR SCIENTISTS

Before we embark on this section, let us agree on some definitions. "*Goal setting*" refers to an agreement between a supervisor and trainee or employee about that person's objectives over some defined period of time. Goal setting is a collaborative undertaking and is typically done infrequently. "*Feedback*" refers to guidance given to the trainee or employee about how they are doing both in accomplishing their goals and in other areas such as behavior, interaction, and collegiality. Giving feedback generally involves a face-to-face dialogue and is done frequently. It is focused on what is being done well and should be continued and on what can be improved. "*Performance review*" is a formal documented process in which performance is assessed. Performance reviews are done periodically (once or twice a year) but done consistently. Like feedback, they are focused on what is being done well and what can be done better. Goal setting, feedback, and performance review comprise a three-legged stool of performance management. Take one away and the structure collapses.

Performance management elements

All scientists need to have goals. If they do not, neither they nor you have any way of answering the question in this chapter's title, "So, how am I doing?," and you are not in a position to give them the feedback that they need to improve performance.

Although the setting and acceptance of goals in science is common if not universal in the private sector, we sometimes encounter uneasiness from those in academia. We hear comments like "How can you set goals if you are doing basic research? You can't set a goal to discover something, it just happens." Also, if the goal of your project is to cure pancreatic cancer or to sequence an entire human genome in 10 minutes, those might not be things you want to get evaluated on at the end of the year.

A better way to think about goals is that they come in two flavors. The first is what we call "*aspirational*" goals. Aspirational goals include goals such as "Our goal is to cure pancreatic cancer" or "Our goal is to discover novel life-forms on the ocean floor below 10,000 feet." These are things you aspire to but that you may not realistically expect to be able to achieve within any predefined time. Nonetheless, aspirational goals inspire and motivate everything you do. Without them you are just doing experiments.

Next come "*operational*" goals. Operational goals are the things you need to achieve in order to make progress toward your aspirational goals. Because aspirational goals are often so "high level," you might need more than one level of operational goals before you can start planning an experiment. For example, for the aspirational goal of curing pancreatic cancer, one first-level operational goal might be "Determine the role of TGF-β in modulating the severity of pancreatic cancer" and a second level goal below that might be "Within six months, compare levels of TGF-β mRNA with median survival in 100 pancreatic biopsies." This last goal is actually one you could hold someone accountable for, whereas the aspirational goal is not, nor is the first-level operational goal, but for other reasons discussed below.

Operational goals also enable you to manage discovery-based labs. Although it is true that you cannot put discovering something on a timeline, you can definitely make a list of activities that you believe will advance you down the path of discovery, and you can hold people accountable for those activities even if you cannot hold them accountable for the aspirational discovery. So, although I probably would not want to get held accountable for discovering new life-forms on the ocean floor, I might agree to something like "Collect 300 water samples at ocean floor depth from a 10 square mile grid 150 miles east of Bermuda within 9 months." and "Prepare samples for light and electron microscopic analysis within 12 months."

Creating operational goals from aspirational goals is called "projectizing" your work. Even the highest-level, most lofty aspirational goal can be projectized by breaking it down into specific, achievable activities for which you can hold people accountable. Those activities in turn serve as the basis for your frequent feedback and periodic review.

Of course, nothing ever goes as planned in research, so you need to be flexible and willing to change goals or modify timelines as you encounter bumps in the road. The quote (attributed to various sources) "No battle plan survives contact with the enemy" can be restated as "No experimental plan survives the first experiment." When the real world intervenes, you need to adapt. But changing goals too often, especially aspirational goals, can be demoralizing and confusing to scientists. There is no reason to continue down a path that you believe is a dead end, but do not imagine dead ends where there are merely obstacles.

SMART goals

SMART goals are possibly the most hackneyed tool in the Human Resources professional's tool kit. Yet every time we present the concept to a group of scientists and ask how many know what they are we are amazed to see virtually no hands go up.

SMART Goals	
Specific:	What exactly will get done?
Measurable:	What measure will be used to assess progress or completion?
Attainable:	Goals are challenging and not unrealistic.
Relevant:	Goals are relevant to the person or group's project and objectives.
Time-specific:	Goals have a specific complete-by date or time frame.

In the previous section we gave examples of a couple of operational goals for which we thought scientists could be held accountable. They may not seem all that sophisticated to you, but they have something in common. When you read those goals, you come away with a clear sense of exactly what needs to be done by whom and when. Without those features, goals are of little use. If instead of "Compare levels of TGF-β mRNA with median survival in 100 pancreatic biopsies by July, 2018" we had written "Measure TGF-β in samples from patients with different disease profiles as soon as possible" you wouldn't know which or how many samples, exactly what to measure (message, signal, degradation products?), when you needed to do this by, or what we meant by "disease profile." In short, this is not a SMART goal. What is a SMART goal? Look at the figure above. SMART goals answer the all-important questions of exactly what you want done (Specific), in a way that you can measure or quantify (Measurable). They also are Attainable (goals can be ambitious but not impossible) and Relevant (don't saddle people with goals that have nothing to do with their jobs or projects) and, perhaps most important, Time-specific (what is the target date for the completion of this goal?). We believe that the operational goals we created in the previous section have these characteristics (see Appendix 2 for more on SMART goals).

Scientists are notoriously skittish about associating times with their goals. It feels too much like a deadline and scientists know that very few things in science regularly happen according to a deadline. There are too many uncertainties, unknowns, honest mistakes, and incorrect assumptions. Although these concerns are legitimate, timelines are necessary elements of goals and expectations. Timelines are targets to work toward and help keep people focused, motivated, and maybe just a bit anxious. If a timeline is not met because of unexpected scientific complexities that you had no way of anticipating, you need to change the timeline or your approach. If it was not met because the person responsible for the work just was not putting enough time into the project, you need to give them guidance on how to better allocate their time and monitor them for improvement. If you do not have a time target, you are not working on a project, you are engaging in a hobby.

Match the goals to the person

Goal setting should be a collaborative undertaking to a degree that matches the experience and role of the employee or trainee. No one, especially self-reliant postdocs, likes to be told what to do. They would much rather decide for themselves or, at the least, play a role in the crafting of goals. Junior people may need to be "given" goals until they get the lay of the land. But even if you are giving someone one or more goals you will always be better off asking the trainee to come up with some of their own as well. You do not need to accept these at face value and can instead use them as learning tools to help the trainee become adept at setting challenging and useful goals. The idea of matching the goals to the person is illustrated below.

Match goal specificity and management style to the person

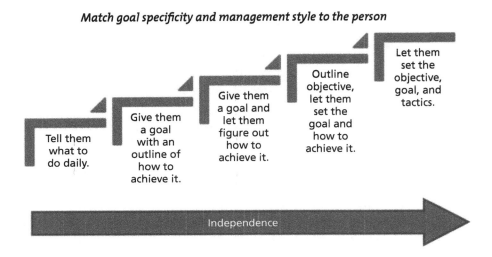

The point is that different people need different types of goals and different levels of guidance. You need to be deliberate and thoughtful about how you manage the people in your team. At one of Carl's workshops we were discussing this topic with a panel of scientists. One of them, Adam Siepel of Cold Spring Harbor Laboratory, said something that stayed with us. He said (I'm paraphrasing), "One of the most profound realizations I made after I had been managing a group for several years was that, without realizing it, I was managing everyone as though they were each a younger version of me." Only when Adam realized that he was doing this was he able to step back and calibrate his management style to the needs and situations of each person.

"CAN I GIVE YOU SOME FEEDBACK?"

Once you have set and agreed to goals with your team members all you need to do is sit back and wait for the spectacular results to roll in. If only.

The next step in the performance management process is to monitor progress and provide ongoing and frequent feedback. Feedback, the second leg of the three-legged stool of performance management, is your most effective tool to ensure that employees are productive and effective. It falls into two domains: science and behavior. Feedback on science includes whether the employee is making progress toward their goals, and if not, why not. Feedback on behavior focuses on whether the employee is meeting your expectations for interacting with others, collaborating well, producing results on time, and being responsive to input and feedback. We often find that although scientists have no trouble giving feedback on science, they are less likely to give feedback on behavior unless it has become so extreme that others are complaining. Becoming adept at giving feedback on behavior will limit the likelihood that simmering interpersonal matters will erupt into full blown crises.

Who among us upon hearing the words "Can I give you some feedback?" doesn't shudder with dread? Why? Because often when someone says this they really mean, "Sit down and let me tell you what you did wrong." And worse, those are the *only* times we get any feedback. It is true that sometimes people do things wrong. They did not do the right experiment or did not do it the way they were supposed to or they forgot to control for some crucial variable or they are being obnoxious to everyone else in the lab. Or all of the above. So, at times, we definitely feel like telling them what they did wrong. But what does the intended victim of our so-called feedback hear after we utter these words and start on our harangue? They hear static. When people are hearing feedback that is accusatory, demeaning, insulting, and unhelpful, they mentally shut down and go into a defensive mode that limits their ability to actually hear and respond to what you are saying. Instead they are either mentally focusing on marshaling rebuttals to your accusations or are overwhelmed by feelings of despair and shame. In either case, they are no longer listening. So, when they start pushing back, or when they just look at you blankly, your first response is "You are not listening to me. You didn't hear anything I just said." And you would be right.

Just as employees may have an aversion to hearing feedback, we as managers have our own aversion to giving it, especially on behavioral matters. Reasons for this aversion may include fear of hurting or insulting the other person or of fracturing the relationship. In the following, we show you how to give feedback that is helpful as opposed to hurtful.

Effective feedback focuses on
• What is being done well and should be continued.
• What can be done better (not what is being done poorly) and can be improved.

For feedback to be useful, it needs to be heard. The best way to ensure that your feedback will be heard is to use the Feedback Golden Rule: *What was done well and should be continued and what could be done better or could be improved.* That is, the most effective feedback is framed in terms of improvement rather than as criticism. Most people are more than willing to hear how they can do better but understandably averse to hearing how they messed up. The beauty of this approach is that it actually accomplishes your objective—improved behavior or performance.

The second most important facet of useful feedback is that it is framed in terms of what the person did rather than in terms of why you think they did it. Stated another way, effective feedback is focused on behaviors and not on what you may think is the underlying cause of the behavior. Compare "It looks to me like your work has slowed down considerably recently. Is there something we need to discuss about this project or the work to get back on track?" with "You seem completely unmotivated recently." The first describes observable behavior (less work), whereas the second provides what you think is the underlying cause of the behavior (lack of motivation). Although that may be true, you cannot know it in advance without a conversation with the person. Or consider this: "I'm having a hard time reading through your lab notebook when we sit down together. I'd like to suggest that you try to take your time when writing and perhaps do your calculations on a separate page to reduce clutter. Does that make sense?" versus "Your notebook is a mess. Why are you so careless?" The first is focused on the behavior and the output (the illegible notebook), whereas the second describes what you think is the underlying personality trait responsible for the behavior (carelessness). Telling someone they are careless may be what you feel like doing, but if you want their behavior to change you will focus on the observable outcome (cluttered notebook) and how it can be improved (write slower, move your calculations).

Feedback that is focused on what can be done better is more than just telling them something and sending them on their way. It includes the two of you jointly creating and agreeing on a plan to improve the performance or the behavior.

Giving positive feedback is usually easy, although most of us do not do it nearly often enough. As we noted above, one reason people get so skittish when they hear the word "feedback" is that it is usually used in the context of "what you messed up most recently." So, if you are in a leadership position, giving out more positive feedback is a great way to desensitize your team from breaking out into a sweat at the sound of your voice. A good rule of thumb is to give out two to three times as much feedback focused on what was done well than on what needs to be done better.

Not only is feedback that is focused on needed improvements hard for the receiver, it is also hard on the giver. We all have a certain degree of discomfort when sitting facing another person getting ready to tell them something that they do not want to hear. Such situations can feel highly charged and can evoke strong emotions in both parties. When this happens, when we are feeling anxious about delivering uncomfortable news or guidance, we may easily lose track of exactly what it was we wanted to say in the first place. One way to counteract this natural response is to have a script for the conversation. We do not mean that you should write out exactly what you are going to say word for word, but rather that you have a mental map (or an outline on a pad of paper) of how you want the feedback conversation to go.

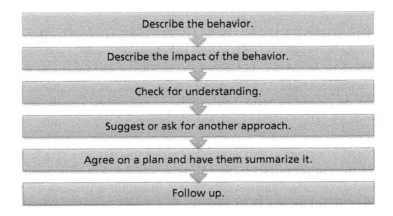

Now we will review how such a conversation might go.

Step 1. Describe the behavior or incident.

"We agreed that you would check with me before you moved on to the next phase of the project. That didn't happen. I would have told you to get the cell sorting data before proceeding." (Versus personality focused feedback "You deliberately left me out of the loop so you could do it your way.")

Step 2. Describe the impact of the behavior.

There are multiple impacts of the behavior described. It could impact the science, the project, and the team. It could also affect your view of the person if the behavior suggests that they do not follow instructions or keep agreements.

An example of feedback addressing the impact on the project or the science is:

"The absence of the cell sorting data leaves a big hole in our analysis and puts our scientific rationale on shaky footing."

An example of feedback focused on the effect of the behavior on the group is:

"If we don't have that data for next week's investigators' meeting it's going to look like our group doesn't follow up on our commitments."

An example of feedback focused on how this behavior affects your view of the person is:

"You have lived up to our agreements in the past and I need to trust that you will to do that in the future. Can you commit to that?"

Which type of feedback should you focus on? There is a school of thought that says that you ought to focus on the impact that will resonate most with the person receiving the feedback (Briggs Meyers and Meyers 1995). But figuring out what kind of focus will have the most impact depends on your knowing a lot about this person's personality. In our view you should cover all bases.

"This may have been an oversight but if we don't have that data for next week's investigators' meeting it's going to look like our group doesn't follow up on our commitments. Also, the absence of that data leaves a big hole in our analysis and puts our rationale on a shaky scientific footing. Can you commit to living up to our agreements in the future?"

Step 3. Check for understanding.

Make sure they heard you, that they understand what you said, and that they agree with your characterization of their behavior. They do not need to agree with your assessment of the impact, but they do need to understand your view point.

You: "Have I got this right? Am I missing something?" and "I'd like to hear your view of this. Do you have any thoughts or reaction to what I just said?"

Wait and listen for a response. There may be an innocent explanation or a misunderstanding. If not, move on to step 4.

Step 4. Solicit or suggest alternative behaviors.

You have described the behavior or incident and told them why it is important. You verified that they understood you. Now it is time to involve them in making sure it doesn't happen again. As in most problem-solving discussions, the more they are involved in crafting the solution or path forward, the greater will be their ownership of the solution. So "This is the way we're going to do things in the future" is not the approach you want to take.

You: "How can we make sure that this doesn't happen again?" or "I'd like to hear your ideas about how we can prevent this in the future. Do you need some time to think about it?"

If they do not have any ideas and seem stuck, you can try, "I have some ideas about making sure this doesn't happen again. Would you like to hear them?" Asking for permission to present your ideas shows that you respect their agency in the situation. Offering options rather than commands puts them in a position of having to buy in rather than having to passively accept.

Step 5. Agree on a plan and follow up.

Here you summarize or, even better, ask them to summarize what you have agreed to.

You: "So, how about summarizing for us what you'll do differently starting next week?"

This and the follow up is important. When you have given someone feedback you need to reinforce positive movement toward the new behavior with more feedback. "Thanks for getting that data to me before the investigator's meeting. It made a big difference in our ability to decide on the next steps to take." If your initial feedback was not helpful and the behavior or incident is repeated, then you need a second meeting. As in the first, you describe the (now-repeated) behavior or incident using the same approach as the first time.

We cannot emphasize enough the importance of involving the employee in and even giving them primary responsibility for the process of finding a solution or new behavior. As noted above, without involvement there is no ownership. Examples of other types of feedback using this same paradigm are given in Chapter 5.

Now or later, public or private?

The best time to give feedback is at or as close as possible to the time that the behavior or incident happened. The worst time is when you are feeling angry or frustrated with the person. If you are angry, you can be sure that your feedback will come off as a reprimand or worse and that you are likely to say something that you will regret later. Sometimes it takes only a short time for anger or frustration to abate and other times you may need to wait a day. Remember, your goal in giving feedback is not to vent your feelings on the recipient but to provide them guidance and input that helps them improve their performance. That is not to say that you need to hide your feelings completely. Expressing happiness or appreciation for something done well is appropriate and useful. Even mild expressions of frustration can be useful in order to make a point. As in negotiation, however, expressions of emotion, especially "negative" ones, are best when used strategically. So, for example, you can consciously decide to show some frustration in order to make a point or to grab the attention of someone who doesn't seem to be taking what you are saying seriously. When used strategically, expressions of emotions are tools that are under your control. Alternatively, when you allow emotions to spill out unfiltered you are under their control and the consequences may be regrettable.

If you have feedback that is not positive, such as a helpful critique of an experiment or presentation, it is best to do that one-on-one in your office. We have espoused this view in our workshops and sometimes get the following question, "Don't you want to critique in public so people in your group see what your standards and expectations are? Do you want people who have just sat through a lousy presentation to see you just sitting there?" Fair enough—you do want people to know what your standards are. But you do not want to do this at the expense of a team member who may feel humiliated if critiqued in public. That is not to say that you should say nothing. When the leader makes a comment like, "It looks like this presentation needs some work. I think it could be better organized and shortened. Why don't we meet in my office after the lab meeting and go over it?," everyone in the room will get the message loud and clear.

We provide a list of the characteristics of effective feedback and ineffective feedback and short examples of each.

Effective feedback is	Ineffective feedback is
Specific "It seems to me that things have really slowed down in the lab. I'm wondering if you're spending enough time on this project." Or, "I've noticed you have been out of the lab six days in the last month. That's affecting your work output. Is there some problem we need to discuss?" Or, "That was a really well-designed experiment. The way you anticipated the problem of bacterial contamination was especially clever."	**Nonspecific** "You need to work harder." "Good work!"
Direct "Hi, Ramesh. Your report was due yesterday. It's holding up the grant submission. Can you meet me in my office in 30 minutes to discuss?"	**Indirect** "Hi, Ramesh. I guess I'll be getting something from you later today?"
Helpful and supportive "It looks to me like the experiment needs to be repeated. I know it's a lot of work. Let's go over how to make sure it comes out better next time." "Can I show you where I think your report could be improved? If you reorganize the first section... ."	**Judgmental and unhelpful** "This experiment is a mess. I'm very disappointed." "Your part of the progress report is the weakest. You need to re-do it."
Focused on observable behavior "Your experimental records are very difficult for me to read. For example, I can't distinguish the raw data from the calculations. What do you think you could do to make this easier to read?" "It seems to me that your work has slowed down. You haven't accomplished any of the goals we get for your project in the past three months."	**Focused on the personality** "You need to stop being so disorganized." Focused on what you think is the underlying cause of the behavior: "You seem to be completely unmotivated."
Thoughtful "Let's go over your presentation tomorrow morning after we've both had time to get our thoughts together."	**Impulsive** "That was just about the most disorganized presentation I ever heard."
Timely (given at or close to the time of the incident or behavior) "The report was due last week. In the future, I really need them to be on time. Can you commit to that for the next progress report? Is there something that will prevent that?"	**Long after the fact** "Let's not have a replay of last year when your report was so late."

(Continued)

Effective feedback is	Ineffective feedback is
Just enough "I'd like to go over your talk, particularly with a view to condensing it and clarifying how you treated the data in the second part."	**Too much** "Your talk yesterday was way too long and then you started late because you didn't have the handouts ready. Also, I didn't understand your data analysis in the second half at all, which should have been a separate talk. Mohan said that, and then you totally ignored him and made a sarcastic comment. By the way, your lab bench is a total mess."

PERFORMANCE REVIEWS

The third and final leg of the three-legged stool of performance management is the performance review. Performance reviews have gotten a bad reputation in the popular press, and for good reason. If you are getting feedback on your performance once a year, it is likely that the things your manager is going to remember or comment on are the most recent mistakes you have made. As a result, and as we have noted earlier, what was billed as a performance review feels more like "What you've messed up most recently." When done by managers who do not make it a habit of giving frequent and useful feedback, mandatory annual performance reviews get to be treated as a burden. The result is a perfunctory "fill in the blanks on the form" exercise that is unhelpful to the employee and possibly demoralizing as well. Yet, as we showed at the beginning of this chapter, postdocs whose training includes structured oversight, including performance reviews, may do better in a number of quantifiable outcome measures (including publications) than those whose training has less such oversight.

The performance review offers you the opportunity to step back from day-to-day issues and to focus instead on more global themes. Is the project on track? Are the individual's goals being met? Are the goals still relevant or should they be modified? What is your view of this person's interactions with others? Are they collaborative and helpful? Are they gaining the skills they will need for the next step in their career? The discussions that these questions engender are not casual conversations to be had at the lab bench or in the corridor. They require a dedicated time and a quiet space. Our experience shows us that employees who see that their manager has thought about them and their goals, who gives them thoughtful feedback, and who cares as much about their career options as about their completing one more manuscript before they leave are more committed and engaged in their work than those who do not benefit from such an environment.

Like effective feedback, performance reviews focus on the feedback golden rule: "What is being done well and should be continued. What can be done better and what could be improved."

Performance reviews should focus on both what the person did and on how they did it. That is, on progress toward the goals that you have previously established with the person as well as on the person's behavior and interactions with others. It will be no surprise that the behaviors that we are going to be monitoring in the performance review are closely related to both the technical competencies and "personal attributes" that we look for when interviewing candidates for the lab or group and that are listed in the Face-to-Face Interview Form in Chapter 4. As with giving feedback, we suggest that you use a playbook or script to guide you through performance review interaction and we provide both below.

We have created a document that combines the performance review with the individual development plan (Appendix 2) that you can use if your institution does not provide one. You can use the performance review sections and the IDP section of this document together or separately. Some institutions may require you to use their performance review form or format. In many cases scientists find those forms lacking because they can be rather generic and it can be a challenge to shoehorn the review of a scientist into a format designed for a wide spectrum of job types (because of inclusion of such categories as customer focus, cost-consciousness [try that in a research setting!], and exclusion of science-specific elements). If you do not have such a requirement or have the flexibility to use another form, you can use the one that we provide in Appendix 2.

Start the review process by forwarding the form to the employee and having them fill out Sections A, B, and C. Section A contains the list of the goals the two of you have previously agreed to. For each goal the form asks the employee to assess their progress at a high level (Completed, In progress, or Not started). Any goal that was not completed will serve as the basis for a discussion during the face-to-face review.

The form also asks the employee to answer questions regarding their views of their mentoring relationship with you and to describe any new skills they may have learned. Section A also has a place where they can describe any special circumstances that may bear on why particular goals weren't completed or started.

Section B asks them to make a first pass at goals for the upcoming review period that will serve as the basis for a conversation with you during which these draft goals may be edited or revised. The employee should send you their completed Sections A, B, and C before the scheduled face-to-face meeting, so you can review them in advance. Section C is the mentee's IDP (discussed further below), which should also be reviewed during the same meeting or at a follow-up meeting.

During the performance review meeting you will discuss their view of what was accomplished on each goal and then give your own view. If you have different views of what was accomplished, now is the time to come to an agreement. A flowchart mapping how this conversation might go is shown below.

Once you have agreed on what was accomplished and have provided feedback on scientific progress toward each goal, it is time to agree on which goals were completed, which ones they will continue to pursue, and which ones should be eliminated or put off. Also, new goals may be added at this point, building on what the mentee has written in Sections B and C of the form. Most important is to end the conversation by having the employee summarize for you what goals the two of you have agreed upon for the upcoming period. Never forget this last step. You will be surprised by how many times you thought you agreed to one thing and the other person remembers something completely different. Having them repeat your agreement ensures this doesn't happen and reinforces the agreement. You might encourage trainees to take notes during this discussion so they do not have to rely on their memory.

The second element of a performance review focuses on behaviors. The combined review form in Appendix 1 lists a dozen behavioral characteristics in Section D that you, the mentor, should review and comment on as appropriate in advance of the review meeting. Some of these are of a technical nature, whereas others relate more to personal characteristics that derive from the list of personal attributes we used in our face-to-face interview hiring form in Chapter 4.

You don't need to spend the same amount of time on all of these items but there may be some areas that you will wish to highlight as needing specific attention. For these, the flow of the performance review feedback is identical to that for situational feedback discussed earlier and is shown in the following flowchart.

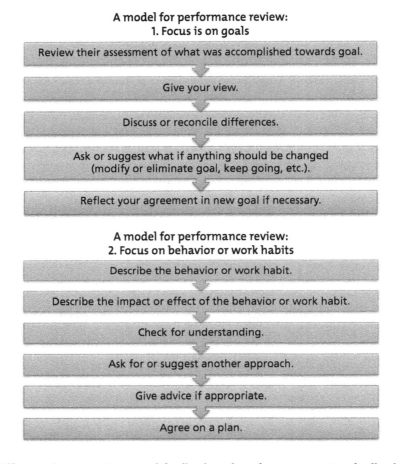

**A model for performance review:
1. Focus is on goals**

Review their assessment of what was accomplished towards goal.

Give your view.

Discuss or reconcile differences.

Ask or suggest what if anything should be changed
(modify or eliminate goal, keep going, etc.).

Reflect your agreement in new goal if necessary.

**A model for performance review:
2. Focus on behavior or work habits**

Describe the behavior or work habit.

Describe the impact or effect of the behavior or work habit.

Check for understanding.

Ask for or suggest another approach.

Give advice if appropriate.

Agree on a plan.

The difference between situational feedback and performance review feedback is that the former is typically incident-focused ("I noticed some friction between you and Ashok at today's lab meeting."), whereas the latter is of a more general nature ("I've noticed that your relationship with Ashok has improved since we met a few months ago. I think that's really helping the project," or "It seems like you've been missing lab meetings a lot these last few months. Is there some reason for that we should discuss?"). Don't limit your behavioral feedback only to the (hopefully) few areas in need of improvement. As you go through the checklist in Section D of the form and use the flowchart above, also highlight areas where the individual is doing a great job and let them know using specific examples of why you think this.

One final note about performance reviews is how often they should be done. Most HR departments require they be done once a year, and if you are a good manager and are giving frequent feedback throughout the year that may suffice. However, during times when performance is changing, and the employee needs more guidance than can be delivered in a brief feedback session, or during the first year of a new employee's tenure, it may be best to schedule a review at six-month intervals. Once things have stabilized, you can revert to every 12 months.

Technical and support personnel need reviews too

Running a lab requires more than ensuring your professional personnel and trainees are doing the right experiments. It also requires supervising support personnel as well. Technicians, lab managers, and administrative assistants provide essential services without which your lab might not be able to function well or even at all. If the people providing these services are not motivated and managed well you may be sabotaging yourself. Just as with your professional staff, your support staff need feedback and performance reviews to help them stay on track and to ensure that they are providing the quality of support you need. Although the focus of feedback for professional staff starts with progress toward goals, the focus for support staff typically starts with what we call performance indicators. Performance indicators are the elements of that person's job that are listed in their job description. For a lab technician, performance indicators might include

- maintain supplies of commonly used lab consumables, chemicals, and enzymes
- train new lab members in the use of shared equipment
- maintain accurate lab biosafety and radioisotope records
- ensure that the lab procedure manual is kept up to date

During the performance review, you will review their performance indicators and jointly assess how they are doing with each one. As with professional staff, it is typically better to start this process by having the employee do a self-assessment in advance of the meeting by rating themselves on each performance indicator. Those who do this are often surprised to find that the employee is often more critical of themselves than they would have expected. As with reviewing goals with professional personnel, this is the time to review indicators for continued relevance and to delete obsolete indicators or add new ones. With some minor modifications you can adapt the form in Appendix 1 for this purpose.

Although support staff are typically thought of as focused on routine or support activities, we find that setting goals for support staff can be highly motivating. Such goals could include learning a new skill or technique or taking on a new administrative task or special project. Remember that people in support roles very often see these roles as stepping-stones to something else. Today's technician in an academic research lab may aspire to go on to graduate school or a higher-paying job in industry. Providing such individuals with learning opportunities is highly motivating and makes for enthusiastic hard-working employees. Sure, they are going to leave at some point, but better to have them motivated and appreciative than disgruntled while they are working for you.

"TAKE MY ADVICE..."

Sometimes you need to provide more than feedback; you need to give advice. Some of us in managerial roles have a love–hate relationship with giving advice. On the one hand, we see ourselves as the technical experts so it is only natural that we have a lot of advice to give. On the other hand, we also see ourselves as mentors helping to develop independent thinking in those who work for us. So, we may seesaw between worrying that, on the one hand, too much advice will sap initiative and make our staff too dependent on us, whereas, on the other hand, giving too little advice risks having them go down the wrong path and waste precious time and resources. As with most of our management advice, the answer to how much advice to give is "It depends." It depends on the level of seniority and experience of the person in front of you, on their degree of independence, and on the importance of the topic you are giving advice on. There are certain areas where you might be inclined to give less advice and let them make mistakes as a learning tool and other areas that are so important that you might be inclined to give a bit more advice to help ensure a timely and successful outcome of a critical project.

As with feedback, how you give the advice will play a major role in how readily it is accepted. Advice is different from a command because it is a suggestion rather than a directive. On the other hand, most of the time when we give advice to an employee it comes with the expectation that it will be taken seriously. If you want more certainty than that you should probably be giving an order or a directive instead. Our guidelines for giving advice are

1. The best time to give advice is when it is solicited by the recipient, "I'd like your advice on how to proceed... ."

2. Whether or not advice is solicited, frame your advice as an option ("I have some thoughts that might be helpful, would you like to hear them?"). If it's not an option, it's not advice, it's a command.

3. Make your advice concrete ("Why don't you try using the modified Pforzheimer technique instead?" vs. "You might try a different approach").

4. Get the recipient involved in crafting the solution ("That's my advice—does that make sense? How do you see it?" "I'd like to hear your thoughts.").

INDIVIDUAL DEVELOPMENT PLANS

Earlier in this chapter we cited a study (see footnotes 4 and 5 on p. 131) that showed that the use of individual development plans for postdocs in particular was highly correlated with productivity measured by publications, as well as with a positive view of the PI. In fact, the use of IDPs was more highly correlated with these outcomes than any other single parameter tracked in that study. There is good reason to believe that postdocs who see that their mentors have their future and best interests at heart and who see how the work they are doing will directly impact their career (both provided by IDPs) will be more motivated and productive than those who feel manipulated and are uncertain of their future. Postdocs who gave clear goals and IDPs know what is expected of them. Written and agreed upon career goals and expectations as they appear in IDPs may make it more likely that postdocs take them seriously and adhere to them. Finally, written goals and IDPs encourage postdocs to take responsibility for their own actions and career path.

IDPs have received support from numerous quarters, especially in the academic sector (see Vincent et al. 2015). Both the NIH and NSF strongly encourage or in some cases require that grantees use IDPs for trainees funded by them. In addition, the NIH requires trainees in its intramural programs (scientists who work at the NIH) to create and use IDPs. Finally, the recent National Academy of Sciences, Engineering, and Medicine report *Graduate STEM Education in the 21st Century* (Leshner and Scherer 2018) recommends that the creation of IDPs be a requirement in all graduate and postgraduate training programs in the sciences.

IDPs come in different forms and formats. Some focus principally on career goals and needed professional and technical skills, whereas others add more detailed elements such as specific research goals and objectives for the coming year. Examples of both used at various academic and research institutions can be found readily on the internet. Although IDPs are different from research plans, and although review of IDP progress is distinct from a performance review, the two categories (research goals and progress vs. IDP goals and progress) are closely related and have significant overlap. Because both IDPs and research progress reviews need to be reviewed periodically (semiannually or annually), it makes sense to us that they be combined in a single document so that both can be reviewed by the mentor and trainee at the same time. As noted earlier, we have created a combined Performance Review and IDP document for use with postdocs or other mentees (Appendix 2 and online in a downloadable editable Word document at www.sciencema.com/downloads). This document draws heavily on the excellent IDP used by the University of Pennsylvania.

YOU TRIED ALL THAT, NOW WHAT? EMPLOYEE PERFORMANCE ISSUES

Your responsibility does not end at giving an employee feedback or even with following up to make sure the feedback is having the desired effect. If your feedback is one in a continuing series of such conversations with an individual, you need to document it for your own records. Rules differ from institution to institution and even state to state but if you think this individual may need to be let go at some point, your documentation may make the difference between a straightforward case and a very complicated one.

Repeated instances of the same behavior or incident may eventually lead to more significant steps, including but not limited to the use of a "performance improvement program" that can be used as the first formal step toward dismissal.

No matter how diligent you are about setting great goals, following them up with thoughtful feedback, and being diligent about performance reviews and IDPs, there will inevitably be some scientists or employees whose performance fails to meet your needs or standards. We have read (and this is consistent with our experience) that about one in seven of new hires across many job sectors will be poor fits no matter how much time or diligence you spend on the interview and reference-checking process. The rubric "Hire slow, fire fast," suggests that the sooner you separate poor performers, or worse, toxic personalities, from your lab or group, the better off your lab will be. A participant in one of our Cold Spring Harbor Laboratory workshops in Leadership in Bioscience made reference to such lab members as "dominant negatives" from the genetic term for individuals with deleterious traits who pass them on throughout the population.

Given the investment you have made in an employee, it makes sense that you will invest additional time to help bring poor performers up to your standards or needs. This could involve once- or twice-a-week meetings, close supervision, reviewing experimental plans, and an open-door policy for questions.

If after a reasonable time, none of these strategies works, you may need to consider the possibility that the individual is not right for the position. In that case you have several options. The first will always be to speak with someone in your organization's human resources office, generally the sooner the better. The outcome of such a conversation will likely be that you will put together a performance improvement program (PIP) for the employee. This is a formal, documented process in which you will keep a written record of all conversations and agreements with the employee. It typically begins with a frank conversation with the employee letting them know that despite your extra guidance and coaching, their work is still not up to your needs and that unless things improve you will both need to consider whether this is the right position for this person.

We recommend that you stress that this is not an outcome you desire and that you will do whatever you can to continue helping them to improve. However, if improvement is not forthcoming in a reasonable time period you will have to consider other options. During this conversation you will set, document, and agree upon specific performance or improvement goals for the person. For example, becoming proficient in a technique, accomplishing certain technical goals, or changing their behavior in specific ways. You will set specific timelines for these that the employee must agree to. You will also schedule a follow-up meeting in three to six weeks, depending on the complexity of the performance goals, to review progress.

At the follow-up meeting you will review progress, and if improvements are apparent and remain so, you can both breathe a sigh of relief. It is often the case that employee performance issues are time-limited, perhaps because of some difficult personal matter being dealt with, and that all that is needed is a bit of time and patience. If, however, improvement is not apparent, then you may need to go through this process once more but before you do you must speak with your HR staff to get their buy-in and input. They will know whether there are any legal or institutional policies that may affect this plan. Once you get their agreement, you will make it clear to the employee that failure to improve during this second time interval will lead to a recommendation of separation from the organization.

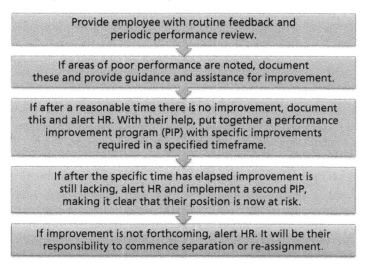

We strongly recommend that you meet with and befriend someone in the HR group before you even think about hiring (let alone firing) your first employee. They can be a tremendous asset to you in good times as well as difficult ones.

Although each organization will have its own policies and procedures, we have illustrated a typical process for dealing with and ultimately separating with a poorly performing employee. We cannot emphasize enough that the number of formal warnings or improvement programs you give the employee and how you manage a separation with an employee depend on the policies of your organization and their HR group. Do not wing it with employee performance problems. Get HR involved, and the sooner the better.

There are some organizations such as governmental agencies or those with strong unions where separating an employee can be next to impossible except under the most egregious situations. In such situations you may need to make the best of a difficult situation and to live with the problem employee until their contract runs out or until the funding for their project expires.

It may seem obvious, but it needs to be said that the last thing you want to do with an underperforming employee is to let your frustration build up to the point where you rush into the HR office and demand that they be fired. Unless the demand is in response to an overt, egregious, or illegal activity (such as intimidation, physical violence or threat thereof), without the kind of documentation and preparation we have outlined above you are not going to get the kind of speedy action you desire. Become familiar with your institution's policies on managing poorly performing employees.

A separation can be a traumatic event for an employee and there is no reason to make it more so. Chances are that if you are unhappy with them, they are not happy in the position either, so helping them move on is to both of your benefits. Help the departing employee see that just because this particular job did not work out does not mean that they are a failure. Help them see their strengths and be positive about their being able to capitalize on those strengths in their next opportunity. This is especially important for younger employees who may still be struggling to find their strengths, passions, and, in some cases, good judgment. One of us (Carl) was fired from an early job as a teenager for a youthful prank that mortified his corporate employer. Fortunately, the incident left no mark on his employment record and only a minor dent in his ego, but it could have turned out differently.

PERFORMANCE MANAGEMENT SUMMARY: AN ORGANIZED WAY TO PROVIDE STRUCTURED OVERSIGHT FOR YOUR TEAM

Soon after a trainee arrives in your lab or group there are six things you should do.

1. Orient them to your lab and its rules and practices.

2. Review with them and jointly sign an advisor/advisee compact (Appendix 1). This document defines high-level expectations for both mentor and trainee in matters of interactions, behaviors, performance, and ethical matters. By signing it, you tell the trainee that you take your job as mentor seriously and that they can expect certain kinds of support and guidance from you. Conversely, by signing it, the mentee commits to taking their job and role seriously and to work and perform according to certain reasonable expectations.

3. You and the trainee should jointly create and agree upon their goals for the first six months in your lab (use the document in Appendix 2 or something like it). After the first six months

this same document will be used to review progress toward the goals you set and to determine whether they need to be changed or amended.

4. Once they start work in the lab and for as long as they are in your lab, give them frequent, specific, and helpful feedback on how they are doing (scientifically, relationship-wise, responsibilities, etc.).

5. Within six months of their joining your group, and preferably much sooner, have them create an individual development plan (IDP; contained in Appendix 2). Review the IDP periodically. When things are moving swiftly every six months may be best, but at other times once a year may suffice.

6. Six months into their stay in your lab use this same document in Appendix 2 to conduct a performance review. Repeat this review every six months to one year.

If you do all these things, your trainees will likely achieve a peace of mind they never thought possible. You will have shown that you care about their career and are willing to work hard to help them define their path, track their progress, and assess and help them improve their skills and performance. Trainees who experience this kind of respect and treatment are motivated, loyal, hardworking, and creative. What more could you ask for?

REFERENCES

Briggs Myers I, Myers PB. 1995. *Gifts differing: Understanding personality type.* Davies Black, Mountain View, CA.

Evans TM, Bira L, Gastelum JB, Weiss LT, Vanderford NL. 2018. Evidence for a mental health crisis in graduate education. *Nat Biotechnol* **36:** 282–285.

Leshner A, Scherer L, eds. 2018. *Graduate STEM education for the 21st century.* National Academies Press, Washington, DC.

Vincent BJ, Scholes C, Staller MV, Wunderlich Z, Estrada J, Park J, Bragdon MDJ, Rivera FL, Biette KM, De Pace AH. 2015. Yearly planning meetings: Individualized development plans aren't just more paperwork. *Mol Cell* **58:** 718–727.

APPENDIX 1

COMPACT BETWEEN POSTDOCTORAL TRAINEE AND MENTOR[6]

This Compact between Postdoctoral Trainee and Mentor presents guiding principles intended to support the development of positive mentoring relationships between postdoctoral appointees and their mentors. It establishes a basic set of expectations that each agrees to and defines some but not all of the behaviors and support that each can expect of the other.

Both mentor and trainee should receive a copy of the final signed document.

Commitments of postdoctoral appointees

- I acknowledge that I have the primary responsibility for the development of my own career. I recognize that I need to explore career opportunities and follow a path that matches my individual skills, values, and interests.

- I understand that there are tools such as the Individual Development Plan (IDP) that I should use to help me define my career goals and develop my training plan.

- I will develop with my mentor a mutually agreed upon research plan that includes well-defined goals and timelines. Ideally, this plan should be developed early in the appointment period and reviewed at least annually.

- I will seek regular feedback on my performance and career planning and ask for a formal evaluation at least annually. I will use this feedback to seek opportunities for development and to build on my strengths.

- I will perform my research activities conscientiously, maintain complete and accurate research records, and catalog and maintain all tangible research materials that result from the research project.

- I will respect all ethical standards—including compliance with all institutional, state, and federal regulations—as they relate to responsible conduct in research, possible conflicts of interest, privacy and human subjects research, animal care and use, laboratory safety, authorship, peer-review guidelines, data ownership, reporting, and sharing. I recognize that this commitment includes asking for guidance when presented with ethical or compliance uncertainties and reporting on breaches of ethical or compliance standards by me and/or others.

- I will show respect for and will work collegially with mentors, faculty, trainees, staff, and other individuals with whom I interact. I will contribute to an environment that is safe, equitable, and free of harassment. I will be an active, contributing member to all team efforts and collaborations and will respect individual contributions. I am also committed to communicating the value of biomedical research to advancing the public good.

[6]Adapted from the American Association of Medical Colleges compact, 2017. An editable Word version of this document can be found at www.sciencema.com/downloads. The original can be found at https://members.aamc.org/eweb/upload/Compact_Between_Postdoctoral_Appointees.pdf.

- I will endeavor to assume progressive responsibility and management of my research project(s) as it matures. I recognize that assuming responsibility for the conduct of research projects is critical to my career path.

- I will have open and timely discussions with my mentor concerning the dissemination of research findings and the distribution of research materials to third parties. I will also work with my mentor to disseminate research results in a timely manner.

- With respect to data ownership, I acknowledge that original notebooks, digital files, and tangible research materials belong to the institution and will remain in the lab when I finish my appointment, in accordance with institutional policy. Only with the explicit approval from my research mentor and in accordance with institutional policy may I make copies of my notebooks and digital files and have access to tangible research materials that I helped to generate during my postdoctoral appointment. I will discuss data ownership with my mentor and reach mutual agreements on future access to tangible research materials and ideas.

- I recognize that I have embarked on a career requiring lifelong learning. To meet this obligation I must stay abreast of the latest developments in science, especially in my specialized field. I will do this by engaging in activities such as reading the literature, participating regularly at relevant seminar series, attending scientific meetings, and interacting with leaders in my field and collaborators. In addition, I will apply for appropriate fellowships and awards that support my transition to independence.

- I will actively seek opportunities outside the laboratory (e.g., professional development seminars and workshops on oral communication, scientific writing, collaborative research, and teaching) to develop the full set of professional skills necessary for success in my chosen career.

- I recognize that the relationship with my mentor continues after my formal training period, and I will commit to being a supportive colleague throughout my professional life.

Postdoc signature   Date

Commitments of mentor

- I acknowledge that the postdoctoral period is devoted to advanced training intended for the development of skills needed to promote the career of the postdoctoral appointee. I will ensure that the postdoctoral appointee has sufficient opportunities to acquire the skills necessary to become an expert in an area of research investigation. I will work with the postdoctoral appointee as the appointee creates a documented individual career development plan, which I will use as the basis for periodic discussions. I will respect the appointee's individual career goals.

- I will ensure that a mutually agreed upon research plan with well-defined expectations and goals is established early in the postdoctoral training period. I will review the plan's progress regularly.

- I will provide regular feedback on performance and career planning and provide a formal evaluation at least annually. I will be accessible to give advice and feedback on career planning and the postdoctoral appointee's individual development plan to help define career goals and identify training milestones.

- I will strive to maintain a relationship with the postdoctoral appointee that is based on trust and mutual respect. I will provide an environment that is intellectually stimulating, emotionally supportive, safe, equitable, and free of harassment. I acknowledge that open communication is essential.

- I will demonstrate respect for all postdoctoral appointees as individuals without regard to gender, race, national origin, religion, disability, or sexual orientation, and I will cultivate a culture of tolerance among the entire laboratory.

- I will promote all ethical standards for conducting research—including compliance with all institutional, state, and federal regulations—as they relate to responsible conduct in research, privacy, human subjects research, animal care and use, laboratory safety, authorship, peer-review guidelines, data reporting, ownership, and sharing. I will clearly define expectations for the responsible conduct of research in my lab and make myself available to discuss ethical, safety, and any related concerns as they arise.

- I will provide the appointee with guidance and mentoring and will seek the assistance of other faculty and departmental/institutional resources when necessary. I will also encourage the postdoctoral appointee to seek input from multiple mentors. I recognize that I must serve as a role model for the postdoctoral appointee and provide access to formal opportunities/programs in complementary areas necessary for a successful career.

- I will provide a supportive training environment to facilitate the postdoctoral appointee's personal and professional growth. I will encourage the postdoctoral appointee to progressively increase levels of responsibility and independence to ensure a successful transition to an independent career.

- I will ensure that the research performed by the postdoctoral appointee is submitted for publication in a timely manner and that appropriate credit is given to the appointee for work done. I will acknowledge the appointee's contribution to the development of any intellectual property.

- I will clearly define future access to tangible research materials according to institutional policy and will discuss this with the trainee and reach mutual agreements that support the trainee's transition to independence.

- To foster career development, I will encourage and assist the postdoctoral appointee to apply for appropriate fellowships and awards that support the transition to independence. I will encourage and facilitate the interaction of the postdoctoral appointee with fellow scientists both intra- and extramurally, including the appointee's attendance at professional meetings to network and present research findings.

- I recognize that there are multiple career options available for postdoctoral appointees and will provide assistance in exploring appropriate options. I recognize that not all postdoctoral appointees will become academic faculty. To prepare postdoctoral appointees for a variety of career paths, I will direct them to the resources that will allow for exploration of various careers, and I will be available to discuss these options.

- I will commit to be a supportive colleague to postdoctoral appointees as they transition to the next stage of their careers and, to the extent possible, throughout their professional lives. I recognize that the role of a mentor continues after the formal training period.

_____   _____
Mentor's signature   Date

COMBINED POSTDOC INDIVIDUAL DEVELOPMENT PLAN AND PERFORMANCE REVIEW

Introduction and overview

As discussed in this chapter, a performance review is done annually or semiannually in conjunction with frequent and routine feedback on specific performance issues. The IDP is a communication and assessment tool aimed at improving the mentoring relationship between postdocs or other trainees and their advisors. When done together, the IDP review and performance review can be powerful tools to define short-term research goals and long-term career goals as well as to give the postdoc guidance on their performance and progress. The following document contains both a performance review section and an IDP section.

Each *newly* appointed postdoctoral fellow should complete Sections B and C of the following document (initial research goals and initial IDP) within six months of appointment. Those sections should then get reviewed (by filling out Sections A and D) every six months to one year as part of the postdoc's routine performance review.

The specific goals of this review process are to

- encourage constructive dialogue between postdocs and their mentors
- clarify responsibilities and performance expectations from both sides (mentor and the postdoc)
- record information about performance and accomplishments and promote more effective performance
- provide postdocs with feedback on their performance and accomplishments during the previous review period
- identify goals for the upcoming period and specific plans to help postdocs meet those goals

Goals of the performance review sections

The performance review is different from routine feedback on performance in the lab or workplace. Routine feedback gives the postdoc real time information on how they are doing relative to the goals the postdoc and mentor have set and the mentor's expectations relative to performance and behavior. The performance review allows the mentor to step back from day to day issues and provide a global assessment of both scientific and behavioral performance. The format of the review is to focus on what is being done well and on what needs to be improved

Goals of the IDP section

The IDP section will change with time since the postdoc's needs and goals will almost certainly evolve. The specific objectives of the IDP section are to

- articulate the career objectives of the postdoc
- identify specific skills and strengths that the postdoc needs to develop in order to attain their career objective based on discussions with the mentor
- identify approaches to obtain specific skills and gain strengths (e.g., courses, technical skills, teaching, and supervision) together with anticipated time frames

Outline of the IDP process

The process is an interactive effort between the postdoc and the mentor.

Outline of steps for postdoctoral fellows and mentors

Steps	For postdocs	For mentors
Step 1	Conduct a self-assessment.	Become familiar with available career opportunities for the postdoc.
Step 2	Prepare the IDP; share IDP with mentor and revise.	Review IDP and help revise.
Step 3	Implement the plan; periodically update and revise the IDP as needed.	Establish regular review of progress; help revise the IDP as needed.

Postdoctoral fellow responsibility

Step 1. Define your goals and conduct a self-assessment.

Outline your long-term career objectives. Ask yourself:

- What type of work would I like to be doing?
- Where would I like to be in an organization?
- What is important to me in my career?

   Assess your skills, strengths, and areas that need development in order to attain those goals. Take a realistic look at your current capabilities. This is a critical part of career planning. Ask your peers, mentors, family, and friends what they see as your strengths and your development needs.

Step 2. Prepare your IDP. Share with mentor and revise.

- This takes place before and during the Performance/IDP review meeting.

Step 3. Implement the plan.

After you and your mentor agree on your IDP, start working on its objectives. Review your progress on and adjust these objectives at the next Performance/IDP review meeting.

   Before you prepare your IDP, review career opportunities you have identified with your mentor. Ask for their ideas.

Mentor's responsibility

Step 1. Become familiar with available opportunities for the postdoc.

By your experience you should already have knowledge of some career opportunities, but you may want to familiarize yourself with other career opportunities and trends in job opportunities.

Step 2. Discuss opportunities with the postdoc.

This takes place in a private meeting distinct from routine research-specific meetings but can be incorporated as part of a periodic performance review. There should be adequate time set aside for an open and honest discussion.

Step 3. Review IDP and help revise.

Provide honest feedback to help the postdoc set realistic goals. Agree on a development plan that will allow them to be productive in the laboratory and adequately prepare them for their chosen career.

COMBINED POSTDOC INDIVIDUAL DEVELOPMENT PLAN AND PERFORMANCE REVIEW FORM[7]

Each *newly* appointed postdoctoral fellow should complete Sections B and C of the following document (initial research goals and initial IDP) within six months of appointment. Those sections should then get reviewed jointly by the postdoc and mentor (by filling out Sections A and D) every six months to one year as part of the postdoc's routine performance and IDP review.

Postdoc Name:
Mentor/Adviser name:
Date of this review:

Postdoc section

Sections A, B, and C are to be filled out by the postdoc and discussed during the review with their adviser/mentor. Once the postdoc has filled out these sections they should forward them to the mentor, so they can review them prior to discussing them at the face-to-face meeting.

Section A. Review of past period

1. Research goals for the period under review (these should be the same as the goals you set with your adviser/mentor at the beginning of this period). For each goal, indicate whether it was Completed, In progress, or Not started. Skip this section if this is the first time you are setting goals.

---

[7] Available for download in Word format from www.sciencema.com/downloads.

| Goal 1.
Check one:	___ Completed	___ In progress	___ Not started
Goal 2.
Check one:	___ Completed	___ In progress	___ Not started
Goal 3.
Check one:	___ Completed	___ In progress	___ Not started
Goal 4.
Check one:	___ Completed	___ In progress	___ Not started
Goal 5.
Check one:	___ Completed	___ In progress	___ Not started

2. Mentoring. Describe your interactions with your mentor. Are you getting enough time with your mentor? How frequently do you meet? How could your mentoring experience be improved?

3. Skills. Describe what new scientific skills or techniques you learned during this review period.

4. Describe any unusual or unanticipated challenges you experienced this year in trying to accomplish your goals. What actions have you taken to meet these challenges? How can your mentor help you?

Section B. Research goals for next period

To be filled out by postdoc and discussed and revised with mentor during review.

1. List your scientific research goals for the coming period including specific research and publication objectives. The postdoc should make a first pass at this section; completion of this section is a collaborative effort between the postdoc and their adviser and can be done during the review meeting. Make sure each goal is SMART (see Appendix 3).
   a)
   b)
   c)
   d)
   e)
   f)

2. Development or training goals (scientific and extra-scientific) for the coming period. Describe or list what new scientific or extra-scientific skills you plan to learn.

Section C. Individual Development Plan: Goals for next period

To be filled out by postdoc and discussed with mentor during review.

1. Describe your current career goals and objectives.

2. Describe your career development or training goals including both scientific and extra-scientific goals for the coming period. You can include new techniques or skills to be learned and publication objectives.

3. Will the development program described in 2 get you to your career goals? If not, what is needed or missing? What additional skills or training will you need to accomplish your career objectives?

4. Are there opportunities that will assist you in reaching your career objectives (e.g., meetings, courses or workshops)? Identify specific events if possible.

5. What assistance can your mentor provide to you in accomplishing the above objectives?

Mentor section

To be filled out by mentor and discussed with postdoc during face-to-face meeting.
Process for mentor:

1. Review the postdoc's assessment of their progress toward goals in Section A1. Indicate any differences of views here. Revise the list of goals if necessary and make sure you both have a copy.

2. Review postdoc's responses to items A2, A3, and A4. Comment and/or amend as needed.

3. Review postdoc's goals for the coming period (Section B) and edit, amend, and discuss as needed.

4. Review postdoc's IDP section (Section C) and advise, clarify, and assist as needed.

5. Review your overall performance ratings (Section D below) and provide feedback and guidance for improvement in relevant categories.

Section D. Overall performance

To be filled out by mentor and discussed with postdoc during review.

The following performance indicators define the expectations for the postdoc's work habits and behavior. For each entry check: Meets standards, Can be improved (add a descriptive comment or suggestion), or Not applicable. Frame your feedback as (A) what they are doing well and should be continued and (B) what can be improved.

a. Takes a thoughtful and rigorous approach to their scientific work.
　　Meets standards
　　Can be improved: (details)
　　Not applicable

b. Seeks and responds to feedback.
　　Meets standards
　　Can be improved: (details)
　　Not applicable

c. Manages conflicts and differences of opinion with skill and sensitivity.
　　Meets standards

Can be improved: (details)
Not applicable

d. Able to manage emotions in tense or difficult situations.
Meets standards
Can be improved: (details)
Not applicable

e. Knows how and when to seek help and assistance in difficult scientific situations.
Meets standards
Can be improved: (details)
Not applicable

f. Is supportive of others in the group by being helpful and by sharing knowledge and resources.
Meets standards
Can be improved: (details)
Not applicable

g. Meets commitments and takes responsibilities (including lab responsibilities) seriously.
Meets standards
Can be improved: (details)
Not applicable

h. Demonstrates ability to think independently.
Meets standards
Can be improved: (details)
Not applicable

i. Keeps abreast of relevant literature.
Meets standards
Can be improved: (details)
Not applicable

j. Possesses appropriate technical skills and abilities.
Meets standards
Can be improved: (details)
Not applicable

k. Demonstrates appropriate speaking and writing skills.
Meets standards
Can be improved: (details)
Not applicable

l. Skilled at managing or supervising others (if relevant).
Meets standards
Can be improved: (details)
Not applicable

m. Other (list)

6. Greatest strengths include:

7. Areas of focus for improvement:

APPENDIX 3

SMART GOALS

Specific: Goal objectives should address the five Ws: who, what, when, where, and why. Make sure the goal specifies what needs to be done with a timeframe for completion. Use action verbs (create, design, develop, implement, produce, etc.). Example: Create cDNA library within two months.

Measurable: Goal objectives should include numeric or descriptive measures that define quantity, quality, cost, etc. How will you and your staff member know when the goal has been successfully met? Focus on elements such as observable actions, quantity, quality, cycle time, efficiency, and/or flexibility to measure outcomes, not activities. Example: Complete all HPLC sample analyses of pond water samples by June 2019.

Achievable: Goal objectives should be within the individual's control and influence; a goal may be a "stretch" but still feasible. Is the goal achievable with the available resources? Is the goal achievable within the timeframe originally outlined?

Relevant: Goals should be relevant to the individual's or group's project. Why is the goal important? How will the goal help achieve objectives? Develop goals that relate to the person's objectives or link with group or organizational goals.

Time-specific: Goal objectives should identify a target date for completion and/or frequencies for specific action steps that are important for achieving the goal. How often should the person work on this assignment? By when should this goal be accomplished? Incorporate specific dates, calendar milestones, or timeframes that are relative to the achievement of another result (i.e., dependencies and linkages to other projects).

CHAPTER 7

Team Meetings:
Who's in Charge Here?

Meetings in a scientific setting take many forms—informal meetings of a few scientists to review experimental data, regularly scheduled lab meetings, and formal meetings of collaborators in a large multi-institutional project. As someone responsible for organizing or leading one of these meetings it is your responsibility to ensure that the objectives of the meeting are clear to all participants and that the meeting is run in a way that these objectives are accomplished or at least addressed in a productive manner. In this section we present approaches that can be used to structure and run such meetings.

Some meetings may have so few participants and such a limited scope that imposing a formal structure on them may appear unnecessary. Reviewing new data with your postdoc, meeting with two collaborators over lunch, or even a weekly meeting of a four-person lab may seem so routine that thinking about structuring the meeting or enumerating the objectives of the meeting seems like unnecessary added work. Yet we can all recall leaving just such meetings feeling that we never got around to addressing one or even any items we hoped to discuss or resolve. The result was that yet another meeting had to be scheduled to cover the items missed in the first meeting. You have better things to do with your time than to attend two or three follow-up meetings, when one well-planned and -managed meeting could have sufficed.

<table>
<tr><td>▸ The role of the leader</td></tr>
<tr><td>▸ The basics of meetings</td></tr>
<tr><td>   Do you need this meeting?</td></tr>
<tr><td>   Lay the groundwork</td></tr>
<tr><td>   Let the agenda work for you</td></tr>
<tr><td>   Beware of decision fatigue</td></tr>
<tr><td>   Minutes and action items</td></tr>
<tr><td>   The rules of interaction</td></tr>
<tr><td>   Focus criticism and feedback on problems, not people</td></tr>
<tr><td>▸ A simple decision-making scheme</td></tr>
<tr><td>   Frame the meeting</td></tr>
<tr><td>   Generate ideas</td></tr>
<tr><td>   Home in</td></tr>
<tr><td>   Closure</td></tr>
<tr><td>▸ Recognizing and addressing impediments to meeting flow</td></tr>
<tr><td>   Procedural impediments</td></tr>
<tr><td>   Individual impediments</td></tr>
<tr><td>   What makes groups effective? Psychological safety</td></tr>
<tr><td>   Group impediments</td></tr>
<tr><td>▸ Tools to manage conflict during meetings</td></tr>
<tr><td>   Be self-aware: Are you comfortable with conflict?</td></tr>
<tr><td>   Monitor and regulate the temperature of the discussion (T_d)</td></tr>
<tr><td>   When the temperature of the discussion increases...</td></tr>
<tr><td>▸ If you are focusing more than 50% of your attention on content or data during a meeting, you aren't doing your job as leader</td></tr>
<tr><td>▸ Final words</td></tr>
<tr><td>▸ References</td></tr>
<tr><td>▸ www resources</td></tr>
<tr><td>▸ Exercises</td></tr>
<tr><td>   1. Case study: The National Institute of Biology</td></tr>
<tr><td>   2. Using silence</td></tr>
<tr><td>   3. Helping others to participate</td></tr>
<tr><td>   4. Dealing with people who dominate meetings</td></tr>
<tr><td>   5. What would you do if...?</td></tr>
</table>

In the following section, we present an approach to running meetings that can be used in formal meetings where ideas need to be generated and decisions need to be made but can be condensed for use in small meetings where a formal structure might seem unnecessary.

THE ROLE OF THE LEADER

Put most simply, the role of the leader in a meeting is to get the group from "here" to "there." The following figure illustrates what we mean.

The leader or facilitator owns the process

Process 1

Problem or agenda   Solution or objective

Problem or agenda → Process 2 → Solution or objective

The team leader is expert not because they know the solution but because they know the process by which the solution is found.

As the leader of the meeting, you have multiple responsibilities. The most important responsibility, however, is to ensure that the meeting addresses the problem or the agenda items in a way that achieves a solution or accomplishes an objective. In the above figure, the leader of the meeting on the top failed to do so, and the result was that solutions were not found or objectives were not accomplished.

Although the point this figure makes may seem trivial or self-evident, in our experience it is anything but. In scientific settings, we find that team or lab leaders very often fail to see that their role goes far beyond serving as scientific expert, fount of knowledge, or arbiter of technical disputes or questions. Scientists in team leadership roles feel comfortable in the role of expert and may believe that because of their expertise their principal role in a lab meeting, for example, is to focus on the science or technical matter being discussed. Let's look at what can happen in meetings run by such leaders.

1. The leader gets immersed in a technical discussion about a particular experimental result and loses track of time and the meeting runs on for two hours instead of the expected one hour. As time goes on the meeting participants become more and more anxious to get back

to the lab and stop participating in the discussion, but the leader fails to notice this. Some lab members with valuable input say nothing in the hope that the meeting will end sooner.

2. During a departmental meeting about recruiting a new faculty member, the discussion gets stuck on whether cancer stem cells should be a new area of focus. The chairman has strong feelings about this and discussion about the science behind cancer stem cells ensues. The discussion takes up so much time that no decision is made and other agenda items fail to get the group's attention.

3. During a meeting of neuroscience team leaders in a pharmaceutical company, the head of the imaging core laboratory unexpectedly makes an impassioned appeal for a new multi-photon imaging system. The vice president for neuroscience allows the appeal to go on for 45 minutes because they like the idea. But others in the room, having no advance knowledge of the request and limited familiarity with the technique, quickly lose interest. Nothing gets decided.

4. A drug development team in a pharmaceutical company is under great pressure to approve a new candidate drug for the next stage of development because the company's pipeline is thin. This is a highly cohesive group that has been working together closely for more than two years and is viewed as a model of cooperation and efficiency by everyone in senior management. Last year the group approved a controversial drug candidate despite strong misgivings by a new member of the group. Now the candidate drug has failed for exactly the reasons the new group member predicted.

5. At a weekly lab meeting, two postdocs get into a heated argument about the meaning of an experimental result. One of them, Ramesh, says to the other, Linda, "Your explanation makes no sense. I don't think you understand the biology of this system. Let me explain it to you." Linda interrupts with, "Ramesh, I've been working in this field longer than you have so don't tell me what I do and do not understand!" Ken, the head of the lab, lets the argument continue because he is thinking about something else. Ramesh and Linda leave the meeting angry and mentally resolve to limit their communications with each other to avoid future clashes.

Each of these examples illustrates a specific error of omission made by the group leader that resulted in the group not getting its work done, not doing its work well, or hampering much needed communication among group members. In examples 1 and 2, the group leaders got immersed in a technical discussion and lost sight of the other matters needing discussion. In example 1, the leader failed to notice that he had lost everyone's attention. In example 3, the leader allowed an unscheduled item to be introduced into the discussion. Because there was no agenda, and because no one knew about this topic in advance, the participants were unprepared to discuss the new equipment, and the discussion will likely have to be repeated at the next meeting. In example 4, the leader failed to be alert for the symptoms of "group-think," which we discuss in more detail below, but which is characteristic of groups that are highly cohesive agreeing too readily about complex or potentially divisive questions. And finally, in example 5, the leader failed to intervene in a contentious discussion that escalated into an interpersonal clash that will likely hinder future communication and scientific discussion.

In each of the above examples, the leader failed to appreciate that their role went beyond being the technical expert and included such important roles as establishing an agreed-on agenda, being responsible for the progress of the meeting, monitoring the quality of the team's decision-making methods, and ensuring that interpersonal conflicts do not sidetrack the meeting or create barriers to open discussion. For the leader to play these important roles they must maintain multiple levels of awareness—scientific, procedural, and interpersonal—during the meeting and be prepared to intervene when challenges arise that can undermine productivity or decision quality. In the following sections we introduce tools and guidelines to help team leaders lead more effective meetings.

THE BASICS OF MEETINGS

We are indebted to Sharon Anderson, MD, Ohio State University for her PowerPoint presentation, "How to run a meeting" (Anderson 2006), for some of the ideas presented in this section. Other ideas for this section come from *The Citizen's Handbook* by Charles Dobson; *Meeting Skills for Leaders: Make Meetings More Productive* (Haynes 2009); *Effective Meetings, The Complete Guide* (Burleson 1990); and *How to Make Meetings Work* (Doyle and Straus 1982).

Do you need this meeting?

It's always a good idea to ask if you need to have a meeting at all, even a meeting that is scheduled to happen on a routine basis. Weekly lab meetings are examples of meetings that we often feel need to be held, regardless of whether there are new data to be presented or decisions to be made. Often, a weekly lab meeting simply serves the purpose of creating a social bond among lab members who may work independently for the rest of the week. In other cases, for example, an ad hoc committee working on a specific project in which there are no new developments, or if the group is awaiting some as yet unavailable information, it will be much appreciated and you can win the weekly popularity contest if you cancel the meeting. Another good reason to cancel a regularly scheduled meeting is if the leader knows in advance that key participants or decision-makers will not be in attendance. Sometimes, the most important action a group leader can take is to cancel or postpone a meeting that they deem will be unproductive. If the meeting is "informational" in nature, ask yourself whether you can accomplish the same thing by simply sending the information to the participants without having to meet.

Lay the groundwork

If you know that one of the items for discussion at a meeting will be contentious you may want to prepare people for the discussion before the meeting. Having a preliminary "heads-up" discussion with people who you expect to have strong views about a topic or who are likely to be personally affected by a decision to be made can make for a more thoughtful discussion during the meeting itself. It can avoid the deer-in-the-headlights phenomenon when meeting participants hear for the first time an important piece of news that affects them and are then expected to immediately come up with ideas or solutions.

People with strongly held views or with a personal stake in an issue can react defensively or aggressively if they feel blindsided by an important discussion that they were unprepared for. Going out of your way to prepare them for such a discussion decreases the likelihood that the discussion will be hijacked by defensive or resentful behavior. This is especially true for meeting participants who are in roles of leadership and authority. It is the clueless meeting leader who chooses the monthly project review meeting with the company's CEO to let it be known for the first time that the latest toxicology study spells doom for a high-profile drug program. Put yourself in the leader's position—would you rather first hear bad news in your office where you can think through an appropriate response or in public with your employees scrutinizing and possibly misinterpreting your facial expressions? Finally, premeeting discussions can also alert the leader to information or viewpoints that they may not otherwise know and that will enable them to develop strategies to manage the discussion in advance.

Let the agenda work for you

Having a meeting agenda distributed in advance is the Platonic ideal. The agenda alerts people to what to expect and can even, on rare occasions, motivate participants to familiarize themselves with an issue before a meeting. But even if the agenda is distributed just at the start of the meeting, it is still important. The agenda is the leader's most important tool for ensuring that what needs to be discussed or decided gets attention and, just as important, that extraneous items do not sidetrack the discussion. It's a lot easier to put off something for discussion by saying, "It's not on our agenda. How about we put that off until the end or until our next meeting?" than to say, "I don't think we should talk about that now."

If you create the agenda, you have the opportunity to set the pace and flow of the meeting.

- In a meeting with multiple agenda items, it is often a good idea to start off with an easy or uncontentious item. The idea is to get the group engaged and, hopefully, feeling good about making a decision together.

- After one or more such "wins," the leader then introduces a difficult discussion item. Another reason to sandwich a difficult item in the middle is that by then, latecomers have arrived, and early leavers haven't yet left.

- Finally, if possible, address one or more "easy" items.

- At the end of the meeting, take a moment to review or summarize what remains to be done, what responsibilities may have been assigned for work following the meeting (sometimes referred to as action items), and, most important, what the group accomplished. It is also useful to briefly recognize individuals who have made any especially significant contributions to the meeting.

By charting such a meeting trajectory, the leader primes the group for hard work, facilitates the hard work, and then reinforces the group's accomplishments, ending the meeting on a positive note. In our experience, group members appreciate this effort by the leader.

By providing this type of group guidance and oversight, the leader can create a bond among group members where none existed before and in this way increase the likelihood of productive group communications.

Beware of decision fatigue

We have all been in meetings that seem to drag on beyond human endurance. Sometimes there seems to be a good reason for a meeting running over its allotted time—there may be multiple important decisions that need to be made about matters that are complex, nuanced, or divisive. When decisions get made near the end of a long and tiring meeting, we often have the feeling of "Let's just decide *something* and get it over with." When this happens, the quality of decision-making may deteriorate. Psychologists refer to this tendency as decision fatigue (Baumeister 2003; Tierney 2011), and there are three principal tools at your disposal to combat it. The first and simplest is to make sure that complex decisions get made early in the meeting (see preceding section). The second is to provide refreshments during long (more than 90 minutes) meetings, which has been shown to improve decision-making quality for reasons possibly related to blood (and brain) glucose levels (Danziger 2011; Tierney 2011). The third is to take control of how long your meetings run by managing the agenda before the meeting, keeping to the agenda during the meeting, and scheduling follow-up meetings to cover items that need thoughtful deliberation.

Minutes and action items

Meeting minutes are the bane of every meeting organizer. Someone needs to record them and that requires thought and attention, preventing the recorder from fully participating in the meeting. Meeting minutes written by an administrative person sometimes fail to capture the essence of technical discussions, requiring the leader to review and edit the minutes. Finally, honestly, how many meeting minutes are actually read?

Certain types of meetings require that minutes be taken: for example, meetings mandated by regulatory requirements, those in which decisions require a recorded vote, and those regarding disciplinary actions. Other types of meetings may not require the taking of minutes.

One effective alternative to taking minutes is listing action items. This is a running list of responsibilities or "deliverables" that are assigned during the meeting, listed under the names of those responsible for them. This list is managed by the leader and updated at each meeting. New items are added to the list at each meeting and completed items are removed. This allows the leader to maintain a running list of outstanding responsibilities. Importantly, because the list is distributed before each meeting, everyone's responsibilities and deliverables are publicly visible. Unlike meeting minutes that are easily ignored, nothing gets people's attention more than seeing their name followed by a list of actions they need to take.

The leader needs to pay attention to action items that get carried forward week after week with no resolution, because this gives the impression that the list is nothing more than a parking lot for actions that you don't ever expect to be accomplished. Find out why these items are not being addressed. It may be that you need to reevaluate why they were put on the list or that perhaps the task should be assigned to someone else.

The rules of interaction

As the leader, you set the behavioral norms for how your team interacts. If you are polite and considerate, your team will tend to be this way. If you are impatient, snide, and demeaning, they will be that way. You may feel frustrated by the slow pace of a project or by a team member who makes the same mistake three times in a row, but, as we discussed in Chapter 2, as leader you need to ask yourself whether behaving as your feelings dictate is productive or counter-productive. If your goal as a leader is to get the work of the group done, then insulting, berating, or scolding a group member in public is probably not your best course of action. These behaviors are demotivating to the person involved and will likely cause others in the group to withhold information that they fear might kindle your ire. Here are some simple guidelines.

The rules of interaction
• Listen respectfully to all ideas. Don't interrupt or make cynical or dismissive comments.
• Ask clarifying questions without being judgmental.
• Challenge assumptions, not integrity, intelligence, or motives.
• Critique ideas or interpretations, not individuals.

The last item especially is worth expanding on.

Focus criticism and feedback on problems, not people

One of the most common types of meetings in a scientific context is the weekly lab meeting, during which lab members present data and interpretations. As leader, you have the opportunity and responsibility to show the team how to critique the science without attacking fellow members.

We saw in the Technical Turf Wars case (Chapter 5) that the scientists dealt with their discomfort with Andrew and the project by venting their anger and frustration on one another. Yet even in circumstances without such overriding issues, scientists often address scientific matters in ways that come across as insults or attacks.

Ideally, team meetings offer opportunities for members to share the results of their work, stimulate one another's thinking, and challenge results and conclusions. Teams and their individual members vary enormously in how they interact. For example, some may choose to ignore questionable data or analyses, whereas others may challenge them emotionally. For many technically minded people, sloppy work or poor experiments are annoying or worse.

The key to getting the most out of team meetings is to keep the focus on the scientific problems and not on the individuals. Both the leader and the participants must take responsibility for this. If the participants do not understand how damaging attacks and sarcastic aspersions can be, even a savvy leader will have difficulty keeping the discussion focused. Changing behavior to channel skepticism toward the problem and away from the person will enhance the likelihood that a critique or suggestion will be listened to and heard. Moving the discussion out of the personal arena increases the chances that challenging questions will be dealt with substantively rather than being felt as personal attacks.

The following two examples show how the leader can model appropriate critical behavior for the team.

Situation 1. Koshi is presenting the results of his assay. For the third time in a row, the assay has not worked. You, the leader, are frustrated and you speak without thinking.

- **Person-centered.** "What is your problem, Koshi? I used to do this assay all the time. It's simple. Just figure it out."
- **Better.** "I'm frustrated by our lack of progress with this assay because it's really holding us back. Koshi, I need you to figure out the problem by the end of the week, all right?"

This response, although it is no longer an attack on Koshi, may still suggest that you are angry, especially if your tone of voice and facial expression don't match the seemingly neutral language. It also places the entire onus on Koshi. Here is another alternative.

- **Problem-centered.** "The lack of progress in this assay is really holding us back, Koshi. Let's talk about how we can address this together."

Situation 2. You are in a project meeting. John, a peer from the drug screening group, is describing an assay that the group developed for your project. You told him last week that you believed this assay to be inappropriate.

- **Person-centered.** "I can't believe you went ahead with that assay! I told you that I didn't want to use it and why, and you just completely ignored me."
- **Problem-centered.** "John, we discussed this assay last week and I thought we agreed not to go ahead with it. Has something happened in the meantime to change your thinking on that?"

Science can be an intensely personal endeavor and like many professionals, scientists can develop a deep personal connection with their experiments, results, and data. At times, scientists identify so closely with their work that it feels like an expression of, or part of, who they are as human beings. Thus, it is no surprise that on hearing their work questioned, criticized, or even impugned, scientists may react with exaggerated sensitivity. In fact, such reactions are so common that we think they are almost the rule and that we should simply expect that people will react with great sensitivity to critiques of their work, even when delivered thoughtfully as outlined above. In recognition of this, we suggest that you adopt the following "behavioral rule."

Behavioral rule #2
Always expect that a criticism or attack of an idea or data will be felt as a criticism or attack of the person Corollary: Expect that you will react this way too.

If you use this rule, you will be careful to phrase your questions, critiques, or criticisms to avoid creating a perceived personal affront. Remember, your primary goal is to critique the data and results in a way that engages, not enrages the other person. If you expect by default that there will be sensitivity to the words that you use, your tone of voice, and your body language, you will be doubly careful about what you say and how you say it.

The corollary, suggested by our daughter, Phoebe A. Cohen, PhD, is just as important. That is, when your data or ideas are criticized, you may feel personally attacked or criticized. Knowing this, you can mentally prepare yourself for this reaction and keep your mind focused on the substantive issue being addressed. Not everyone will have read this book (unfortunately) and many will therefore be unaware of how aggressive or even insulting their comments feel to you. But even well-chosen words can be painful to us when it is our work, our "scientific baby," that is being criticized. Knowing that this is the case and mentally anticipating your feelings can help you to maintain your focus on the work and not on your feelings of hurt. Begin preparing yourself when you hear someone start their question with, "Well, that certainly is an interesting experiment, but... ."

A SIMPLE DECISION-MAKING SCHEME

Often, a group needs to generate ideas and decide between one or more proposals for dealing with an issue. Examples of meetings in which this happens might include a meeting to develop an outline for a collaborative grant application, a meeting to decide on how best to manage limited departmental resources, or a lab group meeting to work on developing a plan to pursue a new experimental discovery. In these cases, the leader needs to solicit ideas from the group, guide the group in evaluating the ideas, and finally ensure that the group decides on one or more proposed courses of action. In the following, we present a general outline of how this process can be accomplished in an orderly manner. The scheme envisions breaking the meeting into four conceptual phases: framing, idea generation, idea evaluation, and closure or decision-making. For each phase of the meeting, we provide key phrases or scripts that the leader can use to keep that phase of the meeting on track.

Frame the meeting

It is the leader's responsibility to alert the meeting participants to the topic of the meeting, even if they think everyone already knows. Consider that there may actually be people in the room who don't know why they're there or (perish the thought) didn't read the agenda. Nothing is more disconcerting than to be 10 minutes into a meeting to decide on which new high-throughput DNA sequencer you are going to buy and having one of the participants pipe in with, "I thought we were here to discuss the holiday party."

While framing the meeting, the leader also has the opportunity to briefly define how the idea generation process will proceed. Will the group start by brainstorming or will members make specific proposals?

It also helps to define for the group whether there are external stakeholders, and if so, who they are. That is, to whom is the group presenting its findings or to whom is the group responsible?

Finally, let the group know how you expect a decision will be made. Will it be by consensus or vote, or do you plan to take a list of ideas to a higher authority for decision?

Having these preliminaries defined up front by the leader provides the group with clarity about its mission and ensures that there are no false starts down the wrong procedural path. This whole framing process need take no more than a couple of minutes, but it will be time well invested and greatly appreciated by the participants.

Frame the meeting		
Action or activity	**Description**	**Example**
Buy in	Get agreement on the problem to be discussed.	"We're here to discuss the integration of preclinical services across our sites; is everyone clear on that?"
Define the method	Ensure agreement on how the group will attack the problem.	"How about we just start with some brainstorming [list ideas with no evaluation]?"
Define the stakeholders (optional)	Clarify to whom you are responsible and what is expected.	"As you know, we've been tasked by the Merger Integration Committee with developing a recommendation…."
Define the end game	Let everyone know how the decision will be made.	"Once we've reviewed ideas, we'll vote" or "We'll take our three best ideas to the review committee."

Generate ideas

Once everyone understands the overall objective of the meeting, it is time to begin idea or proposal generation. There are many ways to accomplish this, and we review three of the simplest.

First, the leader can make an initial proposal. In a group in which the meeting leader is also everyone's boss, this is not the best approach. Junior group members may feel the need to agree with the boss, even if they have different ideas of their own. If you are the boss and you have what you already think is the best proposal that you want to push for, then you do not need a meeting; you need an announcement. Do not bring people together under the pretense of making a decision that you have already made.

In addition, because idea generation and decision-making is hard work, group members will always be glad for an easy and quick resolution. If the leader's initial suggestion is a reasonably good one, group members may feel excused from the hard work that they were dreading and may be happy to reach agreement quickly. Although everyone will leave the meeting feeling relieved, you will have lost valuable team input and, possibly, novel ideas.

"I joined my new department a number of years ago. Our chairman held monthly department meetings to discuss administrative matters relating to the department—things such as shared services, equipment, and space. Whenever a decision needed to be made, he would introduce the issue and then say something such as, "I'll tell you what I think we ought to do… ." You could feel the air being sucked out of the room when he said that. Most of us were fairly junior, so after he said that it was hard to even think about an alternative idea. In addition, because this was usually the first time we even heard about the topic, how were we supposed to have good ideas on the spur of the moment? The result was that usually we went along with his idea. Over time, as we became more comfortable working together—and I mean years, not months—some of us started to pipe in with our own ideas even after the chairman said his. It turned out that he was actually quite receptive to our ideas, and frequently they got adopted. So the problem wasn't that he wanted to dominate the meeting, it was just that he was trying to get things started by making a suggestion. He had no clue that what he was doing was making it uncomfortable for the more junior members to speak up."

Better ways to kick off idea generation are for the leader to solicit ideas from the group as a whole or, if suggestions are not forthcoming, to ask specific people for one or more ideas. Brainstorming, in which the broadest possible spectrum of ideas, even those that seem patently improbable, are generated, is another alternative. The leader manages the process to ensure that all ideas are recorded for later evaluation.

Scientists in particular have great difficulty seeing ideas written on a white board that they *know* with moral certainly lack merit. They would much rather argue the merits of each idea as it is generated so that what they see as a bad idea isn't dignified by being juxtaposed with *their* great idea. The leader needs to be vigilant during the idea generation process to make sure that this does not happen and remind the group that idea generation is just what the name implies and that evaluation and critiquing follow.

The reason to stick to this framework is that critiquing and evaluating ideas can be a lengthy process. If you critique ideas as they are generated, you may never get to list all of the ideas people have. If you first list all ideas without evaluation, it may be easy for the group to see which the best ideas are without spending a lot of time arguing about the weak ones.

Generate ideas		
Action or activity	**Description**	**Example**
Make or solicit a proposal—most specific	To get things started or to set up a straw man	"Andrea, could you make a proposal to start us off?" "Who'd like to start with a specific proposal?"
Solicit a list of ideas—broader input	To make a short list to define boundaries of discussion	"OK. Jim, could you go to the board and list about five ideas from the group?" "Here are three ideas from last week. Can we add three more?"
Brainstorm—widest input	To get the broadest possible scope of ideas; no evaluation at this point	"Let's list as many ideas as we can—no evaluations at this point."
Review	To ensure everyone understands the ideas; again, no evaluation, just clarification	"Let's quickly review these ideas and make sure that we understand each one."

Home in

Once a list of ideas or proposals has been generated, it's time to review and refine them. The list may be too long, and it may contain redundancies or proposals that lack any merit beyond the fact that that someone suggested them. Leading the group in evaluating the initial list for overlaps or repetitions is the easiest way to shorten a long list of proposals.

Next, it may be useful to have those who proposed or support specific ideas clarify them or say something in support of them. Once this has been done, it may be possible to rank the listed ideas by having the group vote for which ones should be included in the top three or four. If there are still too many ideas go through the campaigning and ranking process again.

Home in		
Action or activity	**Description**	**Example**
Look for overlaps	Eliminate duplicates and redundancies.	"Are any of these similar or substantially identical to one another?"
Campaign 1	Have proposers describe and advocate for their ideas.	"Ramesh, you suggested this one—tell us about it please."
Rank	Gauge group's sense of the relative merits of each idea.	"Let's see a show of hands for the ones we like best. Just vote for three [if there are, say, nine alternatives]—who thinks number one is in the top three?"
Campaign 2	Solicit further support or critique for highly ranked ideas.	"Let's have someone say a few final words in support of each of these. Tom, would you like to start by supporting one idea?"

Closure

Once a short list of ideas has been generated and each one has been discussed by advocates and opponents, you're ready to make a decision. If you need to come up with a single idea and you have three, the leader needs to help the group winnow the list down. Eliminating the least popular idea can be achieved by asking how many would be opposed to removing each one from the list. In addition, modifying one of the alternatives may make it stronger or give it wider appeal. The final decision can be made by a vote or by consensus. The leader always has the prerogative to make a final decision if no agreement or majority is possible.

Closure		
Action or activity	**Description**	**Example**
Winnowing	Eliminate least favored ideas.	"How many are opposed to taking number two off the list?"
Strengthening the survivors	Combine or modify the best ideas (into a smaller number, if possible).	"Is there some way we could make numbers two and three into one idea?" "How could we make this idea stronger?"
Review	Check for agreement.	"Let's review our final list for recommendation. Are we in agreement?"

For some relatively formal meetings, you can use the above scheme exactly as presented. For more informal or small meetings, you can eliminate specific elements—most likely, in the Generating ideas and Home in sections. However, we strongly suggest including in a meeting—even between two people—most of the elements in the Frame the meeting section and concluding with some form of summary of what you have accomplished or have agreed upon. If you assume that "we both know what we are here to discuss," you may find yourself talking about the new DNA sequencer while the other person is sweating bullets thinking they're here for a performance review. If you assume that "we both agree on what we decided," you may write out a requisition for the FastLane X7000 sequencer after the meeting while the other person is getting quotes for the Maxi-Gene Pro.

RECOGNIZING AND ADDRESSING IMPEDIMENTS TO MEETING FLOW

No matter how carefully you prepare yourself and your team for a meeting, there will inevitably be impediments and challenges to accomplishing what you hope. We recognize three categories of impediments to productive meetings. The first category is what we refer to as procedural. These interfere with the process or flow of the meeting itself, an example being the attempted introduction of an off-topic or off-agenda discussion item. The second category is individual—group member behaviors, such as uncollegial or inflammatory comments, which interfere with productive discussion or meeting progress. The final category is group behavior—behaviors or thinking patterns of the group as a whole that reduce or limit decision quality.

Just because we refer to the following as "impediments" doesn't necessarily mean that the points raised are without value. Within bounds, and especially in science, there is always value to dissenting viewpoints, even including questions such as, "Why are we even discussing this?" It is just such questions that can get a group to think in a different way or to challenge a previously unquestioned hypothesis. The approaches below can enable the leader to manage the process so that the business of the group is accomplished while at the same time allowing for the consideration of "rogue" ideas or questions.

Procedural impediments

Off-topic discussions

"The real problem we need to be discussing is...."
or
   "How can we discuss next year's program when we haven't even finalized the budget? Let's work on the budget now."

Meetings can be thrown into disarray by members introducing discussion topics that are not legitimately part of the agenda. In our experience, such occurrences are by far the most common source of meeting mismanagement. You can be guaranteed that at any given meeting, someone will attempt to start a discussion that doesn't really belong in that particular meeting.

As the leader, the manner by which you keep the meeting on track is all-important. Abruptly interrupting an off-topic discussion with, "That's out of order" or "We're not talking about that now" can sound hostile and generate unnecessary ill will. Instead, we suggest a general approach that we refer to by the acronym "AVID," for dealing with many types of meeting disruptions.

Use AVID to manage interruptions
A—Acknowledge the behavior
V—Validate the individual
I—Intervene or
D—Defer

Acknowledge the behavior. When someone interrupts a meeting with an off-topic item, the leader first needs to recognize what's going on explicitly, rather than ignoring the interruption.

"Jim, that's not on our agenda for this meeting."

Validate the individual. Let Jim know that the reason you're not going to address this item isn't because you don't respect his ideas.

"I agree that this is an important topic and that we need to address it."

or

"Thanks for bringing this up. It is an important point."

Intervene or defer. By intervening in Jim's attempt to introduce this item, you have kept the meeting on track. You can also defer the discussion to a later time.

"Let's put this on the agenda for next week."

or

"How about you and I talk about that after the meeting?"

A useful tool in such situations is the "parking lot." This is a list of discussion topics for future consideration that is kept either by the leader or by a designated group member. In some meetings, the parking lot list is kept in a prominent location on a white board or flip chart where everyone can see it during the meeting. Having the list prominently displayed makes it easier for people to accept deferral of their ideas for future discussion.

Getting bogged down in a technical discussion

This is another common procedural roadblock in scientific meetings unless the agenda for the meeting is to address that particular technical issue. Most often, scientific meetings have multiple items to be addressed; some technical, some not. It's very easy for scientists to get pulled into a deep technical discussion that might not be appropriate for this specific meeting. Let's look at how Ann, the leader, could use the AVID approach to address this.

"This is an important issue that needs to get resolved and I think that Shoshanna and Miguel need to work on it together. Can you two continue this discussion off-line and get back to us at the next meeting?"

The leader acknowledged the importance of the discussion and validated the roles of Shoshanna and Miguel in bringing it up. She then deferred the discussion to off-line status, thus keeping the meeting focused.

The leader also used another tool, to which we return shortly, of authorizing team members to resolve issues outside of the meeting itself. This has dual benefits—first, of taking a topic that doesn't require everyone's participation outside the meeting, and second, of the leader showing confidence in team members by authorizing them to discuss or even resolve an issue for the group as a whole. This latter benefit can counteract possible hard feelings arising from the leader's halting a discussion that the two or three members involved see as important.

Evaluating solutions during brainstorming

We touched on this previously in our discussion of how to generate ideas during a meeting. Many scientists have a hard time letting an idea written on a white board stay there when they are certain they can demolish it if given only a few minutes *right now*. Whether or not this is true, demolishing the idea right now gets the group into the next phase of the meeting before everyone has contributed their ideas. Moreover, no matter how certain someone is that they can torpedo an idea there will almost certainly be someone else with an unanticipated counterargument, so it's best to wait for the evaluation phase of the meeting.

Using the AVID approach:

"That may be a very good point Richard, but right now we're just listing ideas. You'll get your chance to shoot this one down once we've completed the list."

The leader acknowledged Richard's interruption, validated that it might indeed be a good point, but intervened by deferring the discussion until later in the meeting.

Individual impediments

Another class of meeting challenges is what we have called Individual. Here are three examples.

The naysayer

Ariel says,

"This solution will never work. We don't have enough time to implement it."

It is especially important to recognize and confront naysayers during a meeting. Otherwise, their negativity casts a pall over the meeting like a big dark rain cloud and can unconsciously sap other's energy. By commenting on Ariel's statement, the leader can validate Ariel's concerns without necessarily agreeing with her. Here are some possible AVID responses.

"OK Ariel, good point. Let's ask Guido to put that in the parking lot list, OK?"

or

"Thanks Ariel. I know you have been working hard on the timetable. We will be discussing that shortly, so just hold that thought."

Nonparticipants

Getting maximum involvement of your team in group discussions is important for informed decision-making. However, we have all held meetings in which team members seem unprepared for the discussion or reluctant to participate for one reason or another. Here are two strategies to encourage full participation.

- Distribute the agenda and background information in advance. This of course guarantees nothing, but at least people won't have the excuse that they didn't know what was going to be discussed.

- Assign individual group members the responsibility for specific agenda items. This is a powerful tool and could involve asking specific people to provide the background for an issue or for leading the discussion on an issue. Putting participants' names next to an agenda item (with their advance knowledge, of course) has a motivating effect on their involvement in the discussion. It creates what has been called "active worriers" (Gilmore and Shall 1996), that is, people—other than the leader—who are actively thinking about the topic and the meeting in advance.

Nothing is more frustrating for a team leader than to observe a team member sitting quietly in a meeting and knowing that they have important information or a viewpoint that is not being shared. Rather than trying to figure out why the team member is being quiet, the leader can address the behavior directly:

"Jim, you haven't said much this morning. Tell us what you think."

or

"Kathy, I know you've done some work in this area. What are your thoughts?"

Sometimes, this is all it takes to get a quiet member talking. Often a team member is quiet because they fear that what they have to say will go against the prevailing views of the rest of the team (see the section on Groupthink below). Being invited to express their opinion by the leader authorizes them to speak up, possibly to the benefit of the discussion. As a leader, you don't have to have a degree in psychology to read the facial expressions and body language of a quiet team member and get the message that they are troubled by or skeptical of what is being said.

Remember also that not everyone processes new information at the same rate. Later, or even the next day, some team members may come up with an idea that relates to something that happened in a meeting. Encourage this type of out-of-group thinking as well as discussion, but be sure that the group eventually processes the new ideas.

Monopolizers

At the other end of the spectrum, groups often have at least one person who sees it as her task to make the first suggestion or the most detailed critique of any topic under discussion. If their comments were off-topic or weak, it would be easy to simply move the discussion forward by asking for alternative views. Many times, however, such people are extremely knowledgeable and experienced and their ideas have genuine merit.

Nonproductive individual behaviors in a group/Interventions

• **People who criticize everything**
  ▸ Keep criticism confined to the appropriate phase of the meeting, not during idea generation.
  ▸ Ask them to be specific. Ask if others agree.

• **People who dominate the discussion**
  ▸ Point out what they're doing. "Your points are appreciated, now let's hear a few other viewpoints."
  ▸ Deal with them after the meeting.

• **People who don't say anything**   ▸ Reach out. Ask them for their opinion or input on specific matters.
• **People exhibiting negative body language or facial expression**   ▸ Call them on it. "Jim, I see you looking skeptical. Please share your thoughts with us."
• **Interrupters**   ▸ Intervene. "Joanne, please let Sam finish what he has to say, then we'll get to your point."
• **Cynics ("Nothing we do here will make any difference.")**   ▸ This attitude can sap enthusiasm. "Let's not be cynical. We have a job to do." Address further after meeting.
• **People who get the group off topic and on to a side topic**   ▸ Pay special attention, as this can hijack a meeting. Refocus discussion.
• **Know-it-alls—people who think they already have the answer and who try to force it on the group.**   ▸ Acknowledge their expertise. Remind them that this is a group decision and everyone's input is needed before a decision is reached.

• **Latecomers**
  ▸ Suggest they just sit down and try to come up to speed.
  ▸ Deal with them after the meeting—don't confront them in the meeting.

The problem is that their speaking first or loudest may inhibit or suppress other useful ideas from less confident group members. As a leader you can say (with a smile),

"Thanks Alexi. Let's hear some ideas from other group members."

The leader notes that Bipasha is about to speak and uses the AVID approach.

"Bipasha, I know you're familiar with this issue. Let's let someone else speak first this time. I promise to come back to you."

Without the intervention of the leader, such people will dominate the meeting and can be de-motivating to others. If you know that certain people in the room have a tendency to speak first or to monopolize the discussion, you can also specifically ask someone else for the first comment.

"Renee, let's hear your thoughts on this experiment to kick off the discussion."

What makes groups effective? Psychological safety

We have discussed how to create a climate for productive meetings and to counter or minimize the most common elements that limit group effectiveness. The tools we discussed can be divided into two categories: structural and behavioral. The structural tools include ensuring that meetings have agendas, that the discussion is kept on track, and that meetings end with appropriate direction and mutual expectations. The behavioral tools include techniques to manage individual behaviors that can get the team sidetracked (AVID) and managing group behaviors that can subvert informed decision-making. We think that the behavioral tools are especially important and influential, and we're not alone.

Teams and their effectiveness rank high as topics in the popular press and in psychological publications; a few of our favorite books on this topic are referenced at the end of Chapter 5. One topic on everyone's mind is whether there are measurable (and by extension, controllable) factors that make one team or group more effective or productive than another. Psychologists have studied something called "group collective intelligence," operationally defined as the ability of a group to perform a variety of tasks that is distinct from the abilities and intelligence of the members of the group as individuals (Wooley 2010). Wooley found that of all the variables they studied that might influence collective intelligence, there was only one that showed a strong positive correlation, a characteristic they called "social sensitivity," or the ability of group members to perceive, understand, and respect the feelings and viewpoints of others in the group.

How could social sensitivity contribute to group effectiveness? If you are socially sensitive, you probably already know. In their 1965 classic "Personal and Organizational Change via Group Methods" (Schein and Bennis 1965), Edgar Schein and Warren Bennis, two giants in the field of leadership and organizational dynamics, discuss the concept of team psychological safety, defined as a shared belief that the team is safe for interpersonal risk taking. In a study of the role of psychological safety on team effectiveness, Edmondson (Edmondson 1999) defines it further as "a sense of confidence that the team will not embarrass, reject or punish someone for speaking up…," and "It describes a team climate characterized by interpersonal trust and mutual respect in which people are comfortable being themselves." Edmondson's work showed that teams that had a high degree of psychological safety performed better on a variety of tasks and learned more than teams that did not. Perhaps teams with members who have high social sensitivity create environments that have high psychological safety.

In 2016 an article by Charles Duhigg in the Sunday *New York Times Magazine* caught our attention, "What Google learned from its quest to build the perfect team. New research reveals surprising truths about why some work groups thrive and others falter." The article described Project Aristotle, which was undertaken by Google starting in 2012. The objective of the project was to determine what made one Google team more effective than another and the investigators had access to psychological and performance data from hundreds of teams within Google. They found that among all the variables and measures that they attempted to correlate with team effectiveness and productivity, only one stood out as a significant contributor. You guessed it, psychological safety. Duhigg writes "The behaviors that create psychological safety—conversational turn-taking and empathy—are part of the same unwritten rules we often turn to, as individuals, when we need to establish a bond. And those human bonds matter as much at work as anywhere else. In fact, they sometimes matter more."

The interpersonal tools we provide in this chapter and throughout the book show how to manage people in meetings and one-on-one in ways that promote psychological safety. We have emphasized that the leader, by modeling the kinds of respectful and empathic behaviors that they want to encourage in the group, plays a pivotal role in this process.

Group impediments

Some of the most subtle but important influences on group process or decision-making come not from disruptive individuals or off-topic diversions but rather from the behavior of the group as a whole. Despite the importance of "group" behaviors, they are typically the most difficult for the leader to detect, most often because the leader is so deeply immersed in the group's behavior that they cannot see it for what it is. An expanded discussion of many of the types of group decision-making biases discussed here can be found in *The Psychology of Judgment and Decision Making* by Scott Plous (1993). The following is a case study that illustrates some of the themes we will be discussing below. We will return to the case after we have reviewed the different types of group behavioral biases below. Read the case and make notes about what you see going on between the group members.

▪ *Case Study: Contaminated*

Hanah runs a research lab at Big University focused on gene therapy. The lab has six postdocs and three technicians and has made a novel virus for delivery of therapeutic gene constructs in humans. The virus needs to be propagated in primary chick embryo fibroblasts derived from fertilized eggs. Three months ago in May, the lab technicians started having problems with bacterial contamination in the fibroblast cultures that got worse as the summer wore on. The lab is now at the point where most of the work has been put on hold and everyone is frustrated. Hanah is concerned that two publications will be delayed and final data for a big gene therapy meeting may not be available. The postdocs are on edge because their work is being held up with no definitive resolution in sight. Hanah has called a meeting with the whole team to discuss what their options are.

Scene 1 Four of the postdocs, Rakesh, Ning, Fred, and Samantha, are sitting in the lab's lunchroom the day before the meeting.

Rakesh: "I was watching Juan (one of the cell culture techs) yesterday—man, that guy's sterile technique is a joke. We need to get rid of these useless techs and I guarantee you that then our problems will disappear."

Ning: "Right, we should just do the primary cultures ourselves until we can train new people. It would be worth the extra work just to get going again."

Samantha: "Whatever. We need to do something right now or this is going to drag on forever."

Fred: "I'm with you guys. Let's do it ourselves until we can find competent help."

Scene 2. The next day, at an all-lab meeting.

Hanah: "Does anyone have any new ideas about our contamination issue?"

Rakesh looks at Ning, Fred, and Samantha and is about to speak but before he does, Juan speaks up.

Juan: "I called a tech in the Smith lab, where I used to work. They make primary chick fibroblasts all the time. They said that we should be washing the eggs with a disinfectant before we use them."

Rakesh: (skeptically) "No way—they already do that at the supplier."

Juan: "They do, but my friend says that we need to do it here too."

Hanah: "Juan, that's very interesting. Thanks for taking the initiative to do that."

Rakesh: (looking at Ning, Fred, and Samantha) "Look, we've being making fibroblasts for two years—I never heard of this. Something must have changed in our own lab. It's obvious that we're not being careful enough with the embryo preparation. Ning, Samantha, Fred, and I discussed this yesterday and think we should go back to doing the cultures ourselves."

Juan: "My friend in the Smith lab said they had similar problems last summer, but they went away after they started disinfecting the eggs before they brought them into the culture lab."

Samantha: "I never heard about that and I have a friend in the Smith lab too. This is bogus. It's a very simple problem—it's all about sterile technique. That's it, period."

Hanah: "OK, I know Smith—I'll call her after the meeting and see what she has to say."

Samantha (getting up from table, under her breath but loud enough for Alex to hear): "Waste of time."

They all leave the room with nothing else said.

Confirmation bias

Confirmation bias occurs when group members arrive with a preference for a specific idea or course of action. As a result, they unconsciously bias the information that they take in or the data that they consider in favor of that which confirms their preconceived ideas. It's almost impossible to see this happening during a meeting because you can't tell what people are thinking. But you can suspect that it's happening if

- You know in advance that some or all of the group members have a clear preference for a specific proposal or course of action before the meeting begins.
- People focus their discussion on a specific proposal or idea during the meeting to the exclusion of others.
- Some ideas or proposals are difficult for the group to discuss, as evidenced by reluctant group discussion or prolonged silence.
- The group comes to a decision about a complex issue with little or no discussion.

A tactic that you can use if you suspect that the group is being influenced by preference bias is to explicitly point it out to the group. By calling attention to the behavior, you may be able to nudge people in the direction of a broader discussion and to consideration of alternative ideas or solutions.

"I'm getting the feeling that we have homed in on this particular proposal too quickly. I'm not saying it's a bad proposal, only that there may be other alternatives we're not considering. I know that many of us favored this proposal even before the meeting began and wonder whether we're unconsciously limiting our discussion to only those data that support it. Can someone suggest some other ideas for us to consider? Zvi, could you help us out here?"

Another approach to dealing with suspected confirmation bias is to ask someone to play "devil's advocate" and to speak against a specific proposal. This is especially effective if the person you ask is one who is strongly in favor of the same proposal, because it forces them into a critical thinking mode.

In the case study above, Rakesh, Ning, Fred, and Samantha seem to be engaging in confirmation bias. They entered the meeting having already come to a conclusion. What clues could Hanah have had to suggest that this was the case? First, she would have noticed that the postdocs were quickly dismissive of new information presented by Juan. Second, they immediately focused in on the idea of compromised sterile technique, suggesting that they had discussed the matter beforehand and had already reached a conclusion.

Hanah, to her credit, indicates that she will follow up on this information despite the skepticism of her postdocs. In this case, Hanah chose not to confront the group with an interpretation of its behavior. In fact, it would have been awkward for Hanah to do so during the meeting given the underlying antipathy between the postdocs and the technician. It might be best for Hanah to address this antipathy outside of the meeting.

Sharedness bias

Sharedness bias occurs when group members focus their discussion on information that most members of the group share in common or are aware of, at the expense of new information or data that only one or a few members may have (Larson et al. 1998).

> Groups communicate predominantly about information which all or most group members share before entering the discussion, and neglect unshared information, which only one or a few members have initially. (Klocke 2007)

Sharedness bias is different from confirmation bias in that members may not have reached a conclusion before the meeting. In the case study above, sharedness bias is manifested by the postdocs being dismissive of Juan's new information about how his friend's lab handles eggs. None of the postdocs had heard of this before and therefore dismissed it. The bias was accentuated in this case because Juan was a technician, making it easier for the postdocs to dismiss his new information. As with confirmation bias, it is up to the leader to be alert to this group process and to ensure that minority views or data introduced by one member receive due consideration.

It is questionable whether it would have been useful for Hanah to have said, "You four are ignoring what Juan has to say. You've already made up your mind." Such a statement could be felt by the postdocs as an accusation and might have resulted in defensive behavior (e.g., "That's totally not true; we've been brainstorming about this problem all week."). As with confirmation bias, the best course of action for the leader in the context of this case study might be to use her authority to make sure that Juan's information gets due consideration without interpreting the group's behavior during the meeting.

Groupthink

Groupthink (Janis 1982) is the tendency of a cohesive group to close in on a mutually acceptable solution with limited discussion or without due consideration of all options and consequences of the decision. Characteristics of groupthink include

- high group cohesiveness, leading to
- concurrence-seeking behavior, resulting in
- limited critical thinking.

One classic, historic example of groupthink was the Bay of Pigs invasion of Cuba during the presidency of John F. Kennedy. There, a highly cohesive group of politicians and presidential advisors, all close friends, came to a disastrous conclusion about the likelihood of success of an invasion of Cuba by the United States.

In the case study above, we can also see elements of groupthink. The postdocs seemed to have a close working relationship with one another—maybe they were friendly outside of work, spending time with one another's families, thus increasing the close bond they had at work. Such closely knit groups create a need or desire for their members to form a united front when making decisions and conversely create a psychological pressure against dissent within the group. Thus, even if one of the postdocs had seen merit in Juan's suggestion, it might have been difficult for them to voice that support in the presence of the other postdocs.

As with confirmation bias and sharedness bias, in most cases it may be best for the leader to work around the problem by nudging the group toward a more thoughtful or inclusive discussion, rather than interpreting the group's behavior. In the above case, Hanah did this by committing to follow up on the information that the postdocs wished to ignore. However, if the group leader sees herself as part of the groupthink dynamic, it may be acceptable for her to admit this as a way of helping the group observe its own behavior. Had Hanah herself initially sided with the postdocs, she might have said,

> "I've been watching what we've been doing during the last hour and wondering if we all may be moving ahead with this decision a bit too quickly. I'm sensing that alternative explanations for the contamination aren't being considered in sufficient depth. We seem to have quickly homed in on the idea that we've had a lapse in sterile technique in the culture room. I'd like to suggest that we take a step back and look at Juan's idea again before we move forward."

Because Hanah is admitting that she also may be part of the problem, her comment would not likely be taken as a threat or accusation by the postdocs.

Scapegoating and displacement

When groups are faced with difficult or seemingly intractable problems, frustration builds up, creating internal tension in group members. When this happens, members may unconsciously seek an outlet for their frustration in the form of other group members or people outside the group. In the case study "Contaminated" above, Hanah might have wondered whether Juan and the other technicians were being scapegoated by the postdocs. Other examples might include

- When difficult budgetary decisions need to be made, it's the "bean counters" in the fiscal office who are to blame.

- When your grant is rejected, it's the small-minded, risk-averse reviewers who are to blame.
- When your clinical trial misses its endpoint, it's the fault of the clinicians who insisted on broad acceptance criteria for patient enrollment.

In each case, the group attempts to mitigate its tension and discomfort by assigning blame to an individual or to another group. Scapegoating can be great fun for a closely knit group—it provides a sink for the group's frustrations and seems to build camaraderie and cohesion. Nothing is better than a mutual enemy for keeping a group together. The downside is that focusing on a scapegoat as the cause of the problem ignores the real cause of the problem.

- In the case of the budget, you really do need to come up with creative ways to cut costs. If your group can work with the people in fiscal, as opposed to demonizing them, you may increase the likelihood that you will be successful.
- In the case of a rejected grant, the sooner the group gets over complaining about the reviewers and starts to focus on alternative paths to funding, the better off it will be.
- In the case of the clinical trial, putting together a meeting of the clinicians with the science team may help everyone understand why the trial was designed the way it was and what might be done differently the next time (if there is a next time…).

Of course, in the spirit of the saying, "Just because I'm paranoid doesn't mean that people aren't really out to get me," it may actually be the case that some person or group is, in fact, responsible for an error. However, scapegoating is an easy way to avoid hard work, and it is the leader's job to discern when this is happening and to intervene.

Closely related to scapegoating is displacement. This happens when the group is frustrated by or angry with one person but expresses its frustration or anger to someone else. A common example is when the group is angry with the leader—perhaps he made an unpopular or unilateral decision. Group members may be reluctant to express anger toward the leader and unconsciously transfer their anger to another group member—perhaps one who seems to support the leader's viewpoint—as a safer way to vent their frustration.

The astute leader, observing the body language and facial expressions of the group members will be cognizant of this and will intervene.

"I can see that we're all frustrated by this problem and I know that this is not a popular decision I've made. But it's important that we not let our frustrations spill over onto one another. Let's be respectful of everyone's views and focus on getting our work done."

Rather than interpreting the group's behavior ("Look at yourselves, can't you see that you're really angry with me but displacing your hostility onto Juan?") and being met with blank stares or outright hysterics, the leader reframes the group's frustration as being with the problem they're trying to solve.

Silence

Sometimes the whole group is silent. The following case study is taken from Carl's career. It is an example of how self-awareness helped him to recognize that his own behavior during group meetings was contributing to a dynamic of passivity on the part of some members. Identifying his discomfort during these meetings allowed him to anticipate his behavior and choose to behave differently. His new behavior contributed to a changed group dynamic that encouraged rather than inhibited participation.

> ■ *Case Study: The Power of Silence in Groups*
>
> When I ran a research laboratory at St. Elizabeth's Medical Center, we held weekly lab meetings during which members presented and discussed their work. Sometimes, the discussion was lively and animated, but other times, it felt as though the presenter was simply telling me what they had done while the other members of the group stared off into space. At these times, I was the only one asking questions, and in place of a discussion, only a two-way dialogue took place. I occasionally felt like a prosecutor dragging information from a reluctant witness. At times, this made me feel very frustrated and angry. In retrospect, I am certain that at those times, I felt least secure about myself and my abilities as a leader. I felt that others were avoiding their responsibilities and forcing me to do all the work. I thought this to be unfair. After all, the success of our research efforts would ultimately determine continued funding from the NIH and affect the jobs that these people held.
>
> During this time, Suzanne, a practicing psychotherapist, suggested that I attend a weekend workshop in group dynamics. It was being held in Boston, and on a lark, I agreed.
>
> In one of the "exercises," participants formed small groups of seven to nine people, led by a facilitator. Our "task" was to study and experience the interactions within our group. Period. We were sent into a room in which the chairs were arranged in a circle; when everyone was seated, the facilitator closed the door, sat down, and said nothing. This continued for what seemed like an eternity. Some of the more anxious members (I was perhaps the most anxious) started talking in an attempt to figure out what we were supposed to be doing there, as though this was some sort of a test.
>
> I went so far as to grab a marker and list the objectives for the group on a white board. I felt that we needed to list these so that the group could reach some agreement on how we would achieve them. Some joined in my frenzy of discomfort, whereas others either did nothing or periodically questioned the urgency of our need for clarity. If you read about the "Tea Bag Company" in Chapter 1, my behavior, and the contrast between how I behaved and how some of the others behaved, will sound familiar.
>
> This continued for about two hours that day and occurred again during two successive days. With the help of other members of the group, and some frustratingly minimal observations from the facilitator, I was able to recognize and admit to the intense discomfort that I felt when the group was silent. I also recognized my own need to fix the discomfort by doing something to make it disappear. I could not see that everyone—not just I—was responsible for what was happening. I could not allow the process to work itself out and for the possibility that others might have a different approach to dealing with the "task." I was too focused on my own anxiety and need for control and clarity.
>
> During the course of the weekend, I became more comfortable with silence and the ambiguity that it contains. I became more accepting of the fact that others in the group needed to process what was going on in a different way than I. Some dealt with the discomfort by talking about their experience at that moment—a stark contrast to my approach, which was to try to change the experience into something else.
>
> What did I learn relative to my lab? I learned to accept the times when the group is silent, to let this happen, and let others in the group take some responsibility for it. It was a revelation to realize that if I let the silence continue, which to me was excruciating, often (not always) someone would say something. When I adopted this attitude during my lab meetings, inevitably the comments that were made and the direction in which we went were quite different from where they would have gone in the past. Sometimes, these comments illuminated a facet of the work that might not have otherwise emerged if I had jumped in with my own thoughts. In the parlance of the group world, I had given the others "space" to think and act.

What it took for me to change my behavior was my own experience in the model group. I needed to experience and identify my discomfort and learn from it. I also had to try a new behavior (being quiet) and experience the consequences of that behavior. Most important, I needed to develop the ability to recognize my discomfort at the time that I was experiencing it. This recognition is what gave me the opportunity to decide what to do about it, rather than simply react to it. Finally, I needed to learn that I could live with the discomfort and that others in the group also had responsibility for the group's task. It took trust in the group for me to allow the meeting to flow, rather than to manipulate it.

When a group is dealing with a difficult task (e.g., trying to decide on the next step in a scientific project on the basis of ambiguous data) or having an uncomfortable conversation (agreeing on budget cuts that affect different departments to different degrees), behaviors emerge that are in a very real sense behaviors of the group as a whole rather than of any one individual. The concepts in this section will help you recognize these behaviors and will give you guidance about how to counteract them when they threaten to interfere with effective discussion or decision-making.

An underlying theme we see in group interactions is that during difficult or contentious discussions, members will behave in ways that help them reduce their feelings of discomfort or ambiguity. Thus, if the group is dealing with a thorny problem, such as the contamination issue in the case study above, members will have a tendency to focus on readily available or shared information as a way of coming to a quick agreement. This may not be a problem if the solution is a good one. But if there is some doubt about the quality of the solution and the rapidity with which agreement has been reached, the leader must intervene to help the group manage its discomfort while it arrives at a better solution. Leaders can do this by

- acknowledging that everyone in the room, including the leader, is uncomfortable with the uncertainty of not having a facile solution at hand; and

- that it is OK to live with this discomfort until a solution is found; as long as

- the group does not transfer its discomfort into frustration with group members or scape-goats; and

- as long as the group does not try to manage its discomfort by grasping the first idea that is proposed.

TOOLS TO MANAGE CONFLICT DURING MEETINGS

One of the underlying premises in Chapter 3 is that differences of opinion and even conflict are a natural part of our lives both at work and at home. The goal of becoming a good negotiator and a good leader is not to eliminate conflict—we suggest below that this would actually be a bad thing to do, and likely impossible in any case—but to learn to manage it better. In Chapter 3 we introduced negotiation tools that are useful in managing a conflictual discussion between two people. Many of these apply to groups as well. Here, we introduce additional tools that are especially useful for dealing with conflict in a group setting.

> ### Managed conflict is central to team discussion: Use it
> - Conflict is neither good nor bad; it is a natural consequence of differing viewpoints.
> - Meeting participants are problem solvers, not adversaries.
> - The views of all participants deserve respect.

Many of us are naturally averse to conflict. When a disagreement arises it makes us feel uncomfortable—we may get anxious, we may feel threatened, or we may think that the ensuing discussion will turn into an argument, possibly threatening our relationship with the other person. These attitudes are especially disadvantageous if you are a leader. If you feel uncomfortable with conflict as a leader, you may press for easy but suboptimal solutions to shorten a contentious group discussion. This is particularly dangerous if you're not even aware that you're doing it. A more productive attitude for a leader to have is that the successful management of conflict during a meeting is what enables the complete airing of different viewpoints (Amason et al. 1995).

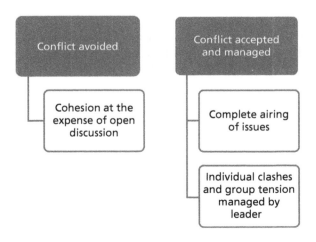

Be self-aware: Are you comfortable with conflict?

The first tool to manage conflict within your team or at a meeting is to become comfortable with conflict yourself. We do not mean that you should be eager to engage in or observe the type of conflict that involves personal insults, shouting, and petulant or passive–aggressive behaviors. These are just the behaviors that you as leader must stanch before they even begin (more on this below). Rather, we mean conflict around genuine differences of opinion and viewpoints. As the chart above indicates, such substantive exchanges are required for legitimate viewpoints to be expressed, even if some of those viewpoints don't make sense to everyone in the room. To learn whether you are comfortable with contentious discussions, ask yourself the following questions.

Use self-awareness during contentious team discussions. Ask yourself:

- Do I sense my body getting tense? Is my left leg vibrating out of control? Is my face tense?
- Do I feel annoyed with the people engaged in this discussion?
- Is my mind wandering away from the discussion?
- Do I have a strong desire to, or do I actually, change the topic to a less contentious one?
- Am I tempted to press for a quick solution to relieve my anxiety?

If the answer to any of these questions is "yes," then you may be conflict averse. Being aware of this tendency is the first step in overcoming it. If you are uncomfortable with conflict, you might try to stop a useful discussion prematurely. Instead, what you should be doing is keeping the discussion going while also making sure the conflict is focused on the science and the ideas rather than on the people expressing them. The tools that follow will help you do this.

Monitor and regulate the temperature of the discussion (T_d)

As group leader, you are really playing two roles. On the one hand you are (presumably) an expert in your field. As such, you will be participating in your group's discussions in an active manner, both as an expert and as a mentor, helping your team work through problems and their solutions. You also play another important role—that of the facilitator responsible for ensuring that the team functions effectively, interacts appropriately, and stays on task. If you're spending all your time focused on the data, you cannot play this second role effectively or at all. So you need to make a conscious commitment that during the meeting you will be jumping back and forth between these two roles.

In your role as facilitator, one parameter you need to be watching is what we call the temperature of the discussion. This means just what is says—as the discussion "heats" up, the quality of people's comments transition from matter-of-fact, through dogmatic, argumentative, aggressive, and finally to hostile.

This is a phenomenon that we have seen often and that we call "affective generalization." Affective generalization happens when people in a group get stuck discussing a challenging topic, for which there may be no easy or readily apparent solution. The members of the group experience what is called "cognitive conflict," or content-related conflict, a kind of mental discomfort associated with trying to accomplish, possibly unsuccessfully, a challenging task. If the cognitive or content-related conflict continues, the members may unconsciously transfer their mental or cognitive anguish to other team members, manifesting what is called "affective conflict"—that is, conflict expressed as negative feelings or actions toward others in the team. This is reminiscent of displacement that we introduced on p. 185 in which group members transfer their frustration from the leader to another group member.

Although cognitive conflict, focused on the substantive issues of the problem, is useful and even necessary for problem solving, affective conflict, focused on or between team members, is counterproductive (Amason et al. 1995). The more affective conflict is present, the less comfortable team members will feel in expressing viewpoints that go counter to the majority view—just the type of behavior the leader needs to encourage if they wish to counteract the insidious effects of groupthink, sharedness biases, etc.

Just as in a chemical reaction that needs to be at a certain critical temperature to provide the activation energy for the reaction, so too do discussions need some degree of "heat" to get people engaged and thinking. Too little heat and your team members are staring out the window in a soporific daze; too much and they're lunging at one another across the table.

If you as leader are too focused on the data being discussed, you may miss the early signs of transition from cognitive to affective conflict, and then it may be too late. Heated words may have already been exchanged and insults or innuendos traded, poisoning the formerly collegial atmosphere of the meeting. If you're too focused on your scientific role at the expense of doing your job as leader, you're going to find yourself running some pretty unpleasant meetings.

Indicators that T_d is increasing
• Team members are starting their comments with "you."
▸ "You're not making any sense."
▸ "Your proposal completely ignores the data that I just presented."
▸ "You don't seem to understand the importance of this issue."
• People are getting stuck in positions rather than focusing on underlying interests (see Chapter 3 for more on this topic).
• Team members are reacting to critiques of their data in a personal manner.

During scientific meetings leaders need and expect a free and open exchange of ideas and viewpoints. This is something we all would agree with, except when the free and open exchange of viewpoints relate to OUR data. As we noted earlier in this chapter, it's perhaps inevitable that scientists will take critiques or criticisms of their data personally. When we're functioning in scientific mind-set and critiquing or questioning someone's data or interpretation, we don't always choose our words carefully.

As leader, it is your job to monitor how people critique others' data and how they respond to being critiqued, and to use and make people aware of the behavioral rule put forth on page 170, "Always expect that an attack on, or criticism of, an idea or data will be felt as an attack on, or criticism of, the person." And its corollary, "Expect that you will feel this way too."

When the temperature of the discussion increases...

Beyond monitoring the discussion for "you" statements, emotional responses, and emerging conflicts, what options do you have once a conflict actually breaks out?

Point out what's happening and reinterpret people's frustrations with one another as frustrations with the problem you're trying to solve

If people are yelling at one another, you don't have to be an organizational genius to figure out that your scientific discussion has gone off-track.

The savvy leader might say, "This is a tough problem and it's going to be hard to reach a consensus. That's making us all a bit frustrated. How can we get back on track? Should we put this issue aside for a minute and come back to it later?"

Here, the leader has essentially called a time-out from a heated discussion and given people a few minutes to cool off. He has also reinterpreted their frustration as being with the problem, thus defusing somewhat any interpersonal tensions that may have developed. This is one of the few instances for which interpreting the group's behavior to the group can be a useful intervention.

Use humor

Sometimes, a clash just happens. You may not have seen it coming, or it may have been building up before the meeting, but there it is. It may not be an outright knockdown argument, but it's enough that people in the room feel uncomfortable enough to stop productive discussion. In such circumstances, humor is often a good tool to use. For example, after a particularly uncomfortable exchange between Marcus and Inder, the group has become silent. The leader says, "Marcus and Inder do this all the time—then they go out for a beer." Or after there has been an especially divisive discussion with no agreement in sight, "OK, now that we're all in complete agreement…," or after Marcus has given an especially harsh and insensitive critique of Inder's experiment, "OK, Marcus, now tell us what you *really* think of Inder's experiment."

> I was at a board meeting where we had a number of proposals to vote on. One of them generated some particularly argumentative discussion that went on much longer than anyone had anticipated. The person presenting the issue got defensive, and I could tell that there was considerable discomfort around the table that might spill over into later parts of the meeting. The very next item for discussion was a trivial administrative matter that everyone knew was going to be approved without discussion. As soon as it was introduced I put on a comically skeptical look and said in a wary tone, "Now wait a minute…," as though I were going to challenge the matter. Everyone laughed and the tension in the room was relieved.

None of these tongue-in-cheek interventions will get you a spot in your local comedy club, but they might provide just the right amount of levity to blunt the awkward group silence that follows an argument. If only a few laugh—even a snicker is enough—it may be sufficient to defuse the tension and get the group refocused. Making a joke out of an argument is the leader's way of saying, "I know neither of you really meant what you said, and we're not going to get distracted by petty disagreements." Of course, if someone has said something especially hurtful or insulting, it's perfectly appropriate to say so during the meeting ("Marcus, let's tone it down a little. I think you owe Inder an apology.") or after the meeting if you want to impress Marcus with stronger advice.

IF YOU ARE FOCUSING MORE THAN 50% OF YOUR ATTENTION ON CONTENT OR DATA DURING A MEETING, YOU AREN'T DOING YOUR JOB AS LEADER

Scientists are always surprised by the 50% figure. Most team leaders tell us that they routinely focus the majority of their attention on the science, the data and the content—after all, they're the experts, right? But if you've read this chapter and the many real-world examples of what can happen when the leader is focused exclusively on the content of a meeting, you'll see the danger in such unitary focus.

We readily admit that the 50% figure is plucked out of thin air. We have done no scientific studies showing that this is the optimal number. Indeed, in some workshops, Carl routinely says that you shouldn't be focusing more than 33% of your attention on content—mostly to get people's attention. It works.

The following table lists just a few of the things that the leader needs to pay attention to during a meeting.

> ### The leader's checklist
>
> - Are we on agenda?
> - Are we on time?
> - Are we spending too much time on this topic?
> - Should we move this discussion off-line?
> - Is the group killing new ideas too quickly?
> - Are interpersonal issues impeding the discussion?
> - Is T_d increasing?
> - Are some team members not participating?
> - Are group impediments appearing (groupthink, sharedness bias, etc.)?

Hopefully, not every meeting you run will present every one of these potential detractors to group effectiveness. But unless you are constantly on the watch for them, you can be guaranteed that one or more will emerge before your meeting is over. Having this list in front of you (written on the palm of your hand?) is one way to keep one-half of your brain focused on the "meta-meeting," while the other thinks about content and data. Another way is to think of your meeting as existing simultaneously in three closely related contexts, as illustrated in the following cartoon:

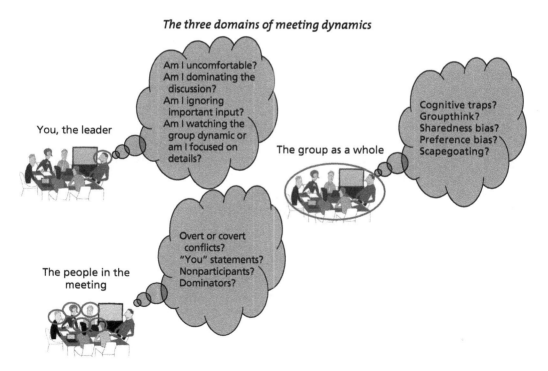

The three domains of meeting dynamics

Thinking of your meeting as existing simultaneously in three separate spheres—the personal (you), the interpersonal (the team members), and the group-as-a-whole forces you to recognize that there's more going on than just science. Left to its own devices, the group will not necessarily behave in such a way as to get the science done. Helping the team stay on track and interact effectively is just as much the job of the scientist-leader as is making sure the science is of the highest quality possible.

FINAL WORDS

Because science is seen as a discipline dominated by "experts," it feels natural to some senior scientists to dominate a discussion or to impose solutions to scientific problems. However, such domination can lead to a culture of passivity among more junior team members, who end up feeling that they have nothing to contribute. More important, leader-dominated discussions may constrain the scope of considered options. Daniel Feldman describes three steps to help overcome this tendency, which we paraphrase here (Feldman 1999, p. 74).

1. **Let go.** Practice letting go of the need to be seen as being right, have the only solution, and control the discussion.

2. **Solicit ideas.** Actively solicit ideas and suggestions from team members.

3. **Be open to input.** Soliciting ideas will do more harm than good if you do not give everyone's input serious consideration. Listen respectfully to concepts and suggestions from those outside of your own area of expertise or with less experience.

None of the above is to imply that actively running a meeting, or making decisions in your role as leader, is not appropriate at times. In fact, in many circumstances, such actions are essential, especially if there is a critical decision that needs to be made urgently. However, in a scientific team meeting the task is not always so well defined.

The explicit task of lab meetings, for example, is to hear and critique the results of the lab member's work. This task involves asking questions about technique, assumptions, interpretations, and all of the details essential to the scientific process. But there is also an implicit task, which is to think creatively or at least disseminate information that would stimulate creative thought later. The leader's role in this process is to ensure that members have the opportunity to think and interact freely; often this simply requires modeling respectful behavior for the group or staying out of the way. Discussion during a lab meeting might stimulate someone to make a novel connection, generate a new idea, or ask a question that could take the group in some exciting, unexpected direction. If you are hearing a lot of comments such as, "We already tried that," "We've been over this before," or "That's a dumb idea," and very few such as, "Hey, I never thought of it that way before," examine your own responses and reactions. You may find that group members are taking their cues from you. If this is the case, experiment with the three behavioral suggestions listed above to see whether members participate more over time.

The case study "The Power of Silence in Groups" shows the way in which Carl learned new behaviors by participating in a seemingly artificial situation. The workshop in which he participated used ad hoc groups to allow participants to experience difficult situations; role playing can achieve the same result. This is called experiential learning. Some of the following exercises use experiential learning to help you to become more effective at managing some of the situations covered in this chapter.

REFERENCES

Amason AC, Thompson KR, Hochwarter WA, and Harrison AW. 1995. Conflict: An important dimension in successful management of teams. *Org Dynamics* **24:** 20–35.

Baumeister RF. 2003. The psychology of irrationality. In *The psychology of economic decisions: Rationality and well-being* (ed. Brocas I and Carrillo JD), pp. 1–15. Oxford University Press, Oxford.

Burleson C. 1990. *Effective meetings: The complete guide.* John Wiley and Sons, New York.

Danziger SL-P. 2011. Extraneous factors in judicial decisions. *Proc Natl Acad Sci* **108:** 6889–6892.

Doyle M, Straus D. 1982. *How to make meetings work: The new interaction method.* Jove, New York.

Duhigg C. 2016. What Google learned from its quest to build the perfect team. *New York Times Magazine*, Feb 25, 2016. https://www.nytimes.com/2016/02/28/magazine/what-google-learned-from-its-quest-to-build-the-perfect-team.html.

Edmondson A. 1999. Psychological safety and learning behavior in work teams. *Admin Sci Q* **44:** 350–383.

Feldman DA. 1999. *Emotionally intelligent leadership: Inspiring others to achieve results.* Leadership Performance Solutions Press, Falls Church, VA.

Gilmore TN and Shall E. 1996. Staying alive to learning: Integrating enactments with case teaching to develop leaders. *J Policy Analysis Management* **15:** 444–456.

Haynes ME. 2009. *Meeting skills for leaders: Make meetings more productive*, 4th ed. Axzo Press, Boston.

Janis IL. 1982. *Groupthink.* Houghton Mifflin, Boston.

Klocke U. 2007. How to improve decision making in small groups. Effects of dissent and training interventions. *Small Group Res* **38:** 437–468.

Larson JR, Foster-Fishman PG, and Franz TM. 1998. Leadership style and the discussion of shared and unshared information in decision-making groups. *Pers Soc Psychol Bull* **24:** 482–495.

Plous S. 1993. *The psychology of judgment and decision making.* McGraw-Hill, New York.

Schein EH, Bennis WG. 1965. *Personal and organizational change.* Wiley, New York.

Tierney J. 2011. To choose is to lose. *New York Times Magazine*, August 21, pp. 32–46.

Wooley AW, Chabris CF, Pentland A, Hashmi N, and Malone TW. 2012. Evidence for a collective intelligence factor in the performance of human groups. *Science* **330:** 686–688.

WWW RESOURCES

https://www.ohsu.edu/xd/education/schools/school-of-medicine/faculty/faculty-development/upload/How-to-Run-a-Meeting-Jan-2016.pdf Anderson S. 2006. How to *run a meeting.*

http://www.citizenshandbook.org/ Dobson C. 2006. *The citizen's handbook. Practical assistance for those who want to make a difference.*

EXERCISES

1 Case study: The National Institute of Biology

Some of the concepts in this chapter are illustrated in the following case study. As you read the case and follow the interactions of the participants, note how the meeting could have been framed to run more effectively.

The National Institute of Biology (NIB) is running short of funds. For the third year in a row, Congress has failed to increase their budget to keep pace with inflation. Greg, the director of the NIB, has asked Antonia, the chief operating officer (COO), to spearhead a session with the senior management team to generate a list of their best ideas on how to cut their operating budget by 20% starting next month. Antonia had her administrative assistant set up a meeting. The participants were told that the topic was "to be announced."

The scene is a conference room with eight members of the senior staff of NIB sitting around the table.

Antonia (COO): "Greg says we need to cut our operating budget by at least 20%."

There is a stunned silence in the room.

Antonia: "I've already taken a hard look at the numbers with our finance people and am convinced we can get most of the cuts we need by layoffs and operating cutbacks in just a few areas including Immunology. That would leave our most high profile genetic screening projects with enough resources to remain operational. I've already spoken to Hiroshi in the Genome Center. Ning (Head of Immunology), I'm sorry you were out of the office."

Ning (head of immunology): "Gee, thanks. Actually the reason I was out of my office was that I was in an all-day meeting and we finalized the contract with our architects for the lab renovations we were planning."

Antonia: "That was totally irresponsible of you."

Ning: "All I know is that you've never really understood what goes on in the immunology branch so it makes perfect sense that you'd cut that first."

Greg (director of NIB): "Let me step in here. Ning, can you explain why you signed the contract with the architects? What are we renovating anyway?"

Ning: "This is new lab space for the stem cell surface marker project."

Alice (director of cell biology): "I thought we never really agreed that this surface marker approach was a good idea. You still don't have the full cell sorting data, do you?"

Ning: "Yes, we have it now. I brought the data with me. See? Look at this bar graph—just the bars on the left."

Everyone, including Antonia, huddles around Ning to get a look at this new data. They spend five minutes trying to understand what it means.

Greg: "OK, let's hold on a second. The purpose of this meeting is to generate ideas for budget cuts—not to review specific projects or data. We're getting off track."

Ning: "Well someone's off track because I didn't even know what this meeting was about until two minutes ago."

Greg: "It's clear that this is a difficult conversation, and it's understandable that we're all somewhat in shock about our financial situation. That's bound to generate some tension.

> But let's all be respectful of one another and focus our frustration on the problem at hand. Ning, is there any flexibility in that contract with the architects?"
>
> **Ning:** "Of course, we have a 30-day escape clause in there."
>
> **Antonia:** "OK, I apologize for jumping on you, Ning. It's been a rough couple of weeks."
>
> **Greg:** "Let's take a step back for a minute. I think Antonia's suggestion about specific cuts was meant to be helpful, but it looks like we just got things off to a rocky start. Let's open the discussion up to any good ideas about cutting our operating budget. How does that sound?"
>
> The discussion proceeds for about 30 minutes more with a lot of ideas being tossed about but no real conclusions.
>
> **Antonia:** "OK, we've got a ways to go yet, but we're off to a good start. Let's plan to meet again in two days and resume the discussion after we've all had a chance to think about this some more."
>
> **Greg:** "We had a few other items to discuss but I'll have my assistant set up another meeting to do that later this week."
>
> A few people groan and everyone wanders out of the room talking amongst themselves.

Before you read the analysis below, list as many things as you can that went wrong with this meeting. What could Antonia have done better? What could Greg have done better? What, if anything, did either of them do well?

Analysis:

- Antonia failed to advise people of the agenda for the meeting. This meant that people showed up for what was clearly going to be an important meeting without having had the opportunity to prepare for it. You don't have to be a management guru to figure out that this isn't a great way to start such an important meeting. It's no wonder that the attendees were annoyed—they were blindsided by the prospect of a budget cut. Antonia should have spoken individually with the participants in advance of the meeting. Not having done this, her meeting starts on a note of annoyance and resentment.

- Although Antonia didn't talk to other department heads, she did talk to Hiroshi specifically. This enhances the feelings of exclusion on the part of the other members, especially Ning, whose department had been singled out for a cut.

- Antonia started the meeting with a proposal of her own, affecting specific departments. Although the proposal may make sense, by starting with her own proposal, she essentially disenfranchised the rest of the group and gave the impression that a decision had already been made.

- Once Ning started to explain the surface marker project, the members all huddled around to review the data. An astute leader would be prepared for this. Focusing on scientific data is the perfect way for the group to relieve the tension generated during the acrimonious exchange between Ning and Antonia. Antonia herself, supposedly leading the meeting, also got caught up in the scientific discussion. But, as Greg was quick to correctly point out, it was not the primary focus of the meeting. Greg intervened to get the meeting refocused, but he was met with further hostility from Ning.

- In response, Greg interprets the group's behavior to them, suggesting that they are all frustrated by the budget cuts and that as a result they're taking out their frustrations on one another. This is appropriate because Greg is also including himself as having fallen prey to the prevailing frustration.

- Despite spending another 30 minutes on fresh ideas, Antonia dismissed the meeting without any summary of what they accomplished and without any explicit instruction for follow-up responsibilities for the participants.

- The meeting ended with several items that were supposed to have been discussed being postponed.

It's not surprising that the people walking out of this meeting felt dispirited and frustrated. Antonia could have run this meeting better without too much effort.

List what Antonia could have done to make this meeting more productive.

2 Using silence

- If you are a controlling manager, start by imagining what it would be like to be quiet for short periods of time in a meeting.

- Note your reaction to this imagined exercise and use this to catch yourself when you feel controlling during a meeting. Take note of what is triggering this reaction. For example, ask yourself if you are talking to reduce anxiety or control the group.

- Try being quiet for brief periods to see how you feel and how people react. Does anyone who is usually silent start talking?

3 Helping others to participate

- At your next team meeting, identify those who do not contribute very much. Make it a point to ask one or more of them their opinion of some matter under discussion, whether they understand what is being said, if they have any questions, or if they have anything to add. Do this at each meeting for different people.

- After several weeks, note whether these people have become more participatory or the group is receiving input that it might not have otherwise.

4 Dealing with people who dominate meetings

The next time that you are leading a meeting in which someone is dominating the discussion, wait for a pause in the monologue to say something such as, "Karen, you're making some good points. I'd like to hear what other people have to say. Jim, what are your thoughts?" Does this approach bring others into the discussion? You may need to have a private conversation with Karen.

5 What would you do if...?

Here are a few scenarios relating to team meetings to stimulate your thinking.

- In the middle of your lab meeting, John, one of your postdocs, tells Alice, another postdoc, that she is spending too much time on Facebook and that he is doing most of the work on their joint project.

- During a team meeting, three people are looking at their smartphones whenever anyone else presents data.

- You are a team member and during your presentation, you notice that the leader and three others are answering e-mail on their iPhones.

- You are a team leader and your meeting has gone on for two hours—an hour more than you intended. People are anxious to leave and are pushing for a decision on an important matter.

- You are a team member in a meeting that has gone on for two hours—an hour more than you thought it would. The leader shows no signs of wanting to end the meeting but you and others are tired and frustrated and have other work to do.

- Lydia repeatedly introduces topics that get the group away from the focus of the meeting.

- You are a team member and the leader never prepares or uses an agenda for her meetings. You never know what's going to be discussed and often find yourself unprepared for important discussions.

CHAPTER 8

A Delicate Art:
Manage Your Boss

Scientists who become managers and leaders of technical organizations are cut from the same cloth as those they manage. Like anyone else, their blind spots and limited self-awareness can prevent them from picking up important interpersonal cues and contribute to conflict aversion. Nonscientists who manage or lead scientists may have strong interpersonal skills, but they can be at a disadvantage if they do not understand scientists or how science is done. This chapter shows how you as a scientist can navigate the choppy waters of interacting with your boss, whether they are a scientist or not. It also shows how you can turn interactions that start out as confrontations into productive problem-solving sessions. Finally, you will learn how to understand your boss's problems, deadlines, and goals; disagree with him or her; and respond to criticism.

- ▸ You and authority: Is there a pattern?
- ▸ Cultural influences in dealing with your boss
- ▸ Hidden boss traps
- ▸ Dealing with an angry boss
- ▸ The three tools
  - *Tactic 1. Manage the boss's anger or frustration*
  - *Tactic 2. Identify underlying interests*
  - *Tactic 3. Be cognizant of your role and responsibility*
- ▸ If your boss is not a scientist
- ▸ Handling feedback
- ▸ References
- ▸ Exercises and experiments
  - *1. Pattern recognition: Your interactions with authority*
  - *2. Misattribution of your boss's behavior*
  - *3. Working with an angry boss*
  - *4. Disagreeing with your boss*
  - *5. Understanding your boss's needs*
  - *6. Criticizing your boss*
  - *7. Think before you speak*
  - *8. Your boss's perspective*
  - *9. Boss traps*

YOU AND AUTHORITY: IS THERE A PATTERN?

Self-knowledge can help you to understand whether feelings of insult or hurt are caused or magnified by your own emotional filters. Self-awareness can help you decide the most appropriate way to behave (or not behave) in response to these feelings. The benefit lies in ensuring that your actions and reactions are appropriate to the circumstance. Nowhere is this benefit more important than in interactions with your boss.

People in authority can elicit strong reactions in us. But in many cases these reactions have more to do with associations to a parent or other authority figure from childhood than with the person himself. These reactions may lead to behaviors that get in the way of our working relationships with those in authority. Becoming aware of your own reactions to authority figures is the first step in deciding whether your interactions are influenced more by the past than the present.

For the great majority of people, preexisting reactions to authority figures originate with a dominant parental authority figure, often, but not always, a father. How might your reactions to your boss be related to relationships with childhood authority figures? Let us look at some hypothetical examples.

1. If you were fearful of a childhood authority figure, you may find yourself avoiding or fearing your boss. Perhaps what your boss does or says does not seem to cause others concern, but feels threatening and anxiety provoking to you. These kinds of reactions may lead you to avoid your boss, or to overreact to comments or actions that others find benign.

2. If childhood authority figures did not command or earn your respect, you may find it hard to respect your boss. You may be skeptical of what they say, poking fun at them behind their back or ignoring their directives. Because you have no conscious intention to disrespect them, you may be surprised when they confront you with your behavior.

3. If authority figures were either absent or paid little or no attention to you as a child, you may find yourself acting as though you do not have a boss, even though you really do. The following brief case study illustrates this type of behavior and its consequences.

▪ *Case Study: Ignoring Your Boss*

Ulla was a senior group director at the Nanotech research center. She reported to Yoko, the vice president for research. Yoko hired Ulla because she was both highly productive and independent. They worked well together until Yoko took a particular interest in the largest of Ulla's projects that was beginning to become very costly. Yoko was careful not to interfere with Ulla's leadership of the project, but several times during the last three months she reviewed Ulla's expenditures with her and recommended alternative approaches to cut costs. Ulla listened politely to her suggestions but did not implement any of them. As the year progressed and budget time approached, Yoko learned indirectly that Ulla was planning an expansion of the project, doubling its budget.

Yoko told Ulla that she had one month to improve her communications with her and show that she could be responsive to Yoko's directives and suggestions. She suggested that Ulla meet with the head of human resources to develop a plan to achieve this.

Ulla was shocked by this turn of events and completely unaware that she was behaving as though Yoko did not exist. However, once she discussed her behavior with the head of human resources she was reminded of a similar problem that she had had with her PhD adviser in graduate school. With the help of human resources, Ulla came up with a plan to improve her working relationship with Yoko. Ulla proposed that she meet with Yoko weekly and that she copy her on any important correspondence related to project expenditures and budget. Ulla felt that routine contact with Yoko would help her to maintain an awareness of their relationship and prevent her from behaving as though she did not report to Yoko. Ulla had to work at reminding herself to touch base with Yoko on a regular basis, but with time Yoko saw an improvement in Ulla's responsiveness and the crisis passed.

Because we knew Ulla well, we can tell the reader that as a child she had very little contact with her father, who was the central authority figure in her family. He was regularly absent from the home, and when he was there, he was often the recipient of her mother's disrespectful comments. It is no wonder that Ulla grew up believing that authority could be safely ignored and she needed to fend for herself if she wanted anything. It was not that authority figures made her angry or threatened her; it was simply that she had no use for them.

Not all of us have, nor do we necessarily need, the kind of insight into our past relationships with authority figures that Ulla had. All we need is an ability to detect patterns in our past and current interactions with authority figures. When Ulla remembered that she previously had a similar problem with her thesis adviser, she concluded that her disregard for authority figures was a recurring pattern to which she needed to pay close attention in the future.

Interactions such as those between Ulla and Yoko are never unidirectional. Yoko also has some responsibility for the events. She encouraged Ulla's independence, perhaps to lighten her own supervisory burden. As long as things were going well, Ulla's independence was not an issue for Yoko. This independence felt very comfortable to Ulla, who, as we have seen, would just as soon have ignored Yoko altogether. Only when budgetary matters caused Ulla's and Yoko's interests to diverge did the problem emerge.

This case study highlights the importance of being alert to behavior patterns from past relationships with authority figures for avoiding those same behaviors in the present. As you become adept at this, you will detect patterns in your relationships with authority figures over time and in different settings—with teachers, advisers, mentors, project managers, CEOs, etc. Moreover, you will be able to foresee or anticipate problems before they arise, proactively managing yourself and your relationship with your boss.

CULTURAL INFLUENCES IN DEALING WITH YOUR BOSS

It comes as no surprise to anyone that there are big cultural influences in how people deal with their bosses. Scientists who come from cultures in which the lab head or department chairperson are held in high esteem by virtue of their position or authority often have a hard time adjusting to the mores of labs in the United States, for example. It is not uncommon for some scientists from Asian cultures, for example, to be reluctant to question or even to ask for clarification from their adviser or supervisor. This can work to their disadvantage in a culture in which a high premium is placed on speaking up and asking questions, even (and in some cases, especially) of authority figures.

If you come from a culture in which people who ask such questions are viewed as being disrespectful, you may find it hard to change your behavior when you move to a different country. Of course, even within a given culture, there are dramatic differences from lab to lab and boss to boss in how people interact with authority figures. Your best course of action is to watch closely how people interact with your new boss and also to ask your peers what the boss's expectations are. Some lab directors thrive on and encourage having their reports or trainees ask them challenging questions. Others may not be as receptive to tough questions, especially in the presence of others. We recently read an anecdote in the obituary of the Nobel Laureate, Baruj Benacerraf. The story (possibly apocryphal) relates that a student approached Dr. Benacerraf and asked him how to pronounce his first name. "You don't" was the reply. You wouldn't need to be a genius to figure out that your interactions with Dr. Benacerraf should at least start on the respectful and cautious end of the spectrum.

HIDDEN BOSS TRAPS

Some problematic behaviors such as being demeaning, overly demanding, inconsiderate, and manipulative are relatively easy to spot in bosses. But other behaviors that seem harmless or even beneficial on the surface can produce consequences every bit as negative. If you are engrossed in your work and have difficulty seeing beneath overt behavior patterns, you may find yourself being manipulated, abused, or misled by your boss. It is in your best interest to learn to recognize questionable behaviors before they get the best of you. Here are some patterns to look for.

- **Boundary issues.** Your boss talks to you about her personal problems. This makes you feel special. But it also leaves you with a vague feeling of discomfort that you brush aside. Pay attention to that feeling; it is telling you something important. Developing a personal relationship of this type with your boss can compromise your professional relationship with her. It can also backfire if your boss reveals too much and later regrets it.

- **Playing favorites.** Your boss plays favorites, and you are the current one. You are put in charge of projects, get to make important presentations, and go to scientific meetings in exotic places. You feel good about this, but others are not treated as well, and you feel uncomfortable about that. You brush aside those feelings and focus on your work. But it is best to pay attention to those feelings of discomfort—it may be that soon someone else will be the favorite and you will be disappointed or angry.

- **Selective communication.** A boss who plays favorites may also disseminate important information to selected members of the organization, making everyone else's job harder and progress slower. If you are on the select list, you feel fine. But if you are on the left-out list, you may not get the information or communications needed to do your job. The result is detrimental to you and the company.

- **The boss as booster.** Because she likes you, your boss promotes you to a position for which you are really not qualified. At first you feel uneasy, but you quickly bury those feelings as you struggle to keep your head above water in your new job. In time, you feel that you have been set up to fail because you lack the experience and skills to succeed. Next time, pay attention to your feelings right away and question the advisability of such a promotion as well as the motives behind it.

If you are disposed to look only at the manifest behavior of your boss, you may be blind to the negative consequences of such behaviors. If you attune yourself to your feelings, doubts, and misgivings, you increase the likelihood that your work and career will not be subverted by your boss's errors in judgment. If your boss is powerful and charismatic or if his questionable behaviors mesh with your own needs (for approval, recognition, acceptance, etc.), you will have a hard time recognizing the downside of those behaviors. As you become more aware of your emotional needs, your ability to make dispassionate assessments of your boss's behavior and its appropriateness will improve.

After recognizing these behaviors, take steps to limit their impact on you and your team, project, and organization. For example, in the above cases, you can say the following:

- "Fred, I'm not sure that I'm the right person for this type of errand. I don't feel comfortable doing it."

- "Fred, I appreciate your confidence. I'd like to share this information with the rest of the team," or "Fred, getting information in this way puts me in an awkward position with the rest of the group. I know that they would really appreciate periodic meetings in which you could update them."

- "I'm really honored by your confidence. I'd like to try this new position unofficially for a month so we can both decide whether this promotion makes sense."

If you're lucky your boss will see the wisdom behind your comments and suggestions and will view you as someone to whom principle, teamwork, and merit matter more than being the favorite.

DEALING WITH AN ANGRY BOSS

The previous sections illustrate some of the ways that limited self-awareness can lead to inappropriate behaviors that result in misunderstandings or clashes with bosses. Moreover, if your boss is a scientist, he may be suffering from the same deficits in self-awareness and interpersonal savvy as you. Limitations such as these, which occur in both employee and boss, can lead to escalating problems and misunderstandings. If this happens, you must defuse the hostility and work toward reestablishing the relationship.

When dealing with your boss, remember that your goal is a productive interaction, not an argument or confrontation. To make this happen, remember to manage your own reactions. If you find yourself getting into a confrontation with your boss, your best tactic is to change the tenor of the discussion to one in which the two of you are allies trying to solve a mutual problem. As in any negotiation, the solutions are defined by the underlying interests of you and your boss.

The following case study illustrates the application of this approach. It involves an interaction between a graduate student and her mentor (it has direct relevance to the topics on mentor–student interactions covered in Chapter 10). Although the mentor–student interaction has some unique characteristics, it also has elements of an employer–employee interaction. The behavior of the employer/mentor in the following may seem extreme but the case, as related to us by a close colleague, is based on actual events. It shows how a junior scientist might have dealt with an infuriating situation.

■ Case Study: When Your Future Is on the Line

A few years ago I was on the thesis committee of a graduate student–let's call her Barbara. Around the middle of November of her third year in graduate school, she e-mailed me to ask if I would discuss a problem that she was having. I knew that Barbara was a member of Professor Aster's laboratory, where she was working on a PhD thesis project. She had been there for about 18 months. My assumption was that this meeting was to be a scientific consultation, because I knew that Aster's scientific interests to some extent overlapped with my own.

The problem turned out to be only peripherally scientific. Barbara told me that she was having a hard time with Professor Aster because she was not satisfied with the amount of time Barbara was spending working on her project. Aster had two other graduate students and two postdoctoral scientists in her lab and it seemed to Barbara that they all worked 12 to 18 hours per day, six or seven days a week. Barbara was married and wanted to have a life outside of the lab. She felt that she put in a reasonable amount of time in the lab, but this apparently was not enough for Aster.

The previous week Aster had called Barbara into her office and expressed dissatisfaction with the progress that Barbara was making in her project. When, during the course of pressing Barbara to spend more time in the lab, Aster learned that Barbara was planning to take a two-week vacation during the Christmas holidays, she got visibly angry. Aster said that she thought Barbara was perhaps not cut out for a life in scientific research. Aster said that some women were happier staying home and having babies, and that maybe this was where Barbara belonged. Barbara was shaken up and distraught by this interaction. As she relayed the story, I could see her becoming visibly upset and overcome with emotion.

I asked Barbara how she had responded to Aster's outrageous comments, and she said that she was shocked and hurt. She was so hurt that she became immobilized; her feelings over-powered her thoughts and she could say nothing. After some moments of silence, Aster announced that she had another meeting to go to and the meeting ended abruptly.

I can still recall the feeling of ice in my veins when Barbara relayed this story to me. Barbara did not know what to do. She had invested 18 months on this project and did not intend to walk away from it. She feared that Aster wanted her out of the lab and was unsure that she could do anything about it. Barbara felt guilty for not working as many hours as the others but was reluctant to sacrifice her personal life or her relationship with her husband for a few more hours in the lab.

I was seething and wanted to tell Aster that her behavior was outrageous and possibly unethical, and that I was going to bring her before a faculty committee for censure. However, as soon as I became aware that I was fuming inside, all my self-awareness alarms went off. I knew from experience that taking action while feeling this way would be a bad idea, both for me and, more importantly, for Barbara.

> I asked myself what was of paramount importance here. Was it to punish Aster for her inappropriate remarks or to help Barbara get through this situation and obtain her degree? I concluded that no matter how much I and Barbara longed to make Aster accountable for her behavior, Barbara's overriding interest now was to complete her project and get her degree in the shortest time possible.
>
> I suggested to Barbara that she needed to weather this storm. I told her that Aster was probably under stress and that Barbara should not take her comments, possibly made in a moment of ill-considered anger, literally. Because I was on her thesis committee, I knew that her work was moving forward, even if it was not at the rate that Aster might have wanted. I told Barbara to initiate a meeting with Aster sometime during the following week for a follow-up discussion and ask Aster to meet with her more regularly to give feedback. I suggested that Barbara make it clear to Aster that she was in the project for the long haul and had every intention of completing her work there. Barbara did this and continued to work hard on her project. She completed a first draft of her thesis 18 months later, and six months after that, she left Aster's lab for a postdoctoral position.

What could Barbara have done differently in this situation? Although she could have made a major issue of Aster's behavior, that would not have been in her best interests. Had Barbara defined her underlying interests as described in Chapter 3, she would have concluded that her chief interest was to complete her PhD work in two more years. It was almost a certainty that seeking vengeance on Aster would have initiated a chain of events that would have had a major impact on this goal. Further, Barbara was not about to quit the lab, because it would result in an unacceptable delay in completing her PhD work.

Just as important, Aster's underlying interest of having a productive lab would not have been satisfied if Barbara left the lab prematurely. After all, Barbara was doing her work and Aster had invested a significant amount of time and resources in her project. Both of them tacitly understood that each would lose if the relationship deteriorated.

In this instance, the faculty member's support and suggestions helped Barbara to negotiate a successful outcome with Aster. Barbara had sufficient self-awareness to know that she needed help in this situation and sufficient observational skills to identify the right person to help her. It is possible that if Barbara had done nothing in response to the meeting with Aster, the situation would have turned out fine. It is also possible that if Barbara had not gone to the faculty member for advice, the situation would have deteriorated and she would have left Aster's lab, setting her education back significantly. We will never know.

But we do know that if Barbara could have done or said something during her meeting with Aster to reduce the chance of a disastrous outcome, she should have. In the following sections, we provide three approaches or tools for dealing with an angry or frustrated boss. Of course, the tools can be applied to anyone with whom you are interacting. We present them here because most of us have greater difficulty dealing with an angry boss or authority figure than with an angry peer or subordinate. If you can get it right with your boss, you are likely good to go with almost anyone else.

THE THREE TOOLS

For Barbara to use the following tools, she first needs to recognize and manage the strong reactions that Aster's comments have engendered within her. Using the techniques described in Chapter 2, Barbara might have been able to recognize what she was experiencing and feeling and anticipate from past experience that her reaction would be to withdraw and be silent. If she can get to that point, she will be better able to apply the approaches listed below.

Tactic 1. Manage the boss's anger or frustration

A productive outcome from a difficult interaction with your boss requires care to avoid exacerbating the situation. To arrive at a resolution of the problems presented by Aster, Barbara needs to shift the tone of the discussion from anger and accusation to joint problem-solving. She must stop feeling like a victim and show Aster that she has heard her and is sympathetic to her concerns. The following guidelines, taken from *The Feeling Good Handbook* by David Burns (1990), work wonders in this and similar situations.

Dealing with an angry person
Remain focused on your task when dealing with an angry or hostile person
Agree, to defuse anger
Empathize, to start a dialogue
Assure or Inquire, to show your commitment

Agree. For Barbara to have a productive conversation with Aster, she must defuse Aster's anger from the outset. It is hard to be angry with someone who is agreeing with you, so Barbara must find something to agree on with Aster. This lets Aster know that Barbara has heard what she has said and will make Aster more receptive to what Barbara has to say. Despite Aster's rather primitive attack, Barbara can still find something with which to agree. She might start by saying one of the following:

- "It's true that I don't work as many hours as Linda or Barry but I do work hard and this project is important to me."
- "This has definitely been a slow month for my project. I'm as frustrated as you are."

Barbara has admitted to no inadequacy, but she has begun to establish a rapport with Aster. By showing that she shares Aster's concern about productivity, Barbara can start to defuse Aster's anger. Once anger is defused, Barbara can build a bridge for communication by empathizing with Aster.

Empathize. Barbara can indicate that she acknowledges Aster's frustration and can understand why she feels that way. Barbara could say that she is aware of the pressures and deadlines to which Aster is subject. Perhaps she has noticed Aster's concern during the past several weeks and understands why she is concerned. She could say that she knows that this has been a slow month for either her or the lab as a whole and understands why this makes Aster uneasy.

- "I know that this is an important project for the lab and I'm sorry that progress hasn't been better."
- "I know you've been frustrated by progress in my project and I'm sorry for that."

Either of these statements sends a clear message to Aster that Barbara cares about what Aster is saying. Once she gets to this point, Barbara is ready for the next step in the process: assurance.

Assure. Barbara can assure Aster that she is committed to doing her part (without saying that she will work 20 hours per day) and understands the need for greater productivity. If these statements are genuine, they will strengthen the bridge that Barbara is trying to build.

- "I want to assure you that this project is very important to me."
- "I am committed to completing this project in a way that you will find acceptable."

Inquire. Barbara is now ready to move to problem-solving, the final stage of the process. She can start by asking problem-solving questions—for example, what Aster would like to see as the outcome of the conversation, what she should be focusing on during the next month, or for ways other than working 20 hours per day to show Aster her level of commitment. Of course, if Aster really does believe that everyone in her lab should be working 20 hours per day, the outcome may be unsatisfactory. The inquiry phase is especially important because it can help uncover the underlying causes of Aster's concerns.

- "Could we meet more frequently so that I have a better idea of how you think things are going?"
- "Can we meet next week so I can show you my experimental plans for the next two months?"
- "What do you think my top priority should be during the next three weeks?"

Tactic 2. Identify underlying interests

As we saw in Chapter 3, identifying underlying interests can be the most important element in a negotiation. The same is true when dealing with an angry person. Aster demanded long hours from Barbara to satisfy some underlying interest. What might Aster's underlying interests be? Without knowing the basis for Aster's specific concern and anger, Barbara is at a disadvantage during the discussion. Consider the following very different reasons for Aster's frustration and the corresponding different ways of addressing the underlying interests.

Scenario 1

- Underlying interest: Aster is under pressure to produce data during the next three months to ensure the renewal of an important Department of Defense grant.
- Way to meet interest: Barbara might be willing to agree to work seven days per week for the next six to eight weeks, with the understanding that this is a temporary arrangement.

Scenario 2

- Underlying interest: Aster is feeling anxious because of a series of setbacks in the lab. Although there are no big deadlines looming she wants reassurance that everyone is fully committed to their projects.
- Way to meet interest: In this case, a sincere expression of understanding, support, and willingness to work harder may meet Aster's needs. Leaders don't always know how committed their lab members are. Hearing the words can be incredibly reassuring.

Scenario 3

- Underlying interest: The other postdocs in the lab are resentful that they work longer hours than Barbara and have complained to Aster. In this case, Aster's underlying interest is to defuse tension in the lab.

- Ways to meet interest: Barbara can ask Aster for suggestions on how she could interact with the other postdocs that would show them her commitment to the project. (Note, however, that it is not Barbara's problem how the other postdocs feel about her work hours.) If Aster is the one dissatisfied then we're back to scenarios 1 or 2. If she's not, it's up to her to convey this to the others in the lab ("I'm interested in results more than hours in the lab.").

Without knowing the specifics of Aster's anger or frustration, Barbara won't know how to respond. Here are some questions that Barbara could ask to help her learn—or guess—Aster's underlying interests or concerns.

- "This is the first concern I've heard about how much time I spend in the lab. Is there something specific that triggered this?"

- "Just so I understand, are you also concerned about my productivity? I think I've been reasonably productive during the last year."

- "Does this have anything to do with the others in the lab? I know they tend to be here later than I, but of course our projects are very different. I put in my time when it's needed and that's not always every day."

Here's another, somewhat simpler, scenario involving an angry boss.

Lab head: "This report is awful! It's disorganized and doesn't make any of the important points we discussed. A two-year-old could have done this better."

Sounds pretty dire. Let's apply the Agree, Empathize, Assure, and Inquire technique.

You: "Well, it may not be the best report I've ever written (**agreement**). I know how important this is to you (**empathy**) and I'm sorry you're disappointed in it (more **empathy**—but don't overdo it). I was under some time constraints (lame excuse, but not untrue). I wanted to make sure that you had this in time (assurance). What would you like me to do (**inquiry**)?"

It would be a truly barbaric boss who wouldn't respond to this series of responses by calming down at least a little. After all, the boss does want the report and it does need to be written. Try it.

Tactic 3. Be cognizant of your role and responsibility

Barbara has at least two roles in this case. One is that of a trainee and the other is a member of a team working on a project. In her first role, she must keep her training program on track so that she can finish within two years. This requires maintaining her relationship with Aster. In her second role, she is responsible for helping advance a group project. Barbara must let Aster know that she understands her responsibility to the group effort and is willing to pitch in (within reason) to help advance it.

Reminding yourself of your role and responsibility in a difficult situation is often an effective way to gain some perspective on an emotion-laden interaction. It forces you to step back and look objectively at your responsibilities. This can be a valuable tool for interrupting knee-jerk reactions until you have had enough time to think through your options. Ask yourself:

- What is my responsibility to the team, or organization?
- Am I acting in accordance with that responsibility?
- Am I contributing to the work of the group or team?

In summary, if all Barbara can think of in response to Aster's confrontation is how she can spend more time in the lab to generate more data, she may have missed the issues that underlie Aster's outburst. If Aster's frustration with Barbara's working hours reflected her belief that Barbara was not committed to her work and was not interested in helping the lab to maintain its funding, Barbara needed to do more than just work longer hours. Judicious use of the "inquiry" tactic during the discussion with Aster could help Barbara look beneath the surface and seek the underlying reasons for Aster's frustration.

IF YOUR BOSS IS NOT A SCIENTIST

One of the hardest challenges facing a scientist can be working for someone who is not a scientist. Scientists tend to be concrete, linear thinkers. They are cautious in their conclusions and may be loath to make the kinds of projections, speculations, and leaps of faith that come naturally to those with more of an entrepreneurial bent. Executives with business or entrepreneurial backgrounds may make connections that seem grossly speculative or even irresponsible to scientists. They may be willing to take risks that seem unjustified or even foolhardy to a scientist. These executives may become easily frustrated by scientists who strive for certainty and thoroughness. Conversely, scientists may act in a condescending manner toward an executive who controls their fate, but who they see as being scientifically naïve.

Scientists must understand that nonscientist chief executive officers (CEOs) may have only a vague understanding of how research is done. Some are skeptical when a scientist implies that they cannot guarantee that a discovery will be made or a promising project completed in 12 months. In a sense, the scientist and the executive are speaking different languages. As a result, a hidden dialogue takes place when they interact. The following is an illustrative case study.

■ *Case Study: Speaking Different Languages*

Pharmex is a successful biotechnology company employing more than 500 people. It makes extensive use of metrics to monitor productivity in its drug discovery division. Last year was a difficult one for the company. Two promising drugs that were either in or about to enter the clinic had been withdrawn because of safety problems. The company's stock had dropped, and industry analysts were skeptical that Pharmex could improve its profitability given its existing pipeline of drug candidates. Alex, the CEO, is convinced that the company is not in as bad a shape as it seems, and he plans to give an upbeat presentation about some of the more promising project possibilities at the forthcoming board meeting. He has asked Luisa, the head of his cardiovascular research division, to meet with him to discuss presenting the progress on a cardiac arrhythmia drug that Pharmex is developing. The overt dialogue follows.

Alex: "Luisa, we have a board meeting coming up in two weeks and I want you to give them an update on the AR 287 project. We've put this off twice and now we need to bite the bullet. Put together some timelines that show when you believe it will be ready for its first trial in man."

Luisa: "Well, that will depend on a lot of factors. We're still trying to find a good biomarker for activity in man. That has been a slow process because our expression profiling group spent six months on something that turned out to be bogus and now we're back to square one. I really don't have anything concrete to tell the board."

Alex: "You need to do better than that, Luisa. Can you just make some projections for us? I mean, you must know how much work you have to invest to come up with a biomarker, whatever that is."

Luisa: "Alex, it is simply not possible to project when discoveries will be made. You just don't understand research."

Alex: "Well, I understand this: Either you figure out how to give us a useful update and timeline for the rest of this project, or I will find someone else to do it."

Luisa's responses to Alex may have been honest, but they did not achieve much more than to create additional anxiety for him and lead to an ultimatum for her. Let us imagine what was going through each of their minds as they had this discussion. Here is the covert dialogue.

Alex: "Here we go again; this is like pulling teeth. How am I supposed to develop a budget and a timeline for the board if these scientists never commit to anything? What can I do to get a straight answer? What do they do in those labs all day, play computer games?"

Luisa: "Here we go again. Whenever I talk to this guy, it feels like a root canal without anesthesia. He just doesn't understand that it is not possible to guarantee anything in research. All he wants is projections and timelines. Am I supposed to make up numbers? I wish that he would just back off and let us get on with our work."

Because Luisa is focused on the science, all she hears in Alex's request is a demand for results. Although Luisa understands management's need to see results, Luisa feels that she cannot tell them what she does not know. By limiting her focus to the scientific impediments to meeting Alex's demands, she has lost sight of what lies beneath these demands. Only by discovering Alex's underlying motivation, interest, and needs will she be able to help him and fulfill her role and responsibility. Let us see how Luisa might use some of the approaches discussed above in this situation.

Manage Alex's frustration or anger (agree/empathize/inquire)

- Luisa can start by agreeing that progress has been slower than any of them would like. She can also agree that it is unfortunate that they have not been able to provide concrete information to the board on this project in the recent past.

"I know we've had to put off reporting on this project to the board more than once. Progress has been slower than we had hoped."

- Luisa can then empathize with Alex's dilemma by agreeing with the need to present concrete information to the board. This will go a long way in reassuring Alex that they have the same objectives. She could say,

"I know that you are under the gun here and that the board would be thrilled to see positive results and projections on this project. I want to help and will do whatever I can to pull together an accurate and realistic picture of our progress. Although you know that I cannot pull a rabbit out of a hat, I can be upbeat and positive because that is how I feel about this project."

Hearing this from a scientist will make most CEOs think that they have died and gone to heaven. Luisa will have made a friend, while also being honest and frank about what she can do.

Finally, Luisa can inquire about the specific type of information that Alex would like to hear in her presentation. Further inquiry will also help her to understand the issues beneath his request. Does he simply need data on AR 287, or is he trying to meet other underlying needs? Whereas it may sound as if he is asking for something that Luisa cannot deliver, he may actually be asking for something much simpler—an indication that she understands his problems and that she is willing to work with him to help address them. If Luisa simply assumes she knows why Alex is asking for this information she may be dead wrong. The following are some possible underlying interests that underlie Alex's request:

- Demonstrate progress on this specific project to the board.

- Get reassurance from Luisa that the project is moving forward.

- Get Luisa to commit publicly to a specific list of deliverables and deadlines for the project.

- Showcase his ability to motivate and inspire research and development in the company.

- Use the board meeting and Luisa's presentation to help him decide between continued funding of this project versus others.

Luisa can get a good idea of the interests that underlie the CEO's request by asking questions.

- On what would Alex specifically like her to focus?

- What should be the underlying theme of her message to the board? Should she show data or just review progress and projections?

- How much technical detail should she include?
- Should she discuss the cardiovascular group's alternative to this project if it fails to progress?

Unless she tries to find out what underlies Alex's request, Luisa is working in the dark.

"OK, just so I know how to frame my presentation, what specifically do you think would be most useful? I won't have any definitive data to show. Would a timeline of our planned studies be useful? How much detail do you think I should include? Do you want me to say anything about our backup compounds, or just focus on AR 287?"

By addressing Alex's underlying interests (as opposed to his positional demand for results), Luisa can deliver something other than, or in addition to, results (that she believes she cannot deliver). If in response to some of her inquiries the CEO says, "Look, Luisa, you need to give us some indication that this project is moving forward and that we have alternative strategies in place if it continues to stall," Luisa can reasonably conclude that Alex is looking for reassurance. This does not mean that she should dissemble about results, but that in this circumstance Alex's demand for results may mean something other than data. Perhaps it signals a review of the project plan, discussion of steps taken to ensure that the project is keeping to its timeline, review of contingency plans, and her assessment of likely outcomes. These all add up to reassurance that all is under control.

Both nonscientists and scientists may use the same words to mean different things. For example, when a senior executive (especially one without a science background) asks for an update on a project, they almost never mean, "Show me your latest data on those polymorphisms in the *GAG 34* gene." They are looking for the big picture: Is the project on track according to the original timeline? Are there any unanticipated roadblocks? Any new avenues of approach? If you present the big picture succinctly, accurately, and honestly, you will be the one this executive comes to the next time they need information. If you open your lab notebook and start flashing micrographs of a rat's liver or pictures of polymerase chain reaction products, do not be surprised if you are never asked again.

Second, executives with little or no science background may have a different conception of time than scientists. Dubinskas (1988) explains that managers with a business background tend to view time in a linear manner, punctuated by well-defined targets and milestones. Scientists, on the other hand, may view time as having an infinite flexibility that reflects the reality of scientific research: You cannot schedule discoveries. These viewpoints need not be incompatible, but both parties must understand that words or phrases such as "as soon as possible," "quickly," and "fast track" may have different meanings for each of them. If not, confusion and disappointment may result.

HANDLING FEEDBACK

At some time or another, if we're lucky, most of us receive feedback from someone in authority, either in a formal performance review or via an informal discussion. Such feedback is one of the most important sources of information you can get about your behavior, effectiveness, and work performance. It is unfortunate but true that most people tend to think of offering feedback only when something is wrong or your performance is not up to expectations. For such situations, knowing and anticipating how you react to negative feedback is the best preparation. Here are some tools that you can use to increase your chances of taking the best advantage of feedback.

- Repeat the feedback in your own words and ask if you have it right. Often we hear something other than what was actually said, especially when it has an emotional impact on us. If your boss says, "Jim, your presentation to the investors who were here last week wasn't at all what I expected. It was over everyone's head and it needs a lot more work." You might respond with, "So you're saying that you expected a less technical presentation and you'd like me to rework the one I gave to make it more suitable for a lay audience, right?" You might be correct or you may get the response, "No, I don't mean that I want you to pitch it for a lay audience but I do think that you need to cut down on the number of equations you show." That response would make you thankful that you clarified his feedback instead of just going ahead and acting on your assumptions.

- Before responding to feedback, especially negative feedback, find something with which to agree. We discussed the rationale for this earlier in the chapter, and the same applies here. Your goal is to bring a positive and collaborative tone to the discussion. So Jim might have said, "Well, I did notice that a few people appeared to be a bit baffled while I was talking," or "I guess my scientist side sometimes gets the better of my communicator side." After this, you can inquire about the feedback and identify underlying interests: "Did one of the participants say something to you?" or "Do you think that they missed the point of what I had to say because I bogged them down in the equations?"

- Once you have done the above, use nonconfrontational problem-centered language to clarify aspects of the feedback with which you might not agree. For example, "I thought I was following the guidelines that your assistant sent out before the meeting. I may have misunderstood them, but I thought they advised showing detailed scientific support for our conclusions, including equations."

- Identify ways to solve the problem together. What do you need from your boss to help you improve or change? It may be that you need increased feedback or more frequent meetings. Keep the focus on what you or your boss needs to do to improve the situation in the future, rather than on what is wrong with the situation now. Jim might suggest, "Perhaps if I had a bit more regular feedback I would do a better job of knowing how you want these presentations structured. What if we meet once a week during the next month or so while we're in a fund-raising mode?"

- Finally, ask your boss to name what you are doing right so that you have a concrete idea of what he wants. In Jim's case he might have asked, "Were any of the presentations I gave last month closer to what you had in mind?" or "The equations aside, was the rest of the presentation on target? What were the strongest parts, so that I can better gauge what it is you wish to see?"

Often, we're not sure what to make of the feedback we get, or we may be so tied up with feelings of resentment or anger that we can't think clearly while we're getting the feedback. Then, the next day after we've had a chance to think about what we heard, we wish we had been able to speak up during the feedback session. If you think this is happening to you try saying, "Thanks for your feedback I really appreciate it. Would it be OK if I process this for a day or so and then meet with you again so I can work out how best to address your concerns?"

Remember, however, that getting feedback from your boss is not a negotiation. Part of their job is to tell you how they think you're doing. What they tell you isn't usually a matter for debate unless you think that they are missing important information or are misinformed.

The exercises that follow help you to gather data about patterns of behavior that you may exhibit when interacting with authority figures. They also include experiments for exploring new behavior patterns and determining whether these can make your interactions with your boss more productive.

REFERENCES

Burns DD. 1990. *The feeling good handbook*. Plume/Penguin, New York.

Dubinskas FA. 1988. *Making time: Ethnographies of high-technology organizations*. Temple University Press, Philadelphia.

EXERCISES AND EXPERIMENTS

1 Pattern recognition: Your interactions with authority

Write down the names of three authority figures from your professional career. These could include your current boss/supervisor/manager and any from the past. Next to each name, write the first four feelings that this person evokes in you. Then, answer the following questions:

- Do your feelings toward these different authority figures have anything in common?
- Do these common features suggest that you might be reacting to feelings (perhaps the ones that you have identified as common to several authority figures above, or others) that are evoked by your boss, but that originate from your past (i.e., are they only loosely related to your boss?).
- If so, do these reactions affect your work or the effectiveness of your interactions with your boss?

2 Misattribution of your boss's behavior

Often, we assume that something a boss says that we find upsetting or objectionable was specifically directed at us or in response to something we did. Sometimes just enumerating alternative explanations can help us to see that these are just as, if not more, likely than the "self-centered" explanation.

List three things that your boss has said or done that angered or threatened you. For each incident, list three explanations that have nothing to do with you and three reasons that have everything to do with you. Evaluate the likelihood of each set. Often, this exercise will enable you to make a more accurate assessment of the situation and identify inappropriate or exaggerated reactions on your part.

3 Working with an angry boss

The next time your boss says something that seems angry or threatening,

- Say something that conveys an understanding of your boss's issue.
- Find one thing with which to agree.
- Ask one clarifying question.
- Check whether these techniques change the dynamic of the interaction.

4 Disagreeing with your boss

- The next time your boss says something with which you disagree, find some element in what was said with which you do agree.
- Start a conversation by talking about that element.
- Transition to what you do not agree with in a way that focuses on the impact of the decision, statement, etc. on the project or company.

- Assess whether this approach helps you to convey an alternative view in a way that is collaborative rather than argumentative.

5 Understanding your boss's needs

- Think of something that you know that your boss needs from you, but has not yet asked for, and do it.
- If you are focused on your work or on technical matters, it is easy to lose sight of the fact that your boss needs information, viewpoints, and support from you. Forcing yourself to think like your boss or putting yourself into his shoes is one step closer to creating a more productive relationship. Make a list of three things that your boss might need from you, especially what they might not know they need, and give them one of them before they ask.
- Take note of whether your new behavior changes the tenor of future interactions with your boss.

6 Criticizing your boss

- List one or two comments or behaviors that you have recently said or done that belittle or denigrate your boss.
- What was the stimulus for these actions? What were you feeling when they happened?
- Review your authority behavior pattern (earlier in the chapter) and ask yourself whether your behavior was motivated in part by feelings unrelated to your boss.

7 Think before you speak

- Catch yourself the next time you feel anger, hostility, or animosity toward your boss. Instead of acting on your feelings, remember them.
- Later, review what you would have done or said if you had expressed the anger, hostility, or animosity and the possible consequences.
- If, as is likely, you decided that you were better off not acting on your feelings, remember this the next time you experience these feelings.

8 Your boss's perspective

It is easy to resent what your boss says or does especially if it negatively impacts you.

However, sometimes resentment is the result of ignorance of the pressures and constraints under which your boss works.

- List the external and internal pressures and constraints to which your boss is subject.
- Pick one decision that your boss has made with which you disagreed or that had a negative impact on you. Take your boss's perspective and write down as many reasons as you can that might have influenced their decision.

- The next time your boss makes a decision with which you disagree, remember the results of this exercise, and list the possible underlying circumstances or explanations.

9 Boss traps

- List what your boss has said or done that has made you feel even mildly uncomfortable, and that was outside the scope of good management practice and behavior. Include comments or behavior that did not seem overtly inappropriate, but that just left you with an uneasy feeling.
- Can you discern recurring behaviors in this list?
- Use this list to identify future situations in which you may wish to exercise caution in your interactions with your boss.

CHAPTER 9

Win/Win with Peers:
Make Allies, Not Enemies

Early in Carl's career, he was one of a three-person team working on an exciting research project. Midway through the project, the group had a disagreement about who would be included as authors on a scientific paper. They went to their lab director, who said, "Work it out among yourselves. I have faith that you can do it." But he was wrong; they could not. The publication of the paper caused great consternation within the lab. The team disbanded, and the project quickly lost momentum. They all lost.

Some of the most difficult interactions that scientists face involve peers and colleagues. Unlike interactions with employees or superiors, during which, if necessary, one party can always make the final call, peer interactions take place on a level playing field. It is here that negotiation skills are especially important because it may not be possible for any of the parties involved to make an autonomous decision. Inability to resolve a dispute may require intervention by a higher authority, such as a director or administrator. When problems escalate to this level, all semblance of civility is frequently lost and the consequences can be disastrous not only for the individuals,

- ▸ Peer problems in the science workplace
- ▸ Key elements of peer conflict avoidance
- ▸ Tools for dealing with peer conflict
  Recognize the warning signs
  Avoid accusations by addressing the problem in terms of its effect on you
  Address objectionable behavior patterns without devaluing or belittling the person
- ▸ Dealing with difficult people
  Angry or hostile people
  Critical or judgmental people
  Pushy or demanding people
  Passive–aggressive people
  Procrastinators
  Complainers and help-rejecting complainers
  Needy people
  Argumentative people
  Are you one of these people
- ▸ How to discuss science without arguing
- ▸ Pay attention to your own body language
- ▸ Pay attention to the other person's body language
- ▸ Consequences of feeling inferior/superior to peers
- ▸ Reference
- ▸ Exercises and experiments
  1. Your body language
  2. Others' body language
  3. Problems with peers: Focus on the problem, not the person
  4. Hidden peer conflicts

but for entire laboratories, departments, and companies.

This chapter shows how you can apply the core skills from Chapters 1–3 to improve your ability to work through difficult situations with peers in the science workplace.

PEER PROBLEMS IN THE SCIENCE WORKPLACE

Whether you are a graduate student, a professor, or a scientist in a pharmaceutical company, chances are that a significant part of your workday is spent dealing with colleagues who engage in behavior that seems inconsiderate at best and downright malicious at worst. Examples of such behavior include

- The lab member who plays loud and distracting music without regard to its effect on others.
- The postdoc who leaves a mess wherever they work. To make matters worse, they do not consider it a mess, but you do.
- An overbearing colleague who dominates group or department meetings.
- A collaborator who shares the results of your work without your consent.

These and similar types of behavior may elicit angry reactions from colleagues. Just as often, the behavior goes unchallenged. Peers of the offender may not consider it their responsibility to confront the offending party or they hope that a higher authority figure will intervene. For many scientists, their previous experiences addressing such situations have been uncomfortable or confrontational. The result is that they avoid the problem or person. Very often, the resentment escalates until a blowup occurs.

The following case study, based on an actual situation, illustrates problems that can arise when we avoid dealing with peer issues.

■ Case Study: Talk about It

Jack was an assistant professor at Wannabe University and a major user of the department's shared two-photon microscope facility. Because operation of this device required specialized skills, Jack had assumed responsibility for training new users. Although the device was open to all department users, Jack made it difficult for others to get trained and schedule time on the device. He subjected new users to a constant barrage of criticism, and they often found it difficult to find open time that Jack had not already reserved for either his own use or "maintenance."

Several users complained that Jack constantly hovered in the background when they used the machine, ostensibly to make sure that they did not damage the equipment. Jack's presence made users feel self-conscious and constrained. He often interrupted their work on the pretext of making sure that they had not altered the alignment or calibration of the device. As a result, department members were reluctant to use the machine and avoided having their staff trained because the experience was so unpleasant and, in some cases, traumatic.

Department members resented Jack's behavior and often complained among themselves. However, no one had ever spoken to Jack about it. Susan, a faculty peer of Jack's, had complained to the department chairman. His reaction was to downplay the issues and encourage Susan and the other faculty to deal with the matter themselves. He said, "You're all mature individuals; I shouldn't have to step in for a minor matter like this." Eventually, the chairman agreed to have a discussion with Jack but it focused on the details of how users scheduled time on the device. Some adjustments were made in the way time was allocated to users, but Jack's behavior remained unchanged and the problem persisted.

Susan finally went to Ellen, the head of human resources, to complain about Jack's behavior. When Ellen asked if Susan or anyone else had ever had a discussion with Jack about his behavior, the answer was "no." Ellen agreed to speak with Jack to help resolve the problem.

The following day, Ellen met with Jack and asked if he had any idea of the resentment toward him regarding the way he acted relative to the instrument. Jack was genuinely baffled; no one had ever said anything to him about this. He professed to have no hidden agenda and seemed hurt that others would find his behavior objectionable without ever bringing it up to him.

KEY ELEMENTS OF PEER CONFLICT AVOIDANCE

Variants of this scenario are widespread, if not universal, in science and technical settings. One element is the scientist who behaves in a self-interested or objectionable manner but whose behavior is not so egregious as to be overtly inappropriate. In Jack's case, he never actually prevented anyone from using the device, and all of his "interventions" were couched in the context of keeping the machine in good working order so that work would not be jeopardized.

The second element is the reluctance of peers to confront the objectionable behavior. When asked why they never mentioned the effects of his behavior to Jack, they responded with, "It's not my job. The chairman should do it," "I felt that Jack could be volatile and I didn't want to become the recipient of his hostility," or "I could never figure out exactly what to say to him. It seemed like he could always deny any negative intent because on the surface everything he was doing was in the context of keeping the machine in good working order." Some used the excuse that if Jack was so oblivious that he did not see for himself the consequences of his behavior and its effect on others, he would never be receptive to hearing about it from them.

All of these reasons for not confronting Jack's behavior are actually excuses for not knowing how to confront or even discuss the matter. The thought of dealing with the issue in a straightforward manner made everyone anxious. The fear that the discussion would end in hostility was a prime factor for many.

The third key element of the scenario is the failure of the department chair to take responsibility when the faculty could not. The chair was in the same situation as the others. He felt anxious about confronting Jack and found excuses for why he should not do it. When he finally did speak with Jack, it was about scheduling, a neutral, only loosely connected component of Jack's behavior. The abdication of responsibility by a scientist in a position of authority is a theme that we revisit in more detail in Chapter 12 on leadership as well as in the case study found at the beginning of Chapter 13.

In this example, Jack had never been told about the effects of his behavior on others. When Ellen finally confronted him, he was baffled and then resentful that others had been talking about him behind his back. This is the fourth key element of such situations: The protagonist is often unaware of the effect of his behavior on others. Perhaps Jack was behaving in a way that was motivated by anxiety about his own work, in which the two-photon device had a major role. Because Jack did not have a lot of interpersonal savvy, and no one had ever taken him to task for the consequences of his behavior, he remained oblivious to the consequences. The result was his feeling that people were conspiring against him.

With some coaching from Ellen, Susan and others learned how to inform Jack of the effects of his behavior by focusing on issues of accessibility and productivity and without impugning Jack's motives. This chapter shows you how to have these conversations. After a year, Jack had learned to watch both himself and others for signs that he was being overbearing or inconsiderate.

Another example of conflict aversion is when someone declares a successful resolution to a problem before it has actually been resolved. Declaring the problem solved alleviates everyone's anxiety, but guarantees that it will reemerge. (Recall our discussion of group behaviors in Chapter 7.) The following case study, taken from an attempted reorganization at a medical device company, illustrates this.

■ *Case Study: Declaring Success*

Ashok was the vice president for strategy at Mattehorn Medical Devices. The chief executive officer (CEO) asked him to oversee a complex integration of their electrical engineering groups at two different sites. Each group was headed by a powerful vice president of its own, each jockeying for advantage relative to the other. Although Ashok had no vested interest in the details of the reorganization, he did have responsibility for ensuring that the process was completed on schedule.

As soon as the reorganization committee started to meet, Ashok saw that the two vice presidents' behind-the-scenes jockeying for position would sabotage the process. He was at a loss as to how to deal with this and felt very uncomfortable with the vice presidents' underlying animosities. With every issue that pitted their interests against each other, Ashok listened to the back-and-forth positioning of the two, but he made no effort to bring their underlying interests to the surface. Ashok felt that to do so would have required him to expose their seemingly irreconcilable expectations for control of the merged division. This would lead to an angry confrontation that he wanted to avoid.

Ashok tried to find small elements of agreement between the two executives. He focused the committee's discussion on only these elements, eventually declaring, "We're all in agreement." Despite the fact that everyone knew that they were not in agreement, they went along with the charade because it released the tension in the room and obviated the need to deal with the difficult underlying issues.

The result was that the two vice presidents lobbied Ashok and others outside of the meetings. Despite this, the reorganization committee came up with a plan that on the surface seemed to satisfy the charter of the reorganization. But in fact, the two executives were secretly dissatisfied and as soon as the formal process ended, each met with the CEO and argued for modifications to the plan that would favor them. Eventually the CEO became so disillusioned with the outcome that he canceled the reorganization altogether, much to the detriment of the company.

Ashok saw his role principally as referee and peacemaker, largely because he was uncomfortable with the underlying competitive issues between the two vice presidents. He ignored the underlying conflict to avoid dealing with it. Although he did manage to find important elements on which they agreed, a number of more significant issues about which they disagreed were never dealt with during the planning process. At the time, everyone, including Ashok, accepted the omission of these issues because everyone was uncomfortable dealing with them.

Scientists who are prone to avoid conflict must be alert for situations in which they deny its existence to maintain the appearance of agreement. If the alternative to denying conflict is to engage in a bitter and destructive battle, denial may seem attractive at first glance. But more often than not, denial leads to an outcome such as that described above. The long-term solution is to negotiate a resolution that satisfies underlying interests. In the following sections, we present ways to help you recognize when you are avoiding conflictual peer interactions, and how to resolve them.

TOOLS FOR DEALING WITH PEER CONFLICT

Recognize the warning signs

Signs requiring attention include feelings of resentment, fear, disappointment, frustration, or anger toward colleagues. Learn to "capture" or notice these feelings at the time that you experience them. Paying attention to your feelings may enable you to determine what it is that is bothering you. If it is something the other person does or says, or does repeatedly, think through some scenarios for discussing the matter with the other person. The sections below provide examples of such conversations, and the exercises at the end of the chapter help you to develop a better awareness of your experiences during uncomfortable interactions with colleagues.

Another warning sign is if you or others repeatedly complain about a colleague. Notice when you do this and reflect on whether it is in your interest to continue complaining or to find a way to discuss the problem with the colleague directly. The next section presents two related techniques for having such a discussion.

Avoid accusations by addressing the problem in terms of its effect on you

This is always your best first approach for addressing behavior that is objectionable, inconsiderate, or insulting. For example, in the case study "Talk About It,"

Instead of "Jack, you're being really obnoxious by peering over my shoulder every time I use this instrument,"

Say "It is hard to concentrate with you looking over my shoulder. Is there some problem with what I'm doing?"

Instead of "You have no consideration for anyone in the lab but yourself" (accusation),

Say "Susan, it's hard for me to work with your music on so loud" (effect of behavior on you).

_____

Instead of "You paid no attention to the suggestions from my team about the project plan,"

Say "Richard, it looks like none of my team's ideas made it into the project plan."

_____

Instead of "You don't seem to have any interest in hearing what I have to say,"

Say "Larry, please let me finish making my point before I answer your questions."

The second type of statement makes the same point as the first, but in a nonaccusatory manner. In each example, the first statements may be technically true, but do they accomplish what you want? If your objective is to start an argument, then they do. Susan can deny that she means to be inconsiderate; Richard can deny that he ignored your suggestions or is being unfair. Larry may not be able to deny that he keeps cutting you off, but making accusatory statements may cause him to become defensive and make a counter-accusation.

Another way to frame a problem in terms of its effect on you is to start with "I" rather than "you." We introduced this concept in Chapter 3 for use during a negotiation, and it applies here as well. When you express your concerns in the form of an "I" statement, the other party cannot deny or refute what you say.

"You" statement	"I" statement
You always play your music too loud.	I have a hard time concentrating with music playing.
You ignored my team's suggestions.	My team feels that its input was ignored.
You keep interrupting me when I speak.	I'd like to be able to finish what I have to say.

Address objectionable behavior patterns without devaluing or belittling the person

This technique is based on tools used in negotiation that teach us to be "hard on the problem and soft on the person" and is closely related to the "I" versus "you" approach. The difference is that, in this case, the irritating factor is not an incident but a behavior or behavior pattern. The principle of this approach is to separate the problem, incident, or behavior from the person. Carl has found in his workshops that this technique baffles some scientists at first, partly because of their tendency to seek cause and effect explanations. The person seems to be the cause of the behavior, so it makes some sense, they reason, to confront the person. But by now you should know that addressing a pattern of objectionable or unproductive behavior by implying that it reflects some character defect or mental deficiency is not likely to be productive.

Example 1

Instead of "As usual, you were really obnoxious in the team meeting today" (personal accusation),

Say "I thought your comments about my work were unfair and I'd like to discuss this with you" (reference the specific behavior and ask for a clarification).

Example 2

Instead of "You're taking up all the time on the two-photon microscope" (attribution of selfishness),

Say "It's becoming difficult for me to find time on the microscope. Can we discuss how we schedule time on it?" (Reference the specific behavior with evidence, e.g., there being only 10 hours per week available for other users and that affects you.)

Example 3

Instead of "Jack, you're being possessive of this equipment and creating an atmosphere of resentment and hostility" (accusations),

Say "Jack, if you have concerns about the equipment, let's talk about it when I'm done here."

Making statements such as these, which address the behavior and its effects on you, requires thought. Because it's another person's behavior that's making you frustrated or angry, the first reaction you have will likely be to attack the person. It will take practice to recognize your feelings before you act on them. Only then will you be able to decide what to say or do so that you're accomplishing your goal of improving a relationship with a peer, getting your work done, or being able to make a point in a meeting.

DEALING WITH DIFFICULT PEOPLE

Not everyone is as reasonable and even tempered as you are. You will at times have to deal with people in the lab who will tax your diplomacy skills. You can use the "Agree, Empathize, Inquire, or Assure" strategy that we introduced in Chapter 8 to manage your interactions with a variety of challenging personalities.

Angry or hostile people

We showed in Chapter 8 in the section on dealing with an angry boss that you first need to create an environment in which what you have to say can be heard. Only after you have shifted the tone of the conversation from one of hostility to one of collaboration will the other person be able to listen to you. The following example illustrates how to apply some of these tools.

■ *Case Study: Think Before You Speak*

Barbara is an assistant professor in biochemistry working on the role of SMADs in signal transduction. Six months ago, she had an idea about a way to use a green fluorescent protein (GFP)-SMAD construct to monitor changes in the intracellular location of a particular SMAD during cell signaling. She discussed the idea with Mohan, an assistant professor of physiology with experience using GFP constructs. Later that year, Barbara sat in on a seminar about SMADs given by Astrid, a faculty member from another institution. Astrid discussed some preliminary work in which she used a SMAD-GFP construct in almost precisely the same way that Barbara had planned. Moreover, she indicated that this work was being done in collaboration with Mohan.

Barbara felt herself getting red in the face during the seminar. She believed that Mohan had stolen her idea, and she was furious. At the end of the seminar, hardly able to contain her anger, she approached Mohan and pulled him aside.

"I told you that I was planning to do almost exactly that experiment six months ago. Now I find that you did the same experiment with Astrid. This is infuriating and amounts to theft of my work." During this recitation, Barbara became increasingly agitated and shouted loud enough for everyone leaving the seminar to hear, "How can you possibly justify what you did?"

Mohan was stunned and embarrassed. People were looking at the two of them. In desperation, he said, "Look, you're way off base here. I never talked about your work. Why don't you calm down? You have completely misunderstood this situation and now you are making a mountain out of a molehill."

To Barbara, this felt like an attempt to brush her off and she became even more furious. "You're an outright liar, Mohan, and I'm taking this to the committee on scientific misconduct," she shouted.

After hearing Barbara call him a liar, Mohan became furious and said, "Go ahead. You're paranoid and everyone knows that."

In this case, we have the advantage of knowing that Barbara did in fact talk to Mohan about her idea. But we do not know whether Mohan knowingly misappropriated it or whether something else happened. Whatever the case, Mohan's reaction simply fanned the flames of Barbara's anger and resulted in her filing a formal charge of misconduct against him.

Mohan did everything wrong when confronted by Barbara. He denied her anger, or that she had any reason to be angry, by telling her to calm down. Telling an angry person to calm down is probably the least effective way to get them to do that. In fact, it is likely to increase their anger. Then Mohan told Barbara that what she is furious about is not a big deal. It was clearly a big deal to Barbara, and hearing Mohan deny that she had something to be angry about did not help. What could Mohan have done differently? Let us rerun the tape with a new and different Mohan.

Barbara: "I told you that I was planning to do almost exactly that experiment six months ago. Now I find that you did the same experiment with Astrid. This is infuriating and amounts to theft of my work. I demand an explanation."

Mohan: "Barbara, I can see that you are really angry. I see how this looks to you, and I'd be angry too if I thought that someone did that to me. Frankly, it should have occurred to me how this would look, and I apologize for not speaking with you sooner. Can we go somewhere and talk about this? I'd really like to explain how this situation came about. The last thing I want is for our relationship to be damaged because of this."

Principles used

- **Empathize.** Mohan acknowledges Barbara's anger and shows that he understands why she is angry.

- **Agree.** Mohan has agreed that Barbara has the right to be angry based on what she *thinks* happened. In doing this, Mohan has not agreed that he has done anything wrong. It's important to note that telling Barbara that he can understand her anger is not the same as admitting that he did anything wrong. In fact, in this case, Mohan did not do anything wrong. But he needs to create a climate that enables him to explain this.

- **Apologize.** Next Mohan apologizes, not for doing anything wrong, but for failing to anticipate how Barbara would perceive the situation. Apologies work wonders with angry people, even if you are not apologizing for precisely what they are angry about. It shows Barbara that Mohan cares about her feelings and is willing to accept some responsibility.

- **Inquire.** Mohan tells Barbara that he would like to hear more about what she believes happened, further showing that he is interested in her perception.

- **Assure.** Finally, he assures her that it is important to him to maintain their relationship.

All of this will allow Mohan to explain the situation from his perspective in a calm setting.

When they met later, Mohan explained that after Barbara initially came to him for advice he thought that was the end of their interaction about the experiment. The very next month, Astrid approached him with a similar request, asking Mohan to collaborate on her project. Because Astrid was a competitor of Barbara's, Mohan felt that he could not tell Barbara about Astrid's work and vice versa. Mohan felt that he was in a bind. For lack of any other alternative, he went ahead with the collaboration with Astrid.

The issue here is not whether Mohan did the right thing. The issue is what the two participants can do to resolve the problem. In the second scenario, Mohan defused the situation by not reacting angrily to Barbara's insults. He created an atmosphere in which Barbara was willing to listen to what he had to say. In the first scenario, Mohan's behavior resulted in a lot of people spending a lot of time adjudicating something that could have been resolved between Mohan and Barbara.

There is another aspect of this case worth noting. Mohan's collaboration with Astrid was not the result of maliciousness. Recall our discussion of the fundamental attribution error. In our experience, most instances in which it seems as though a colleague has behaved in an insulting or malicious manner are in fact the result of thoughtlessness, inexperience, or naïveté. Barbara was prepared to think the worst of Mohan without knowing all the facts. She owns a lot of the responsibility for the misunderstanding.

In Barbara's case, awareness of her hot buttons might have helped. Barbara's strong reaction to Astrid's presentation could suggest a predisposition to feeling exploited. If Barbara had known this about herself from previous incidents, she might have been able to recognize what was happening to her during Astrid's presentation. She might have asked herself whether an angry confrontation was really what she wanted. What she really needed was to better understand the circumstances of Astrid and Mohan's collaboration. Better self-awareness on the part of either of the participants would have prevented an ugly and unnecessary incident.

More tips on dealing with an angry person

- **Never say "you're wrong"**—Try "In my opinion" or "I see it differently." ("I understand why you are upset about the GFP construct. I'd like the chance to tell you how I see it. I respect your view. Can we sit down and discuss this?")
- **Use the person's name**—"Jim, I know you are angry. Will you tell me what the problem is?"
- **Maintain eye contact to let them know you are listening.**
- **Let the anger run its course**—listen, nod, empathize.
- **When you do speak, don't let them interrupt**—"I'd like to finish what I'm saying."
- **Ignore attacks**—Don't take anything personally.
- **Apologize for real or imagined offenses**—It costs you nothing to say, "If I did something wrong, I will apologize. Tell me what you're concerned about."
- **If you really did do something wrong, APOLOGIZE RIGHT AWAY. The longer you wait, the harder it will be.**

Behavioral Rule #3

Apologize for what you did, not for how they feel.

Telling someone "I apologize if I made you feel uncomfortable," or "I apologize that you felt insulted" are not apologies. If you wish to apologize, apologize for what you did, not for how it made them feel. "I'm sorry I did that. It was wrong" is the way to go.

Critical or judgmental people

Some people have the uncanny ability to find something worth criticizing with everything you say or do. Rather than arguing with them, try the following.

Them: "Don't you think your research program is a bit diffuse? It seems like you are spreading yourself way too thin on all these projects. Look at my lab, for example... ."

(Agree) "Well, it's true that I have a lot going on in my lab."

(Empathize) "I appreciate your concern."

(Inquire or assure) "If our roles were switched, which of my projects would you deemphasize?"

(Note that this approach could also be applied in the case study "Talk about It" on pp. 222–223.)

(Agree) "Jack, I know this is a sensitive piece of equipment and that you're in charge of it."

(Empathize) "That's a lot of responsibility and I understand your concern that it be used properly."

(Inquire or assure) "I share your concern because this instrument is important for me too. How do you think we can help users feel more comfortable signing up for and using the microscope?"

Pushy or demanding people

Them: "I need your completed section of the program project grant two months before the filing deadline."

(Agree) "I'm with you, I always like to have things in hand way in advance so I can get the big picture."

(Empathize) "I know how much work it is to assemble a big grant like this and we all appreciate the effort you are making."

(Inquire or assure) "I'm stretched thin right now, but I'm working on this every day. Is there some way I could get you my part in sections? I'm done with the Rationale and Specific Aims. Would it be helpful to send those to you now?"

Dealing with pushy or demanding people

- **Never make hasty decisions—you may regret them later.**
- **Suggest interim solutions** ("I can get you the outline tomorrow with a summary of how much more work I need to do. When can we get together to discuss a timeline to which I could commit?").
- **Tell them that you are feeling pressured and you'd like time to think it over.**

Passive–aggressive people

Whereas angry or hostile people act out their feelings in your face, passive–aggressive people get at you by doing nothing. How is this possible? Here are some examples of what passive–aggressive people do:

- withholding support by not showing up for a meeting
- sitting silently in a meeting when you know they have something to add

- "forgetting" to communicate important information to you
- refusing to discuss or admit that there is a problem

Passive–aggressive people can be the most frustrating group of people you deal with. Because they infuriate you by not doing things, it's hard to confront them with overtly egregious behavior. Here is how the Agree, Empathize, Inquire technique can be used with such people.

In this example, you may have reason to believe that a colleague isn't coming to meetings because the group didn't agree to a proposal he made four meetings ago.

(Agree) "I know we all have a lot of meetings to go to, but I've noticed that you haven't been at the last three joint project meetings."

(Empathize) "I know you have a lot going on in your lab and you've been traveling quite a bit."

(Inquire or assure) "What can we do to make these meetings work for you? Should we change the meeting time? Are they being held too frequently? What about if we hold them in your office so you don't have to walk across campus? Would you be willing to participate by phone if you are out of town? The team really needs you at these meetings."

Dealing with passive–aggressive people

The meeting organizer keeps "forgetting" to include you in meeting invitations.

- **Present factual data** ("I've not received notices for the last three project meetings.")
- **Be nonjudgmental** (Don't say, "You're excluding me from the project meetings.")
- **Ask open-ended questions—avoid questions that can be answered by a yes or a no** ("When can I expect to be put on the list for forthcoming meetings?")
- **Don't force the issue but ask for specific commitments** ("I'd like to be copied on all team correspondence. Which ones will you be sending me?")
- **Don't get caught up in power struggles** (Don't say, "It's not your decision who gets invited to the project meetings," even if that is true.)
- **Involve them in helping solve your problem** ("What would you suggest is the best way for us to make sure all the required people get invited?")
- **Be clear about the consequences** ("It will be embarrassing for both of us if I have to explain to our vice president why I haven't been at the last two meetings.")
- **Don't give up—maintain a positive attitude** ("I realize you have a lot to keep track of when you arrange these meetings.")

Procrastinators

People who have a habit of putting things off can sometimes look like they are passive–aggressive, but they're not. People put things off for a variety of reasons that have nothing to do with you. Procrastinators may be anxious about making a mistake or appearing inadequate. Often, they are perfectionists who fear handing in a less than 100% perfect product, when 90% is all that is needed.

For a colleague who keeps delaying something they need to give you:

(Agree) "I know that getting this program project submission in on time is going to be a real challenge."

(Empathize) "I also know how busy you have been with all of your other responsibilities. I'm not sure how you manage them all."

(Inquire or assure) "I really need the first three sections of your grant component by the end of the week. Can you commit to that? It doesn't need to be perfect at this point, just close to final. Everyone agrees that your section is really important for the success of this proposal."

For someone who keeps putting off an important discussion:

(Agree) "I know this discussion is something that's been put off several times because you're so busy."

(Empathize) "I can only imagine all the things you're dealing with right now."

(Inquire or assure) "If there's a problem between us that I'm not aware of I'd like to know about it. Our ability to work well together is very important to me. How about if we just agree to spend 10 minutes either now or at nine tomorrow morning?"

Dealing with procrastinators

- **If necessary, present factual data without being judgmental** ("I have everyone's sections for the grant submission but yours.")
- **Be nonjudgmental** (Don't say, "You're always late with your grant submissions.")
- **Ask open-ended questions—avoid questions that can be answered by a yes or a no** ("When can I expect your part of the grant application?")
- **Don't force the issue but ask for specific commitments** ("I need your entire section done by next Wednesday at noon. How much of it can you send me today?" not "Soon," or "Next Wednesday," but "Next Wednesday at noon.")
- **Don't get caught up in power struggles** (Don't say, "In my role as principal investigator, I need to insist that you finish your section by the end of the week.")
- **Involve them in helping solve your problem** ("What would you suggest is the best way for us to make sure the grant gets in on time?")
- **Be clear about the consequences** ("If your section isn't completed on time, the team will want to know why.")
- **Don't give up—praise and reassure if appropriate** ("I know your section of the grant is going to be one of the strongest.")

Complainers and help-rejecting complainers

We all complain on occasion, but some people seem do it for a living. As scientists, we might be prone to getting tangled in a complainer's web because of a tendency to seek solutions to problems. Most frustrating are "help-rejecting complainers" who are much more interested in complaining than in finding solutions. Nothing you suggest will make any difference or will be worth trying because the whole point is to complain.

- **Listen.** Some complainers don't want or need your advice; they just need to vent. Ask yourself, "Is this person really asking for my help or are they just venting?" If the latter is the case, you're off the hook; all you need to do is sit there.
- **Empathize** ("That sounds like a real problem.")
- **Don't try to solve their problem for them.** If you want to be helpful, help them focus on solving their own problem ("What are you planning to do about it?" or "Do you have any ideas for solutions?").

Needy people

Closely related to complainers are people who repeatedly ask for help from us regarding a personal or non-work-related matter. An example might be a colleague or even an employee who is going through a difficult life situation such as divorce, separation, or other extended family crisis who wants to tell you about it daily. Your role in such situations will depend in part on the nature of your relationship with the person. You might be more willing to discuss personal matters with a peer than with an employee, for example. You can still show empathy with an employee's or even a peer's circumstance without getting involved with helping them solve their personal problems. This is called "boundary setting," knowing just how far into someone's personal life you feel comfortable going, or that is appropriate in your role.

Dealing with needy people

- Sometimes all you need to do is listen.
- Be clear in your own mind about the differences between work relationships and personal relationships, then
- Set appropriate limits on your involvement with the personal problems of professional colleagues, but
- If you decide that it is appropriate to offer help, focus on solutions rather than on complaints.
- If their personal problems are impacting their work, suggest that they speak with an employee assistance program (EAP) specialist if your institution has them available.

Argumentative people

Remember, it takes two people to have an argument. Just because the other person is argumentative doesn't meant that you have to follow suit.

Them: "That binding data can't possibly be right. You didn't do it in the presence of magnesium and everyone knows that magnesium is required. You did this all wrong."

(Agree) "I do agree that this enzyme is very sensitive to the reaction conditions."

(Empathize) "I appreciate your picking up on that, Fred. I know you are familiar with these reactions."

(Inquire or assure) "Can we go over the reaction mix together after the meeting? I had read that either magnesium or calcium would work, but I could be mistaken."

When dealing with difficult people, remain focused on your task

Your task is to get your work done, not to act insulted or annoyed. It's hard to get work done with a hostile, argumentative person so

- Agree—to defuse anger
- Empathize—to start a dialogue
- Inquire or assure—to show interest in what they have to say

Are you one of these people?

A few years ago, a friend saw the book *Dealing with People You Can't Stand* by R. Brinkman and R. Kirschner (1994) on Carl's desk. He picked the book up, leafed through it, and asked whether Carl had read the sequel *Dealing with People Who Can't Stand **You***. Of course, there is no such sequel, but his point was well taken.

It's lots of fun to talk about how obnoxious other people can be, but has it ever occurred to you that they might be talking about you in the same way? We all exhibit some of the characteristics discussed above, it is just harder for us to see these behaviors in ourselves than it is to see them in others. Using the self-awareness that you will develop as you go through the exercises in this book, take a step back and observe yourself during interactions with your peers. Every interaction, whether pleasant or unpleasant, productive or unproductive, requires at least two people. Take ownership of your own role in difficult interactions. The next section introduces some additional tools to help you ensure that you do not annoy others in the same way that they annoy you.

HOW TO DISCUSS SCIENCE WITHOUT ARGUING

It is not uncommon to observe participants expressing hostile or insulting remarks to one another in a team meeting, especially when discussing their interpretations of data. Nothing frustrates scientists more than having to listen to presentation of data that they believe are flawed or just plain wrong. Carl has seen people become furious listening to someone recount the results of a poorly designed experiment; they squirm in their seats, make pained facial expressions, and interrupt the presentation with aggressive, snide, or hostile questions. Some scientists consider it their obligation to express their frustration overtly and in damaging ways. This was exactly what Fred thought in the case study discussed in Chapter 2.

There often seems to be an unspoken agreement that the importance of good science is so great that objections, criticisms, and destructive comments can and should be delivered swiftly and mercilessly. The corollary is that people are simply the instruments through which questionable data are delivered, and their feelings and reactions need not be taken into consideration when criticizing their results. Unfortunately, the outcome is that important comments, critiques, and suggestions get lost in the noise of hostility. Scientists who display such attitudes fail to recognize that valid scientific criticism has to be heard and accepted by the recipient to be useful. Consider the following example.

Joe says, "That binding data can't possibly be right. You didn't do it in the presence of magnesium and everyone knows that magnesium is required. You did this all wrong." Fred immediately feels attacked and gets defensive. "No, you're wrong. It doesn't need magnesium; almost any divalent cation will do. I've done it this way before and presented the results in previous meetings. You must not have been listening."

This is the beginning of an argument. Both sides now feel insulted and defensive. The ensuing discussion will likely be more about scoring debating points than determining whether the reaction really should have been done in the presence of magnesium.

The reasons for interactions such as this are multiple. Team meetings are often regarded as competitions rather than peer-to-peer discussions. This is the responsibility of the team leader to correct, and it is addressed in detail in Chapter 7. The team leader must create an atmosphere in which science can be discussed without participants feeling the need to score points at the expense of others. They must also model how to question and critique data in a way that is scientifically rigorous and at the same time respectful of the presenter. The following are some examples of how this can be done.

Example 1

Instead of "That binding data can't possibly be right. You didn't do it in the presence of magnesium and everyone knows that magnesium is required. You did this all wrong."
Say "Fred, I thought that this reaction required magnesium. That could explain why your results look different from last time."

Example 2

Instead of "I can't believe that you did this experiment without a control group of untreated rats. That makes the whole experiment useless."
Say "Was there some reason that a control group with untreated rats was not included this time?" or, stronger, "I understand the results, but I'm having a hard time feeling confident about them because of the absence of a control group with untreated rats. What was your rationale for not including one?"

Example 3

Instead of "I don't really understand anything that you have just said. It makes no sense to me."
Say "You're going to have to help me understand this better. I think that I missed the point of what you just said."

Example 4

Instead of rolling your eyes and ignoring what you think is a lousy idea,
Say "Okay, that's one way of looking at it. It's not one that I favor, but I hear what you're saying."

Example 5

Alice is presenting some disappointing results of an assay to measure the power output of a new type of light-emitting diode (LED) that her company is developing. Raul listens carefully and then interrupts to ask what seems like a straightforward question, "Alice, are you sure that you had the photometer calibrated when you measured the output?"

Alice, who has been working with LEDs for about five years, feels insulted that anyone would ask her such a question. After all, she knows perfectly well that she calibrates the photometer with every use. Alice therefore responds, "Yes, I'm sure. I've only been doing these measurements for five years, Raul." End of discussion.

What if instead Raul had said, "Alice, the last time that I made similar measurements, I had a really hard time calibrating the photometer. Could this be part of the reason for your results?"

To adopt this type of moderated approach, first notice whether you have feelings of superiority, frustration, or aggression before you speak. If you have such feelings but are not conscious of them, you may react to them and say something that comes across more as hostility to the experimenter than as a critique of the experiment. The take-home message is that you will usually get a more productive and useful response if you phrase even innocuous questions as requests for help or clarification, rather than as challenges. Most people respond positively to requests for help.

PAY ATTENTION TO YOUR OWN BODY LANGUAGE

As we saw in Chapter 3, you convey your feelings and thoughts not just by what you say and how you say it, but also by your body language. Carl spent many years sitting in meetings with a look of profound skepticism on his face that often turned into a scowl. When someone pointed out this habit to him, his reaction was, "So? That's how I feel. I'm skeptical of everything." This was indeed, how he felt, and it may even be an important attribute; after all, skepticism can contribute to scientific rigor. But his facial expressions often made others defensive and frequently inhibited open communication.

It may seem obvious to you that colleagues will be much more receptive to questions or comments if you deliver them in a neutral tone of voice and with a facial expression that conveys interest and curiosity, rather than hostility or confusion. But thinking this way is a far cry from behaving this way. Practice becoming aware of what you are feeling when you are about to ask a question. If you feel angry, frustrated, disdainful, or disgusted, ask yourself if your behavior, tone of voice, or facial expression is conveying these feelings. Notice what you feel before or as you speak. It will enable you to decide how you want to come across.

None of the above implies that you must never express feelings of anger or frustration. The point is to take control of when and how you do so. Demonstrating mild anger or frustration can on occasion make your point and get the listener's attention in a way that reasoned statements cannot. But the expression of these emotions should be closely monitored and used knowingly and sparingly.

Also recall (see pp. 53–54) that body language that is inconsistent with the verbal message will negate your message and confuse the recipient. If you are verbally expressing support but avoiding eye contact, looking out the window, or leaning back from the conference table, you are sending a mixed message. Remember to especially monitor your body language when you need to say or do something that makes you uncomfortable, or in situations in which you may not be in full agreement. Such situations are common in the workplace, where we sometimes must say or do something that is appropriate and professional but that makes us uncomfortable personally.

For example, if you say, "Sara, would you be willing to collaborate with us on this project? We really need someone with a background in genetics," but if you are gazing out the window, making only transient eye contact, and standing four feet from Sara, she will likely get the message that you are requesting this as a formality rather than out of genuine interest. Let's assume that scientifically, Sara really is the best choice as collaborator for this project. Let's also assume that you are threatened by her or believe that she dislikes you. These feelings make you uncomfortable when you talk to her and result in your sending a mixed message to her—your words say you want to collaborate, but your body (unconsciously betraying your deeper feelings) is saying the opposite. Your professional needs are at odds with your feelings.

If you pay attention to your feelings of discomfort as you approach Sara, you can anticipate that they may affect your body language or behavior. Knowing that, you can pay special attention to your facial expressions and body language to ensure that you are conveying a unified message.

Looking at Sara in the eye, smiling, facing her straight on, leaning forward a bit, perhaps making a friendly open gesture with your hands, you say, "I really want you as my collaborator in this project!"

The first exercise at the end of the chapter helps you to observe and make conscious choices about your own body language.

PAY ATTENTION TO THE OTHER PERSON'S BODY LANGUAGE

While you watch your own body language, watch the other person's as well. The way in which they move, behave, and look at you may be more revealing than what they say. Look for the behaviors listed on page 54 for clues to what is really going on, and follow those observations with questions to ensure that you are interpreting their behavior correctly.

"Have I said something that you disagree with?"
"Am I misunderstanding what you meant?"

In a collegial discussion with a peer, such observations and inquiries can enable you to gauge the other person's reactions and facilitate a productive discussion. If the other person is dissembling, or saying one thing while behaving in a seemingly contradictory manner, you may have reason to question their sincerity or commitment. You need not confront them with your uncertainty, but it would likely motivate you to seek stronger assurances about any agreement. Exercise 1b on page 240 helps you to practice your skills at reading body language.

CONSEQUENCES OF FEELING INFERIOR/SUPERIOR TO PEERS

Feelings of inferiority and superiority are common in scientific settings, where performance and knowledge are constantly being tested and measured. If you are unsure of yourself to begin with, working in a competitive environment can make matters worse. Feelings of inferiority may cause you to behave in a self-denigrating or self-effacing manner and downplay your skills or accomplishments. The more you behave this way, the more you and others will believe in your limitations. The resulting consequences for your career can be unfortunate: Important contributions may be ignored, you may be passed over for promotion, and your ideas and comments may be viewed as unimportant. Your feelings end up becoming self-fulfilling prophecies. Some scientists explain their self-deprecating manner as a desire to be humble. If you think that you are being humble but you feel marginalized, ignored, and bypassed, it would be wise to reexamine your motivations and feelings.

Becoming aware of feelings of inferiority is the first step toward overcoming their consequences. The next time that you catch yourself minimizing what you know or did, remember the situation and write down as many details of the circumstance as you can. Ask yourself whether your contributions were really as limited as you had made them out to be. What would the consequence have been had you spoken in a more positive manner about yourself or what you did? Investigating your behavior in this way may enable you to uncover erroneous assumptions. For example, if you felt that speaking honestly about the value of your work would feel as though you were boasting, ask yourself why being positive about your contributions has to feel that way.

The next time you are about to do or say something that is self-deprecating, stop to think of something positive to say about yourself instead. For example, if someone says, "Alice, your role in this project was incredibly important. I don't think that it could have succeeded without you," instead of saying, "Well, others contributed as much as I did. I couldn't have done this without Mario's help—he's the one who really deserves your thanks" or "I was just doing my job; it's not a big deal," try saying, "I really put a lot into that project and I'm glad that it turned out to your satisfaction," "I was happy with the way it turned out as well," or simply, "Thanks."

The opposite of being humble is being an aggressive self-promoter. We all know scientists who constantly seem to promote themselves, their accomplishments, and their knowledge. In principle, nothing is wrong with this. Sometimes it is what you need to do to survive and advance in a competitive world. Something is wrong, however, when this behavior is at the expense of others, the group, or team. Often, such behavior stems from feelings of inferiority or insecurity.

Just as self-deprecation can have career-limiting consequences, so too can self-promotion, when it is done in ways that minimize or belittle the contributions of others. If you are a self-promoter, it is important that you observe how others respond to your behavior. (Watch their body language—did you suddenly lose their attention when you started tooting your horn?) If you sense that they are reacting with skepticism or annoyance, it may be time to tone it down. Take time to reexamine how you promote yourself, paying special attention to what you may say or do that reflects negatively on others. If you need to self-promote, do it in a way that does not devalue or diminish others. Keep the focus on your accomplishments, skills, and knowledge, not on the deficiencies of others.

REFERENCE

Brinkman R and Kirschner R. 1994. *Dealing with people you can't stand*. McGraw-Hill, New York.

EXERCISES AND EXPERIMENTS

1 Your body language

Notice your body language

a. During the course of your interactions with others, take a moment to observe how you stand, sit, hold your hands, etc. Use the list on page 54 for suggestions about what to look for. See if you can relate what you do with your body to how you feel during the interaction. Ask yourself if noticing your body language helps you to become aware of your feelings at that moment. Awareness of your own body language can alert you to how you might react in a future situation and gives you the opportunity to consciously choose your behavior.

b. Write down several instances during which your body language was consistent and inconsistent with what you were saying. Review each situation and ask whether the interactions during which your body language was consistent with your message were more productive than the others.

c. Make a list of three people with whom you have significant interactions at work in a typical week. For each of these people, the next time you interact, record the following responses:

	Agree	Disagree
1. I am uncomfortable making eye contact	☐	☐
2. I feel tense	☐	☐
3. I fidget	☐	☐
4. I cross my arms or legs	☐	☐
5. I routinely make eye contact	☐	☐
6. I stand close to the person	☐	☐
7. I behave in a relaxed manner	☐	☐
8. I am comfortable touching or making other appropriate physical contact	☐	☐

The people for whom you agreed with statements 1–4 are probably those with whom you have some discomfort or uneasiness. Ask yourself whether this uneasiness is affecting your work or professional relationship in a negative manner.

d. Pick one person from the "agree" list of statements 1–4 as an experimental subject. During your next several interactions with this person, make it a point to use one or more of the positive body language behaviors listed on page 54. After the interaction, ask yourself the following questions:

- Did my changed body language affect the dynamic of the interaction?
- Did it change the way in which I interacted with the other person?

- Did it change how the other person responded or acted toward me?
- Did these changes lead to a more productive outcome?

2 Others' body language

List three people with whom you interact during a typical day. For each of them, watch and record any of the body language messages listed on page 54 during your interactions with them for one day. Does taking note of these messages give you information about your interactions? Write down several instances of body language messages that you observe that are consistent and inconsistent with what the person is saying. Often, a person's body language betrays their discomfort with what they are saying. In one or more cases where body language seemed inconsistent with what was being said, ask questions that encourage the other person to elaborate, clarify, or restate what they said. Does this technique improve the quality of your communication? Does taking note of body language help you to have a more productive interaction?

3 Problems with peers: Focus on the problem, not the person

In Chapter 3, we introduced the concept of focusing criticism or frustrations on problems, not people. This concept is, of course, applicable in almost any situation, and we use it liberally throughout the book. The following exercises can help you apply this principle to interactions with colleagues. Write down your answers to each question. Sample answers for situations 1 and 2 follow.

Situation 1. You and your fellow faculty members have agreed to set aside a room in your research building to be shared by all. However, when you return from a three-week vacation, you discover that one of your colleagues has installed a mass spectrometer in the room as well as two of his technicians, effectively taking up the entire space.

- What would happen if you do nothing?
- What could you say that could be construed as attacking your colleague?
- What could you say to attack the problem?

Situation 2. After having a confidential discussion with a colleague, you discover that someone from another institution knows the details of your discussion.

- What would happen if you do nothing?
- What could you say that could be construed as attacking your colleague?
- What could you say to attack the problem?

Situation 3. Describe a current conflict that you are having with someone in your workplace. Answer the following questions about how you might deal with this problem.

- **Avoidance.** What would you do if you avoided the problem?
- **Attacking the person.** How would you address this problem by attacking the person?
- **Attacking the problem.** How could you focus on the problem?

List as many of your underlying interests as possible for resolving this problem. List the consequences of each of the three approaches. Which approach enables you to satisfy the greatest number of your underlying interests?

Situation 1 sample answers

- **Avoidance.** You do not say or do anything. The result is that everyone feels resentful and that group agreements can be ignored at will.
- **Attacking the person.** "What do you think that you're doing in the equipment room? You've hogged all the space and left none for the rest of us. This is inconsiderate and outrageous."
- **Attacking the problem.** "It looks like your group has taken up quite a bit of space in the shared equipment room. The way it stands now there's not much room for the rest of us. Is this just temporary? Maybe we need to revisit our discussion of how that space is used at the next department meeting."

Situation 2 sample answers

- **Avoidance.** You do not say anything and you never discuss your work with him again. You miss the benefit of his expert advice and this results in a delay of your work.
- **Attacking the person.** "I can't believe that you talked about my project with the people from MIT. Have you ever heard of confidentiality? I thought that I could trust you, but that's clearly not the case."
- **Attacking the problem.** "Did you talk to the MIT people about our conversation? [You first need to establish the facts.] I just assumed that our conversations were confidential. Can we agree on that from now on?"

4 Hidden peer conflicts

Many of the difficulties that we have with peers stem from feelings that have nothing to do with the person herself, may be partly our own fault, or are based on lack of familiarity or contact with the person. Unless we are aware of these feelings and how they influence what we say or do, we are at their mercy. If you are conflict averse, you probably avoid thinking about the problems or uncomfortable feelings associated with these people. Remember that you do not need to deal with every problem that you identify. Focus on those that you think will have the greatest impact on your work and your organization. The following exercise will help you to become more aware of hidden or ignored conflicts that may be interfering with your work.

Make a list of as many colleagues as you can with whom you have a problem, conflict, or just uncomfortable interactions. For each person, answer the following questions:

a. What problem do I have with this person?

b. Is the problem interfering with my work?

c. Is the problem interfering with my responsibilities to my organization?

d. Even if it is not interfering, could my work be improved if the relationship were improved?

e. What could this person do to alleviate the problem?

f. What could I do to alleviate the problem?

g. How could I alleviate the problem without attacking the person?

CHAPTER 10

The Slings and Arrows of Academe: Survive to Get What You Need

This chapter shows how trainees can cope with the sometimes confusing and stressful interactions they can have with their mentors. We also show how mentors can provide training experiences to help trainees, as well as themselves, be more productive. Chapter 11 applies these same concepts to managing in the private sector, which has its own unique set of problems.

We begin this section of the book with an exploration of science management in the academic setting because many poor management practices in the science workplace originate there. The standards, mores, behaviors, and expectations learned in academe remain with scientists throughout their careers. Scientists in training may become imprinted with their mentors' management styles, perpetuating poor or absent management from one generation to the next and from academia to the private sector. Although many academic mentors are skillful leaders, just as many, if not more, are not. The following examines why academic institutions pay no heed to managerial and interpersonal skills in their faculty and what can be done about it.

ACADEMIC RESEARCH: A FOUNT OF CREATIVITY OR A DEN OF DYSFUNCTION?

Your time in academia can be the best of times or the worst of times, and very often is both. On the one hand, you have the freedom to engage in exciting research with intelligent, motivated people. On the other hand, you may end up feeling like an indentured servant with backstabbing mates, working under a mercurial master for a pittance. In the worst cases, scientists in academia are demeaned, humiliated, exploited, and bullied more than at any other time in their careers. How they deal with these experiences determines whether they survive to complete their training.

The academic laboratory is arguably today's most successful paradigm for scientific discovery, technology development, and engineering advances. Such laboratories are run by scientists who are sometimes referred to as "principal investigators" (PIs), whose research is supported almost exclusively by competitive research grants from the government or private foundations. For the majority of PIs, the quest for funds to support their research is never-ending, time-consuming, and a source of considerable anxiety.

Grants mean money, and money translates into the ability to hire scientists and technicians, buy equipment and research supplies, and attend scientific meetings. Difficulty in obtaining a grant or in renewing an existing grant may jeopardize your job. Thus, academic scientists are highly motivated to generate funds to support their research. But this motivation is based not only on the negative consequence of losing one's job, it is also based on the positive reinforcements of obtaining research funds.

Academia rewards individual achievement

The rewards from academic research include professional prestige, recognition, and funding. These materialize largely from outside the scientist's own institution in the form of grants, publications, and invitations to speak at meetings. Scientists are also rewarded by their own institutions through promotions and other perquisites (space, departmental funds, etc.), which are allocated largely on the basis of the scientist's external recognition. Thus, there is a strong and direct connection between the output of the PI's laboratory and the rewards that accrue to the PI. PIs are acutely aware of this connection and therefore behave like entrepreneurs, working for themselves, albeit still a part of a larger organization.

PIs are not the only ones to benefit from their hard work and entrepreneurial drive. It is also in the interest of the institution, dean, and department chair that their PIs are recognized and rewarded. Productive scientists mean more funding for the institution, prestige that facilitates fund-raising, and political leverage for deans and department chairs. For all of these reasons everyone roots for the success of the faculty scientist.

The motivation of members of the PI's laboratory is also easy to define. The trainees clearly expect that their performance will be noticed, recognized, and documented (eventually in the form of professional publications), and that this will have a direct bearing on their future opportunities.

The dark side of the academic reward system

Reward for individual achievement is at once the strength and the bane of academic institutions. It is a strength because achievement is tightly coupled to a powerful and valued system of rewards that confer professional satisfaction, recognition, prestige, power, and advancement. Yet this efficient reward system presents significant challenges to the academic system. Because every research laboratory succeeds or fails based on its own accomplishments, the interests of academics are largely self-interests. Motivation to contribute at a departmental or institutional level is of a secondary nature or even nonexistent. Chairing committees, interviewing prospective graduate students, and even teaching are often viewed more as responsibilities to be fulfilled than as opportunities to help advance the individual, the department, or the institution. Moreover, because the coins of the academic realm are funding and publications, department chairs and deans can find themselves at a loss when trying to reward communal service. Thus, although the academic research enterprise is characterized by high creativity, motivation, and entrepreneurship, it is challenged by a paucity of group cohesion and a relative lack of shared objectives and common focus.

On your own

Because academic institutions benefit from their achievements, faculty in general and scientists in particular are given wide berth for conducting their affairs and running their laboratories. Only in the most egregious cases of sexual harassment, research misconduct (e.g., falsification of data), or financial wrongdoing will a university or departmental official step in or meddle with the way in which a PI runs their laboratory. An almost universal lack of oversight of, or even interest in, the management styles and interpersonal behaviors (within bounds) of PIs exists in academia. The result, and we speak from personal observation and experience, is that management of some academic labs (thankfully, the minority) is characterized by bullying, insensitivity, threats, incompetence, and neglect. We have seen these characteristics in labs run by scientists who were successful by any of the conventional criteria, and in labs run by scientists whose careers were foundering.

This void in oversight can also be attributed in part to the fact that those in a position to encourage improved management and interpersonal practices—department chairs and deans —are often themselves scientists who lack the very skills that need to be promoted.

If you are a trainee associated with a mentor who has less than optimal managerial and mentoring skills, or if you are a mentor and you find yourself challenged by your responsibilities to trainees, the following sections have something for each of you.

AN ILL-FATED ACADEMIC RESEARCH PROJECT

The following case study illustrates issues that can arise between mentors and trainees in an academic setting. The case also illustrates how both scientific and social factors play major roles in the outcome of scientific projects. Following the case, we review what happened and what could have been done differently to improve the outcome. The case is a composite taken from Carl's own first-hand experience and from the experiences of others from whom we have heard first-person accounts.

■ *Case Study: What's Wrong Here?*

Liz is an associate professor of biology at Wannabe University working to discover the mechanism by which certain types of white blood cells move through the body to seek out and engulf pathogens. Fred, one of her senior postdocs, has generated data demonstrating the importance of a previously unidentified protein in the movement of white blood cells. Fred presents this data at the weekly lab meeting during which Liz and her six postdocs, two technicians, and two graduate students review the week's work.

One of the technicians and Sue, the postdoc for whom she works, meets Fred's data and his presentation with disinterest because they have just started an experiment of their own. They are exchanging written notes about when the next sample needs to be taken off ice and placed in the 37° incubator. Fred notes this with mild annoyance because Sue knows a lot about his project and he had hoped for her feedback.

Two of the other postdocs with the most experience, Alan and Richard, ask probing questions about Fred's data in a way that makes him feel defensive. He feels as though he is being given a hard time and that if someone else were presenting the data, they would be readily accepted. Moreover, Alan and Richard are close friends and frequently challenge others over seemingly minor scientific issues (this seems like one of those times to Fred). Because he is distracted by these feelings, Fred fails to take careful note of all of the questions or objections raised about his work. After the meeting, he decides to avoid too much discussion about the most controversial part of the project, again because he is fed up with the nitpicking of his data. He decides to quietly proceed with that portion of the work on his own.

Liz is intrigued by the data and immediately decides that she would like more effort directed to the new protein, to discover its function more rapidly than Fred could on his own. She also begins to consider how Fred's work could be used in a grant application that she is completing. Because she is so preoccupied with her thoughts, she stops asking questions. Fred interprets this to mean that she agrees with Alan and Richard's criticisms and has also lost interest in what he is saying.

After the lab meeting and without any preliminary discussion, Liz calls a meeting with Fred and a more junior postdoc, Zheng. Liz says that she would like Zheng to get involved in the project to accelerate the work. Fred is taken by surprise and reacts with anger saying that Zheng will slow the project down. After the criticism that he felt during the group meeting, he feels that this is an attempt to double-check his work or perhaps pry it away from him. He is so angry that he mentally resolves to start looking for another position as soon as possible, something he has been considering for a while.

The next day, Zheng, having been hurt and angered by Fred's outburst in yesterday's meeting, reminds Liz that he has his own project and does not wish to dilute his effort by working on someone else's project. He also intimates that he doesn't think he would get along with Fred. Liz reluctantly agrees and then tells Harvey, her most junior postdoc, that he must work with Fred. It takes several months to train him because Fred is unenthusiastic and is spending some of his time looking for a new position.

Within three months, Fred finds a position in industry and leaves the lab. Liz assigns Harvey to be lead investigator of the project. During a lunchtime discussion with a colleague, Liz laments the amount of time lost on this project and her concern about her grant application. The colleague notes that another professor, Lee, has a postdoc who has done similar work and may agree to collaborate on the project. Liz says that she does not think this would work because she and Lee had a contentious disagreement several months ago.

After lunch, Liz wonders why so many things have gone wrong in this project, but she cannot seem to put her finger on any one crucial incident. Delays and impediments seem to have cropped up time and again. Maybe, she thinks, it is just bad luck.

What went wrong?

If you have worked in a research lab, you probably will not find anything in this case study to be out of the ordinary or surprising. Liz is not a malicious or inconsiderate person. She is a dedicated scientist trying to be successful and train her staff and students. Yet at almost every juncture, progress was either delayed or hindered by an interpersonal, not scientific, issue. Here are a few of the lost opportunities and their consequences.

Incident	Alternative behavior	Possible effects of the alternative behavior
Liz becomes distracted during Fred's presentation.	Liz may know that she has a tendency to become distracted and distant because this is probably not the first time that this has happened. She could have told Fred what was really on her mind so that he could adjust to her new level of interest.	Fred may not have felt dismissed by Liz's apparent distance and might have been able to better focus on feedback during the meeting.
Sue shows little interest in Fred's presentation.	Liz often multitasks at lab meetings. She brings her computer and reads her e-mail during presentations. Liz has set a precedent that it is acceptable to pay only peripheral attention when someone speaks. She could become aware that this behavior has a negative impact on her lab members. She could change her behavior to show them that it is important to pay attention during presentations.	Lab meetings could become more interactive and productive. Presenters such as Fred would not leave feeling that the meeting was a waste of time.

(Continued)

Incident	Alternative behavior	Possible effects of the alternative behavior
Alan and Richard gang up on Fred.	Liz could have moderated this discussion, instead of becoming distracted by her own thoughts. If Alan and Richard have a record of double-teaming lab members, it is Liz's responsibility to call attention to their behavior and teach them to raise their concerns in a more productive manner.	Alan and Richard may have asked important questions that would have been more thoroughly discussed in a nonconfrontational atmosphere.
Liz springs Zheng on Fred with no warning.	Springing Zheng on Fred without warning showed a lack of awareness of how he would feel and react. Liz needed to prepare Fred by explaining that she wished to help him, not take the project away from him.	Zheng's participation could have helped Fred move the project forward. Instead, a more junior person got assigned, with detrimental consequences.
Liz fails to take advantage of Lee's postdoc because of an unresolved conflict with Lee.	Had Liz been proactive in recognizing and resolving conflicts with her peers, she might have had a better relationship with Lee and gotten the help that she needed.	Liz would have been able to recruit expert help for the project, further accelerating the work.

This case is a good illustration of the complex interplay between technical execution and social dynamics in the lab. By being more attentive to her own behavior and its impact on others, Liz could have changed the outcome of this story. She might have taken steps to defuse animosities and misunderstandings in the lab before they ended up hampering the research. She also could have mended her relationship with Lee, enabling the participation of others in the project. Progress was hampered not by scientific, but by interpersonal matters.

The case contains examples of interpersonal interactions that most working scientists will find unremarkable and even routine. No overt battles took place, no threatening confrontations occurred, and no one stormed out of a meeting in a huff. No one behaved in a reprehensible or censorious manner. Yet the interactions in aggregate resulted in major delays in an important project, lost professional opportunities for the participants, and wasted research dollars.

All of the scientific participants were negatively affected by the outcome. Fred left the lab before key studies could be completed and failed to gain senior authorship on a major scientific paper. Harvey ended up owning a project prematurely, prolonging the time it took to generate important data. Liz, under pressure to generate this data for a grant renewal, failed to do so. Her renewal was turned down, and it took another year to produce the results needed to resubmit the application. The National Institutes of Health ultimately paid for several years of work on a project that yielded fewer results than might have been possible had it been managed better.

The antecedents of the outcome are many. Liz never received any training in managing a group of scientists. Her primary objective was to retain her funding, and she viewed her students and postdocs as instruments for meeting this objective. Liz's department chair did not know the way in which she ran her lab and offered no help or advice, despite the fact that he had invested hundreds of thousand dollars of departmental funds in her start-up package. Liz's colleagues may not be any better equipped than her at discerning the reasons for the misunderstandings, interpersonal clashes, and poor management skills that hampered her work.

IMPROVING YOUR SKILLS AS A MENTOR

There are many reasons why observing, evaluating, and improving your skills as a mentor[1] should be a high priority if you are in the sciences. First, and foremost, is the fact that you have a responsibility to your trainees to provide them with the kind of professional guidance that only a mature and experienced professional can provide. Second, is that your productivity as a principal investigator is directly connected to the productivity of your trainees. If you behave in ways that alienate or confuse trainees, you will decrease their productivity in the lab, thereby sabotaging yourself and your own interests in the process.

One of your most important functions as a mentor is to provide consistent and helpful feedback and advice to your mentees. We discussed this in Chapter 6, which contains specific scripts and other tools you can use to help you perform this important role. Also, in Chapter 6 we discussed the importance of establishing a Mentor/Trainee agreement with your mentees and have provided a sample of such a document at the end of Chapter 6. Finally, if you are mentoring postdocs, refer to the section in Chapter 6 on Individual Development Plans (IDPs) and their importance. We have provided a copy of our own version of a combined IDP and Performance Review document as well as a Mentor/Trainee Compact at the end of Chapter 6 as well as online in editable form at www.sciencema.com/resources.

The many facets of good mentoring are discussed at length in numerous excellent guides and publications. The basic functions of a mentor in a scientific setting are reviewed by Barker in *At the Helm* (2002; pp. 181–230) in the chapter on mentoring, which also contains a comprehensive list of readings and references relative to mentoring in the life sciences. Skills discussed by Barker include helping trainees to establish a reliable methodology for their research, providing scientific oversight, and modeling appropriate behaviors for interactions with peers and collaborators.

It is the last group of skills on which we wish to focus here. In the absence of well-developed self- and interpersonal awareness, you may find yourself inadvertently transmitting ineffective attitudes and behaviors to your trainees.

Trainees learn from your behavior

Trainees pay close attention to what mentors say and how they behave. Trainees may interpret overt behaviors such as belittling others in public and arguing with colleagues as the norm if a mentor exhibits them routinely. Similarly, passive or covert behaviors such as not addressing difficulties in the lab or undermining others in their absence may also be seen as acceptable if exhibited by a trusted mentor.

None of us is perfect, and we all exhibit behaviors that we regret in retrospect, or that we would rather not have a trainee emulate. You can use your own lapses in interactions with others as lessons for your trainees, just as you would use scientific triumphs and failures as teaching tools. Doing so implies an acceptance of the fact that your role as mentor extends well beyond the transmission of technical information and skills. By showing trainees that you have the capacity to examine and learn from your own behavior, you transmit one of the most valuable lessons that a mentor can provide.

[1] For our purposes here, mentor also refers to someone who oversees a trainee's lab work or research.

As illustrated in the table on pages 249–250, the outcome of the case study could have been improved if either Liz or Fred had been more self-aware. As a mentor, Liz needed to pay more attention to the effects of her behavior on those in her group. For example, she needed to understand that trainees place a great deal of importance on what may seem like inconsequential comments, behaviors, or facial expressions.

Anticipate the consequences of your behavior

If you are a mentor with a number of people working for you and you have difficulty spending time with each of them, you may be especially vulnerable to being misinterpreted. The absence of "face time" with you will incline trainees who need feedback or attention to read more into the brief interactions that they do have with you than they might otherwise. Maintaining an awareness of your own state of mind is the first step in knowing how your thoughts and feelings will influence your behavior with trainees.

If you feel pressured, you may know from past experience that you tend to act in a distracted or dismissive manner. If you are elated about an exciting result from one project, you may ignore other projects, leading trainees to wonder whether they are responsible for your lack of interest. If your personal life is a shambles, you may be venting your frustrations on the people in your lab without realizing it. You may be so preoccupied that your lack of interest causes work to slow down. If you are aware of your state of mind, you can anticipate behaviors that impact others and take steps to act differently. The exercises at the end of the chapter may help you to improve your behavioral self-awareness.

Focus on process, not just content

Scientists who are deeply involved in problem-solving may make comments that they realize in retrospect were ill considered, inappropriate, or hurtful. As a mentor, sound a mental alert to yourself when you become deeply involved in a science discussion with trainees. We know from personal experience that it is possible to monitor your affect and behavior without damping your enthusiasm for the science. It is important to be aware of how trainees react to you and what you say. To avert misunderstandings, watch for signs that someone has misinterpreted or had a negative reaction to what you have done or said. Students or trainees may be confused or baffled by your comments, misunderstand instructions, or disagree with your viewpoints. Yet, for any number of reasons, they may not say or do anything in response. For example, they may find you intimidating, be so unsure of themselves that they are terrified of asking for clarification, or come from a culture that frowns on questioning a mentor. In such cases, the mentor must be alert for behaviors that manifest feelings of discomfort, confusion, or anxiety.

A subtle furrowing of the brow or a brief look of confusion can be revealing. Perhaps while you are talking, you notice that your student is furiously drawing pictures of tiny bugs crawling across a page in his lab notebook. In these cases, you might ask, "Jatinder, is there something I just said that's confusing to you?," "Marie, you look like you have a question. Do you?," or "Doug, do you want to talk more about this after the lab meeting?"

The benefit of such attentiveness is obvious. If students or trainees are unclear about or disagree with something you said, or if they misunderstood you, it is likely that they will end up doing something other than what you expected.

SURVIVING AS A TRAINEE IN ACADEMIA

Those not familiar with the politics of academic research laboratories might be tempted to wonder why trainees would choose to work in laboratories that are poorly managed, or for PIs who behave in insensitive and churlish ways. The answer is simple. Most trainees chose their mentors based on their scientific interests and accomplishments rather than on their managerial or interpersonal skills (although other factors do come into play; see the case study on pp. 291–292 entitled "Junior or Senior?"). Despite exhortations by friends and advisors to seek mentors who will be mindful of the trainees' needs, often what really matters is the mentor's research track record and prominence in their field. As a result, mentors with brilliant scientific minds but poor or even disastrous managerial or interpersonal skills can flourish and produce students in their own image. Once a decision is made to pursue a doctoral level research program, academic trainees are, for all practical purposes, indentured to their mentors.

The perpetual cycle of academic training

- All scientists are trained in academia.
- Academic faculty are rewarded for their research, not mentoring.
- Trainees leave academia with technical skills and little else.
- Trainees will tolerate a lot to get credentialed, thus there is no incentive to improve the system.
- Trainees who go on to become faculty perpetuate the cycle.

Trainees have much at stake in this relationship. Most important is that it will lead to a granting of the credentials needed to practice their chosen profession. As a result, trainees have an intrinsic and powerful motivation to persevere and perform even if they are ignored, mistreated, or manipulated. Trainees will put up with a great deal because their objective is clear and their success depends on the mentor.

If you are a trainee, once you have spent four or five years learning both science and, by observation, the management of science from your mentor, you may take away more (or less) than what you bargained for. If you continue your career in academia, you may perpetuate the same managerial deficits as your mentor (see the figure above). Although this may negatively impact your productivity, you may never know it because it is unlikely that anyone will ever call your attention to it. If, however, you leave academia, you will almost certainly run into problems.

Scientists who enter the private sector often take a long time to learn that interpersonal and management skills have a major impact on long-term advancement opportunities (see Chapter 11). This is in stark contrast to their experience in the academic setting, and this shift in expectations is rarely, if ever, made explicit. It is a distinction that young scientists miss at their peril.

As a trainee in science, there may not be much you can do in the short term to create wholesale change in the academic world. But there is a lot you can do for yourself to minimize misinterpretations of your mentor's behavior. You can also learn to identify management styles that are worth emulating from those that are not.

Observe how your mentor manages and relates to trainees and employees

It is easy for a student or trainee to assume that successful mentors know how to manage scientists. After all, there must be some reason that your mentor arrived at where they are today. But perhaps they have their position despite, rather than because of, their managerial and interpersonal skills. Maybe if their managerial skills were better developed they would have been even more productive and more successful.

Our recommendation is that you learn to observe management styles early in your career. The way to learn is to note the management techniques that seem to be effective. We define effective management techniques as those that help meet the objectives of the enterprise (lab, group, company) and that are implemented in a manner that is respectful of others and mindful of the consequences for the members of the organization.

Note that this definition does not mean that everyone will always be happy with every decision made by a manager or leader. Do not fall into the trap of believing that mentors or managers who make decisions that lead to disappointment or disagreement are poor managers. Effective managers often need to make difficult and unpopular decisions. Conversely, do not be fooled into believing that a lab is well managed just because there are no overt arguments or conflicts in evidence and that everyone, including the PI, goes out for beer together and on weekend canoe trips. We once knew a PI who ran this type of lab, but his graduate students typically took two years longer than most to complete their degrees.

One way to assess the quality of a mentor's decisions or actions is to ask yourself the following four questions:

1. What is the objective underlying the decision or behavior?

2. Is the objective known or made known to those affected?

3. Is the objective consistent with the mission of the lab/group/individuals?

4. Was the resulting decision or behavior implemented with respect and consideration for those affected by it?

If you know the answer to question 1, you are at least in a position to understand the rationale for a decision. Decisions or behaviors for which the objective is either unclear or unknown are bound to cause confusion and consternation. If the answer to questions 2 and 3 is "yes," both the objective and its relationship to the (presumably shared) goals of the group are known. Decisions or behaviors that lack a clear and known relationship to the goals of the enterprise or its members will be resisted, resented, or ignored.

Question 4 addresses the area in which many managerial actions fall short. Managers often believe that their work is complete once their staff or trainees understand the rationale behind a decision. The result is that decisions get implemented without careful consideration of how they will impact people. Mentors who make decisions that have a clear rationale consistent with the group's mission, but who implement them in ways that result in bruised egos, hurt feelings, and alienation, should be observed but not emulated.

Learn to take note of your mentor's actions and the outcomes of those actions. By evaluating what you observe as described above, you can identify those managerial behaviors that are worth emulating.

Be aware that your mentor may be oblivious to the impact of their behavior

If your mentor has limited managerial and interpersonal skills, it is likely that at some point you will feel hurt, insulted, or misused by them. Trainees often feel insecure, especially during their early years of training, and readily interpret a dismissive comment or lack of interest as being caused by something they did or said. This was what Fred did when he misinterpreted Liz's distraction for skepticism or lack of interest (see the case study "What's Wrong Here?" on pp. 247–249). A good rule to follow in such situations is what we call the "95% rule:

Behavioral rule #4
Ninety-five percent of anything anyone does or says in your presence has nothing to do with you.

Although mentors often have important messages and advice to give their trainees, remember that not everything your mentor says or does is a direct reaction or response to you. Indeed this rule applies to most people you encounter in most situations of life. Each of us is the most important person in the world—to ourselves. It is an unfortunate truth, however, that we are not the most important person to others. Everyone has their own problems, concerns, and petty annoyances in their lives, and it is largely these that affect their moods, their willingness or interest to listen to our problems or complaints, and their general friendliness when we encounter them. Knowing and accepting this fact of life will save you countless hours of obsessive analysis of people who seem distracted when you talk to them, don't say good morning as effusively this week as they did last week, and who don't return your phone calls. It's almost always about them, not you.

Of course, there's the other 5% that does pertain to you. Here are a few situations that you can be sure are in that 5%: The person is (1) gripping your shoulder with their hand and looking you straight in the eye, (2) giving you the results of your annual performance review, and (3) sitting in your lap (definitely not at work). Beyond these situations, it is difficult to devise a foolproof guideline to help you to decide when a comment or action is meant to convey a message, is simply a manifestation of thoughtlessness or distraction, or has nothing at all to do with you. When in doubt, you will usually be better off inquiring if there is a problem rather than worrying about whether there is one. For example, saying, "I noticed that you didn't have much to say about my project report this morning. Was there some reason for that?" is much better than spending the rest of the week worrying that your mentor thinks your work is a disaster.

Take account of the pressures and deadlines to which your mentor is subject

In the case study "What's Wrong Here?" Fred was oblivious to the professional pressures being experienced by Liz. Fred was only focused on people's reactions to his data in that moment. Had Fred applied the 95% rule he might have been able to temper his annoyance at Alan and Richard's interruptions ("They do this to everyone, I'm not going to let it aggravate me.") and his frustration at Liz's distraction ("I know she's under a lot of pressure right now so it's understandable that she'd be distracted."). Keeping in mind that Liz had a difficult impending grant renewal and was eagerly seeking ways to enhance her chances of success might have alerted Fred to the possibility that any new or exciting data would be welcomed, and that Liz would let him know this in her own way and in her own time. Below is a satiric rendition of the relative importance of your project to you and to your mentor.[2] Although you shouldn't take it literally, after all, many mentors are deeply invested in their mentee's projects and training, looking at this figure now and then should help you calibrate your expectations.

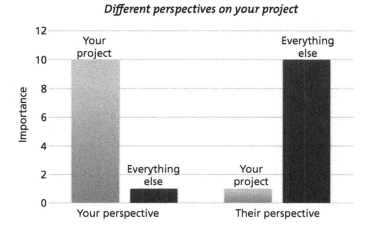

Different perspectives on your project

---

[2]Figure derived from http://phdskills.blogspot.com/2014/01/manage-your-phd-supervisor.html.

The pressures to which a mentor is subject should not excuse insensitive and manipulative behavior. But trainees who are aware of their mentor's pressures will be less likely to interpret distraction as disinterest and aggressive questioning about a key experiment as a personal attack. This point is closely related to the subject of the preceding section: A lack of awareness of your mentor's needs, motivations, and pressures makes it more likely that you will seek personal explanations for their behavior.

Pay attention to your mentor's pressures, as well as the needs and motivations that affect their actions. Be especially aware of grant deadlines, manuscript submissions, acceptances and rejections, upcoming tenure decisions, and other academic responsibilities such as teaching and committee assignments. None of these factors excuses bad or absent management, but being aware of them can help you to place your mentor's behavior in context and reduce the chances that you will misinterpret the occasional thoughtless comment or lapse in attentiveness.

Becoming more aware of the world in which your mentor functions allows you to improve your relationship with them (for more, see Chapter 8). If improving your relationship results in better guidance, more thoughtful input, or increased face time, the time that you spend doing so will be well worth it.

Your needs are not always synonymous with those of your mentor

In the best of all worlds, your mentor's success benefits you as much as it does them. But at times, your mentor may have a different agenda than you. In a strong job market, keeping highly skilled graduate students and postdocs in the lab for long periods of time to support new work may be of great benefit to a mentor but of little or no benefit to the trainee. In a weak job market, it may benefit both mentor and trainee.

It is not uncommon for trainees to feel that their own interest in moving on, either to new projects or a new lab or institution, is being inhibited by an advisor who is attending to their own interests over those of their trainee. Making your own interests and needs known (by using the types of statements illustrated below) is always your first line of approach in such situations.

Let your mentor know the impact of their behavior without being accusatory

Empathy and understanding will take you only so far with an inconsiderate mentor. Remaining silent about feeling abused, mistreated, or ignored will only increase your sense of powerlessness and frustration and provide little help in getting the advice and guidance you need. Mentors who have poor interpersonal or managerial skills or who create unpleasant work environments are almost never malicious—they are simply clueless. The key to getting what you need, whether it be more feedback on your performance, scientific help, or career assistance, is to ask for help without being accusatory. Referring to the tools we introduced in Chapter 9, when asking for help, depersonalize sensitive topics by remaining focused on improving your work and performance. Remember to depersonalize sensitive topics by using "I" statements instead of "you" statements. The table below shows examples of approaches for making your needs or feelings known with your mentor.

Situation	Passive/passive–aggressive	Confrontational	Neutral
You feel that your mentor doesn't give you enough of their time.	Say or do nothing, feel resentful. Become withdrawn and silent at lab meetings.	"You never have time for me. You're always busy with something else."	"I wonder if we could meet more often? I could use a bit more input into this phase of my research."
You think your mentor is giving another student the same project as yours.	Complain to others but don't talk to mentor. Become hostile to the other student.	"This is not right. You're going back on your promise that this was my project."	"I may not be seeing the big picture, but are Gary and I working on the same project?"
Your mentor says something highly critical of your work during a lab meeting.	Feel resentful toward mentor. Say nothing. Start criticizing other people's work.	You get defensive and argue during the meeting.	After the meeting say, "Can we meet to discuss your comments? I'd like to be able to understand and address your concerns."

Dealing with a passive or absent mentor

One of the consequences of having a mentor who is busy or overcommitted is that you, the trainee, do not get the attention and time you deserve and need. Although we are all for respecting the time of a busy lab head, this should not come at the expense of your legitimate needs.

Below are a few tips on how to help your mentor do a better job at mentoring.

- *Schedule meetings, don't just walk in.*
- *Go in with a clear idea of what you want to accomplish at the meeting. Share this with them.*
- *Meet regularly, even if it is for a brief time.*
- *Whenever possible, come with ideas and solutions, not problems.*
- *Take notes and summarize.*
- *It's not all about you. Let them know you understand the pressures they are under.*
- *Respect their time. Be prepared. Be concise.*

If your mentor is busy you can also get their attention, and gratitude, by making yourself useful to them and the lab by volunteering to take on some lab responsibility (within reason). However, do not get caught in the trap that we see graduate students and postdocs fall into all too often. When an adviser is absent or passive it is very easy for some people to feel as though they have a responsibility to manage problems that are legitimately the responsibility of the supervisor. In our Cold Spring Harbor Laboratory workshop on Leadership in Bioscience, we see example after example of this in the case studies of lab leadership that the participants prepare. For example, there might be someone in the lab who is not being a good citizen and leaving the cold room a mess on a daily basis. In the absence of intervention by the leader, others in the lab feel it their responsibility to intervene with the culprit, often with unfortunate results. The intervener has noble motives but because they lack standing and authority they get ignored or impugned and end up feeling that "no good deed goes unpunished." Although it is admirable that some try to be good lab members by trying to enforce rules and respect for others, be careful before you start down this path. If it is legitimately the job of the leader, take the matter to them before you take it upon yourself.

If you do take it to the leader, follow a few simple guidelines.

- *Do not bother your mentor with every petty disagreement. Use your judgment and validate that with another person in the lab.*
- *When you do decide to go to your mentor, focus on:*
  - "We need your help" or "I think you will want to know about this… ."
  - "It's affecting our work… ."
  - "I have a few ideas on how to deal with it, would you like to hear them?"

The bullet about the incident or behavior affecting your work is an important one. We all get annoyed with people at work. Sometimes it is because they are doing something that is genuinely interfering with the group's ability to get work done. Other times it is because the behavior feels annoying or inconsiderate *to us*. Before you go to your mentor to discuss someone else's behavior, make sure it is not just you that sees a problem. If it is just you, then you need to examine what it is about this person that rubs you the wrong way, and more than likely keep your adviser out of it.

If whatever is annoying to you is not actually affecting the work or progress of the group, then you need to think carefully about your motives for bothering your mentor with it. When you do go to your adviser, frame your concerns in the context of the work of the lab or group. Go to your adviser with ideas or suggestions about how to solve or alleviate the problem rather than just complaints.

Remember that you are not alone

Sometimes your best efforts at open mindedness and at understanding your mentor's problems are not enough. Some advisers and mentors are, on occasion or even consistently, inconsiderate or unfair. It is easy to feel that you are at the mercy of your mentor because so much is at stake in your relationship with them. In addition, unless the mentor's behavior is so egregious that it falls outside of the bounds of laws or institutional policies, it is not always obvious to a trainee what is acceptable and unacceptable behavior on the part of a mentor.

As a trainee you need to be aware of the consequences of feeling isolated and at the mercy of your mentor. If you feel inadequate or unappreciated, you may find yourself feeling depressed and hopeless with nowhere to turn.

Thankfully, most graduate training programs are organized such that trainees are assigned to a committee of faculty members who monitor their progress periodically. Some training programs are now being improved by creating more supportive environments for trainees. For example, Harvard University's chemistry department has established multiple approaches for providing graduate students with more opportunities for faculty interaction as well as psychological support. It is unfortunate that this was done after two graduate student suicides rather than before (for a discussion of these events, see Hallowell 1999).

Take advantage of your thesis committee members. In all likelihood, you will encounter at least one faculty member with whom you can establish a rapport. In most cases, simply having someone with whom to discuss problems is all that you will need. In other cases, an ally will be important if you are being treated in an unprofessional or malicious manner.

Do not overuse your faculty ally. It is easy for an ally to suspect that you are chronically dissatisfied if you use your relationship to grouse about minor matters that have more to do with you than your mentor (see above for hints on identifying such issues). Discussing these matters with a peer, especially if they are familiar with your mentor, is a good way to validate concerns that you may have about your mentor.

Yet some scientists may find it difficult to reveal their insecurities to peers, and even more difficult if these insecurities derive from being belittled or harassed by their mentor. It is easy for a trainee to feel ashamed, embarrassed, or inadequate as a result of such treatment, especially if they believe that they truly are inadequate. We have heard others characterize the tendency of scientists to keep matters such as this to themselves as being attributable to a "macho" culture in science, and there is likely some truth to that. This attitude is fostered by the training that young scientists receive, geared toward their functioning as independent investigators in a competitive environment.

It is difficult to break the mold and seek support by opening up to a peer. But if we have learned one thing from our workshops for scientists, it is that everyone has the same problems. When participants in our workshops cite examples and problems with which they are dealing, we see others nod in recognition as one problem after another is discussed. Scientists who have rarely, if ever, discussed their problems, perhaps because they are ashamed of them or believe that the problems are unique to themselves, suddenly see that everyone has the same difficulties. Keep this in mind. Chances are that if you share your uneasiness, fears, and concerns with a peer who seems to be receptive, you will find yourself hearing the same from them.

If you can avail yourself of such peer support, do not fall into the rut of using your peer group only for grousing. Use the guidelines in the sections above to critically examine and evaluate your mentor's behavior. Consider motivations, external pressures, and constraints. Use your discussions to put yourself in your mentor's place and ask what you would realistically do in the same circumstances. In many cases, you will be surprised to discover that decisions that seemed to be unfair or capricious were actually the result of a complex set of circumstances for which there was no easy solution. Perhaps a decision that could have been implemented with finesse and consideration was instead performed in a heavy-handed and inconsiderate manner. Your training experience and your own success as a mentor and manager will be improved immeasurably if you and your peers use these and other types of decisions and behaviors as learning tools.

In addition to taking advantage of peer support, determine whether your institution has resources or personnel to support trainees. For example, the University of Michigan's Rackham Graduate School publishes a comprehensive and insightful guide to mentoring and being mentored (see http://www.rackham.umich.edu/downloads/publications/Fmentoring .pdf).

The free book *Making the Right Moves,* available at http://www.hhmi.org/resources/labmanagement/moves.html, also has a chapter on mentoring that offers solid advice for both mentors and mentees.

In addition, there are several online resources specifically focused on helping trainees to think about and deal with mentoring issues (e.g., http://gradschool.about.com/cs/aboutadvisors/a/mentor.htm).

REFERENCES

Barker K. 2002. *At the helm: A laboratory navigator.* Cold Spring Harbor Laboratory Press, Cold Spring Harbor, NY.

Hallowell EM. 1999. *Connect: 12 vital ties that open your heart, lengthen your life, and deepen your soul,* pp. 117–123. Pantheon, New York.

EXERCISES AND EXPERIMENTS

1 Mentors: Experiments with new behaviors

The purpose of this exercise is to help you to observe and evaluate how you interact with trainees. Below we provide a framework within which to record and assess your interactions. For those behaviors or interactions that were either ineffective or counterproductive, you will be asked later to experiment with alternative behaviors.

Record the incidents. During the next week or two, record two or three interactions with a trainee for which you felt the outcome to be unsatisfactory. For each incident, record

  a. the incident or interaction

  b. your objective in the interaction

  c. your behavior

  d. the effect of your behavior on the interaction or trainee

  e. how your behavior affected the outcome or objective

Look for patterns. After you have accumulated several such incidents, examine them and ask yourself if you see a pattern. Does a specific person continue to come up? What about this person triggers your reaction? What are you experiencing during this reaction? Is a particular type of interaction more troublesome than others?

As an example, many mentors have difficulty giving negative feedback, reprimands, or evaluations to trainees and instead send confusing or ambiguous messages. If you have difficulty deciding whether interactions leading to unsatisfactory outcomes were influenced by your own feelings of discomfort or avoidance, refer to the tools for improving self-awareness presented in Chapter 2 for this part of the exercise. If you have difficulty discerning the effects of your behavior on others, refer to Chapter 9 and the exercises in that chapter on improving your observational skills.

Alternative behaviors. Propose alternative behaviors that might have led to a better outcome. List the possibilities you could have used (refer to the guidelines beginning on p. 254) and describe the anticipated outcome of the new behaviors. You may find it useful to record this information in the form of a chart, such as the one that we used for the case study earlier in the chapter..

Do the experiment. After you have listed several alternative behaviors and thought through their consequences, try one or more of your suggested alternatives in an interaction with a trainee during the following week. Determine whether the outcomes of the interactions involving the new behavior are better than those that occurred with your former behavior.

2 Trainees: Experiments with new behaviors

The purpose of this exercise is to help you to observe and evaluate how you interact with your mentor. Below we provide a framework within which to record and assess your interactions. For those behaviors or interactions that were either ineffective or counterproductive, you will be asked later to experiment with alternative behaviors.

Record the incidents. During the next several weeks, record two or three interactions with your mentor for which you felt the outcome to be unsatisfactory. For each incident, record

a. the incident or interaction

b. your objective in the interaction

c. your behavior

d. the effect of your behavior on the interaction or mentor

e. how your behavior affected the outcome or objective

If you have difficulty discerning the effects of your behavior on others, refer to Chapter 8 and the exercises in that chapter on improving your observational skills.

Look for patterns. After you have accumulated several such incidents, examine them and ask yourself if you can see a pattern. Does a type of interaction continue to come up? Does a specific behavior of your mentor trigger a strong response in you? What are you experiencing during this situation? Is a particular type of interaction more troublesome than others?

As an example, many trainees have difficulty hearing negative feedback or evaluations and instead hear confusing or ambiguous messages. If you have difficulty deciding whether interactions leading to unsatisfactory outcomes were influenced by your own feelings of discomfort or avoidance, refer to the tools for improving self-awareness presented in Chapter 2 for this part of the exercise.

Alternative behaviors. Propose alternative outcomes based on new behaviors. List the possible alternative behavior(s) that you could have used (refer to the guidelines beginning on p. 254) and describe the anticipated outcome of new behaviors. You may find it useful to record this information in the form of a chart, such as the one that we used for the case study earlier in the chapter.

Do the experiment. After you have listed several alternative behaviors and thought through their consequences, try one or more of your suggested alternatives in an interaction with your mentor during the following week. Determine whether the outcomes of the interactions involving the new behavior are better than those that occurred with your former behavior.

If you are a trainee, do the exercises at the end of Chapter 8. These are as relevant to your interaction with a mentor as they are with a supervisor in a traditional work setting.

3 The 95% rule

List three things that your boss has said or done that upset, angered, or threatened you. For each incident, list three explanations for the behavior that have nothing to do with you and three explanations that have everything to do with you. Evaluate the likelihood of each set of explanations. Simply performing this exercise will give you a more balanced perspective on what was said or done and can help you identify inappropriate or exaggerated reactions on your part.

CHAPTER 11

Science, Inc.: Make a Smooth Transition to Industry

A biotech executive recently said to Carl, "New scientists who join our company from academia are like deer frozen in car headlights. They don't know what's coming at them. They have no experience of working to timelines, reporting progress weekly, and having their work critiqued by every Tom, Dick, and Harry who happen to show up at a project meeting. It takes a couple of years for them to adjust."

Just as academic research labs present a spectrum of challenges unique to academia, so too do labs in the private sector. Scientists new to the private sector are rarely prepared for the different norms and expectations there. The culture shock that scientists face when moving from academia is one of the most vexing problems of research enterprises in the for-profit sector. Your ability to sense and respond to changes in expectations will determine how effectively and quickly you adapt to new work environments.

Scientists often have a hard time accepting the loss of both autonomy and the ability to focus on "pure" research without regard for

applicability and are unprepared for the overriding concern with the bottom line (profit). Scientists who tend to become emotionally bonded to their research projects can find themselves feeling disappointed and manipulated in the private sector when projects get terminated.

The following table summarizes differences between academic and private sector research, with one or more personal or interpersonal consequences of each. The hardnosed manager might be tempted to read through the list and respond, "Get over it!" Most scientists do get over it, but the loss of productivity during this transition is substantial and unnecessary.

Characteristic	Academia	Private sector	Consequence of difference
Autonomy	High. Scientific and administrative autonomy is the norm.	Low. The nature of for-profit research dictates that research be consistent with and lead to the advancement of corporate goals.	Scientists may feel manipulated or controlled. May lead to resentment, low morale, and alienation.
Ability to follow own ideas	High/medium. Ideas can be followed within the scope of available time, funding, and manpower. Possibility exists for delving into new research domains.	Medium/low. Projects need to be conducted according to well-defined plans and timelines. Tolerance for exploration of peripheral areas is low.	Scientists may feel constrained and hampered in creativity.
Criteria for choosing a project	Typically chosen for scientific interest and importance to field. Choice may be constrained by interests of funding agencies.	Chosen for consistency with corporate mission.	Scientists may feel that their scientific input into project choice and focus is irrelevant.
Criteria for ending a project	Scientists find it hard to end projects in any setting. In academia, projects can persist for long periods of time, provided funding is available. Often the only thing that can kill a project is the loss of a research grant, and even then institutional funds are often available.	Projects can be killed for any one of a number of reasons including change of company focus, loss of a corporate partnership, or failure to accomplish a predetermined set of objectives.	Scientists who are unfamiliar with the corporate mission or with company politics may feel that project termination lacks justification and that they were not given sufficient time to prove its worth.
Management of lab staff	Laissez faire.	Highly regulated.	Need to learn new skills.
Oversight of scientific work	Episodic and remote. Typically, oversight comes only on submission of a publication or a grant application.	Constant and formalized. Scientists need to stay focused and to deliver results.	Scientists may feel micromanaged.
Periodic evaluation of and input into personal performance	Virtually nonexistent.	Depends on company, but typically is ongoing and explicit.	Scientists need to be receptive to positive and negative feedback.

(Continued)

Characteristic	Academia	Private sector	Consequence of difference
Need to work with multiple parts of organization	Typically of relevance only with respect to support services and administration.	Critical to success of complex projects. Need to understand and work productively with people who have distinctly different goals, viewpoints, and backgrounds.	Scientists arrive with a lack of appreciation of need for input from groups such as clinical, regulatory affairs, pharmacology, etc. They may also lack skills for interacting with nonscientists (e.g., lawyers) on scientific matters (i.e., intellectual property).
Stability of parent organization	Academic institutions are highly stable.	For-profit research organizations vary greatly in their stability. Start-ups may be especially unstable.	Job security may for the first time become a major factor in the scientist's professional life. Research outcomes may impact career and livelihood.

CORPORATE CULTURE SHOCK

Scientists who join the private sector are often dismayed by the constraints to which they are subject. Seemingly simple decisions such as the choice of research tools and scientific approaches are often driven by the strategic focus of the company rather than by a scientific evaluation of the many possible alternatives. Scientists in academic settings exult in the discovery of an unexpected connection between their work and some other apparently unrelated phenomenon. Such discoveries are the stuff of new grant applications, funding, professional satisfaction, and prestige. In the private sector, such discoveries and insights must be sacrificed if they do not fall within the objectives of the organization. Finally, projects may end unexpectedly and for reasons that bench scientists find hard to understand. The following case study illustrates several of these issues.

▪ *Case Study: Terminated*

Nano-Innovations was a three-year-old company with a staff of 30 scientists, most of whom had been recruited from academic labs. Nano had a collaborative project with ExploiTech who provided about $15 million a year in funding to develop a more efficient manufacturing process for the company's nanofibers. This project employed about 15 full-time Nano scientists. The agreement was renewable on an annual basis but ExploiTech had recently told Nano that they were unlikely to renew once the current year ended. This news prematurely leaked to the scientists at work on the project, which still had six months to run under the terms of the present agreement. The scientists had difficulty understanding the termination decision, which had nothing to do with the quality of their work, but more with financial problems and management changes within ExploiTech. These scientists now faced six more months of work on a project that they knew would be terminated and some uncertainty about their own jobs. Moreover, the scientists were frustrated that the project was ending because, with the encouragement of their CEO, they had been working long hours and were very close to making important improvements to ExpoiTech's manufacturing process. Some of the scientists resolved to never become this involved with a project again and to do only enough work to get by.

For a committed scientist, dropping a good idea or project can feel like giving up a child for adoption. Ending a favored project is something that even the most seasoned of research veterans finds difficult to do without much hand-wringing, procrastination, and last-minute reprieves. When Carl worked in academia, he walked into his laboratory on more than one weekend to find a dedicated scientist busily doing "one more experiment" for a project that he thought had been terminated weeks or months earlier.

When a project is terminated for reasons that may have nothing to do with the promise or progress of the work, scientists may become disillusioned. Because scientists become intellectually bonded to the projects in which they are engaged, they often react by feeling frustrated and manipulated. Novice scientists, after being shuffled from one project to another, may avoid fully committing themselves to new projects to minimize disappointment. You do not need to be a scientist to recognize that such behavior is detrimental to any organization.

Such reactions are often attributed to individual personality and adjustment issues or to immaturity. Although there may be some truth to these explanations, these reactions may also be a consequence of the poor preparation that scientists receive for working in the private sector and of the failure of managers to deal with the human consequences of scientific and business decisions.

MASTERING THE ART OF SCIENTIFIC RESEARCH
IN A CORPORATE SETTING

If you are a scientist planning to enter or who has recently entered the private sector, there is much you can and should do to minimize the chances that you will fall victim to corporate culture shock. The following sections will help you to identify aspects of corporate scientific research that we know from experience cause consternation, confusion, and alienation in scientists. Your ability to recognize these problems and anticipate their effects on your attitudes and performance will improve your chances for a successful career in the private sector.

The only science that survives is profitable science

As the previous case study illustrates, scientists need to be aware of the differences in the way that decisions are made in the private sector versus academia. Leaders of science and technology companies, especially those who are or were scientists themselves, have a penchant for motivating their scientific staff by extolling the elegance, power, and beauty of the company's science. They do this because they suspect that this is what the scientists want to hear. Do not be fooled by such oratory.

Be skeptical of companies that recruit research scientists with the promise of unfettered research opportunities. Very few companies live up to this promise, and many of those that try, especially new companies, change their tune after a few years. This is not to say that it is impossible to have satisfying and even exhilarating scientific experiences in the private sector—it is possible, and many do. But remember that the mission of your company is to create shareholder value and this may, at any time, lead to the curtailment or scaling back of your favorite research project. The best way to insulate yourself from disappointment is to be informed about your company and the economic forces that impact it.

Understand your company's mission

Understanding the mission of your company is a great way to keep abreast of which way the corporate wind will blow. But do not expect to learn anything from the mission statement displayed in the main lobby of corporate headquarters. Mission statements convey lofty visions of the company's goals and values and can be useful tools to the skilled leader (see Chapter 12), but most offer little guidance about what's important today or tomorrow. Instead, watch and listen to what fellow employees do and say. In addition, do not believe that your company's mission is unchangeable; after all, the life science research and discovery companies of yesterday are all scrambling to become the drug development companies of tomorrow—after spending a brief interval describing themselves as genomics companies.

Pay special attention to those programs that your company's senior management values most and to the goals on which they are focused. Unfortunately, it is not always easy to discern what senior executives are focused on. They may have a tendency to secrecy, especially when their focus may lead to mergers, layoffs, and downsizings that negatively affect employees.

Understand the market and economic factors that affect your company's business

Look beyond the walls of your company to understand the world in which it exists. Scientists can be myopic in their view of their company and many are oblivious to the external factors that affect its fortunes. For those who transition from an academic setting, the notion that the destiny of their organization may depend on the availability of venture capital and changes in investors' perception of the technology's attractiveness (or "trendiness") may be an alien and even baffling concept.

Arming yourself with information about the external forces that affect your company is the best way to limit feelings of helplessness and surprise when the unexpected occurs. Spend time learning about the market dynamics that affect your industry. During 2001, scientists who followed industry trade publications knew that genomics companies focused exclusively on target identification were falling out of favor with the investment community. These people were not surprised when their companies announced that they would be downsizing their gene discovery efforts in favor of creating drug discovery operations. The more you understand about the market in which your company operates, the better you are able to place decisions affecting your job in the proper context.

One middle manager in a biotechnology company related the following story.

> "Our company started out as a combinatorial chemistry boutique. Our sole focus was to sell our compound libraries to biotech and pharmaceutical companies for use in high-throughput screening for drug discovery. My job was to lead a group developing screening assays. After years of explosive growth and high market valuations for companies such as ours, the investment community became skeptical. The biotech sector moved away from a fascination with clever technologies and more toward products (i.e., drugs). Because neither our company nor any company such as ours had made much headway in actually making or helping to make an actual product, we took a big hit financially. We had to reorganize and we are now much more product-oriented. Because I was watching this trend carefully, in early 2001, I began to volunteer for some of the more product-focused projects. My goals were to learn new skills associated with testing drugs in animals and eventually transition into the preclinical group if circumstances did not change. In fact, that's exactly what happened. Last year, we shut down most of our chemistry and screening operations and added several new preclinical groups. I was made manager of one of the new groups, but many from my old department were laid off."

If you are unsure of how to keep abreast of the industry trends that affect your company, ask the nearest vice president or, better yet, ask the CEO themself. Specifically, ask what to read on a weekly basis to follow industry news. This will get you noticed as someone who cares about the big picture (which immediately distinguishes you from most other scientists in the company) and points you to the same sources to which your superiors refer.

Learn to give up projects

Although it may be difficult to give up a project into which you have invested much of yourself, the sooner you learn to get over it and move on, the happier you will be in the private sector. Companies with multiple projects and programs almost always track progress against predetermined goals or advancement criteria. Meeting these goals qualifies a project for continuation, whereas repeated failure likely targets the project for termination.

We have all heard stories of persistent scientists who argue doggedly and persuasively for the continuation of a project slated for termination. The typical outcome is that the project eventually results in a blockbuster drug or product. I have no doubt that such stories are true, but I also have no doubt that the vast majority of reprieved projects ultimately yield nothing. Those are the stories that you won't hear about.

There is no way to tell early in a project's life whether it will be a success or a failure. If you need to argue for the continuation of a doomed project, focus on scientific merit and supporting data. Be careful using arguments based on "scientific intuition" to justify the continuation of a program that has failed to meet its objectives.

Pay attention to any feelings of resentment toward management for terminating projects or making programmatic changes that affect you or your work. Ignoring or repressing these feelings may influence your behavior and attitudes in ways that have a negative impact on you or your relationship with management. It can be dangerous to harbor resentment when a project in which you had invested much psychological energy was canceled, especially if you have poor self-awareness and self-control. These feelings can result in your becoming sarcastic or argumentative and may lead you to criticize management in ways that do you more harm than good. Refer to Chapter 2 for methods to improve your awareness of such feelings.

Managers may actually find it useful if you express feelings of resentment in the appropriate context in terms of the impact of decisions on yourself and how you feel. Telling your boss, "I was disappointed in how this project was terminated, and I think that others were as well. Morale might be a bit low right now, and I thought that you'd like to know" may help them to recognize and address the impact of their decisions on employees. Telling your boss, "You people don't give a damn about how hard we worked on this project" may feel good at the moment but will not be very useful to either you or your boss.

Do not expect to own your project

In the private sector, individual scientists almost never own a project and often do not even own an experiment. Project ownership is hard to define because so many people are involved. The project may be managed by a program manager who has responsibility for coordinating the timing and execution of the various technical tasks that need to be accomplished. The tasks themselves may get done within specialized technical groups or "lines" managed by senior scientists or directors. Within these lines, individual scientists may be working on more than one project at any given time. In the worst cases, such arrangements can make scientists feel like cogs in a wheel. Scientists newly transplanted from academia, who are accustomed to having their own projects, become concerned that their contribution to a project will be invisible. The result is that they may have difficulty motivating themselves to excel and to expend any extra effort. In fact, those who understand the corporate reward system learn that you do not need to own a project for your efforts to be noticed or appreciated.

Understand the corporate evaluation and reward system

Scientists in an academic setting have a pretty clear sense of how their performance is measured. They do experiments that either work or do not work and participate in or lead projects that either succeed (and get funded or renewed) or fail. They submit papers to scientific journals that get either accepted or rejected. In each case there is a direct connection between the work of the scientist and the outcome or reward. As we show in Chapter 10, this connection fosters a powerful sense of ownership that is an important motivator for scientists in academia.

Project ownership is harder to define in the private sector. This is not to say that scientists are not connected with or committed to their projects, because they typically are. What it says is that the outcome of a specific project will be less tightly coupled to the fate of scientists in the for-profit sector than in the academic sector. This is true not only for the organizational reasons outlined above, but also because in the private sector determining whether a project is a success or not can take an uncomfortably long time. For example, in the life sciences, the value of a drug target or a molecular drug candidate may not be known for five to 10 years after the early scientific research is completed. No one can be expected to wait five to 10 years for their bonus, nor should they. This is true in other technical disciplines as well.

Thus, instead of getting rewarded or evaluated for bottom line success or failure, you get judged on what might be called surrogate performance measures. You may not know whether the new receptor that you found will be useful as a therapeutic target, but you delivered it on time. You may not know whether your remote guidance system will help put a spacecraft on Mars, but it meets all of the criteria set by NASA. Management may not know whether the lead compounds that your team delivered will succeed in the clinic, but you produced 11 in the last year—one more than planned.

In the for-profit sector, scientists are typically evaluated annually. In well-organized companies, scientists and science managers have limited and well-defined goals and objectives for the year, sometimes for the quarter. These goals and objectives are typically divided into multiple categories, among which can be found specific project or scientific objectives. Because projects can last for years, each set of project-related objectives must be finite enough to be evaluated quarterly or annually. Thus, although it is conceivable that a scientist, or more likely a vice president, could have a goal that reads "Develop a new drug against inflammation that acts via newprotein pathway," it is unlikely. The goal is more likely to read, "Develop assays by which compounds binding to newprotein can be screened" or "Validate newprotein as a target for inflammation by Q3." The more complex the project and the longer it takes, the more performance is measured by activities performed rather than results achieved.

In fact, the deeper within the organization one goes, the more removed a scientist's goals and objectives are from actual outcomes or accomplishments. We have seen team leaders struggle to create measurable and meaningful short-term goals and objectives for their scientists. Sometimes the goals include learning a new technique or attending a training course. In a scientific setting, in which important outcomes are the result of the combined efforts of an entire team or combinations of teams, it is often difficult to assign measurable scientific objectives to individuals. Despite this, scientists must maintain their focus on whatever goals they do set with their managers. As noted in Chapter 5, and as we will discuss in Chapter 12, it is the job of the manager or leader to help you tie those goals into the big picture.

Learn to be a team member

In addition to being evaluated by surrogate success criteria, you will also be appraised for your ability to work as part of a team and across or outside team boundaries. The importance of these evaluations in the private sector should not be underestimated. Unlike in academia, where you can sometimes find successful lab directors who have succeeded despite abysmal interpersonal skills, such instances are rare in the private sector. Companies cannot afford to place scientists who alienate others, or are argumentative or hostile, in leadership positions.

Scientists in the private sector ignore this fact at their peril. We have seen highly skilled senior scientists, lacking the most elemental interpersonal skills, end up feeling marginalized and unsupported—without understanding the reasons, either because no one ever told them or because they had been made aware of the problem but were incapable of doing anything about it. If you think you might fall in this category, avail yourself of every opportunity for feedback about your behavior from others.

On a psychological level, being a team member involves relinquishing some autonomy, being exposed to and influenced by viewpoints that may be radically different from yours, and making contributions that benefit the team's work rather than your own. If you do not have ownership and responsibility for your own project, you need to find other ways to stand out or distinguish yourself. This fundamental and natural need for recognition may cause difficulties for scientists in large organizations. The next section suggests ways to gain recognition as a member of a team.

Learn to get recognized for actions that support the group

Scientists have an expectation that individual knowledge, performance, and creativity will be recognized and rewarded (not only do scientists expect this, but so do their managers and supervisors). When Carl reviews a scientist's performance, he considers their contributions in meetings, the quality of their data, the way in which the data were presented, and the knowledge and insight they display in the process. He also reviews how they interact with, help, or seek help from others, along with a variety of other social factors. In many cases, he asks not so much what they accomplished, but rather how they behaved, interacted, and collaborated in the process of trying to accomplish it.

Whereas teamwork is all about sharing, collaboration, and merged effort, the reward structure is all about getting noticed. Schein observes that this dichotomy can be damaging to team efforts: "We create tasks that are group tasks, but we leave the reward system, the control system, the accountability system, and the career system alone. If these other systems are built on individualistic assumptions, leaders should not be surprised to discover that teamwork is undermined and subverted" (Schein 1992, pp. 140–141).

But this need not be the case. As we noted in Chapter 3, the opportunity for individual achievement and recognition and the focus on team objectives are not inimical to one another. If you are a savvy and adept player, you will learn to be successful in both contexts. You can be a team player and get noticed for your own individual achievements at the same time.

Some will never catch on, fearing that they will lose autonomy and power if they share their knowledge and insight with the team. Such scientists try to gain recognition and distinction based solely on their role as independent contributors. These employees become problems in team settings. The difficulty can become serious if their information and contributions are critical to the success of the project.

Because the need for recognition is strong, and because team rewards (pizza and T-shirts!) are seen more as fluff than substance, young scientists may hoard information or ideas and use them as currency to enhance their individual status. In one organization in which Carl worked, the running joke was that you could always tell when a certain scientist got an interesting experimental result because he would be seen literally running out of the lab into the CEO's office. It is not easy for scientists to shed behaviors and needs acquired during the formative years of their education and training.

If you donate your knowledge, data, and insights to the good of the project and the group, you will find that you will be viewed as both a team player and a valuable contributor in your own right. You'll be surprised by how quickly word gets around that you're a team player. A single casual mention by a coworker that you helped them figure out a tough problem (without seeking coauthorship or acknowledgment!) will do you more good than a dozen self-promoting smarty-pants remarks in a lab meeting. You do not need to show your knowledge in secret to the CEO to get noticed nor, when you do so in public, should you do it in a way that embarrasses or demeans others.

The best solution or the first solution?

If your love of science is based on a search for truth and beauty in the natural world, you may have a rude awakening when you move to the private sector. Scientists delving into one scientific phenomenon invariably discover many other related phenomena that may be just as interesting, if not more so. As a basic researcher, you may want nothing more than the opportunity to follow your curiosity from one discovery to the next. However, you will rarely, if ever, have this luxury in the private sector.

Engineers seem to have an intuitive understanding of this reality. Engineering is all about working under a slew of constraints (mechanical, economic, temporal) to create something that works. Basic scientists often feel less comfortable with this mind-set. In many project meetings that I have attended, two implicit agendas were operating. The development group asks, "How can we take what we have now and make it work as a product?," and the research group counters, "We really don't fully understand why this works. As a result, we can't guarantee that it will always work, and so we need to experiment a lot more before we're willing to turn it over to development."

This attitude is less common in established companies, especially in pharmaceutical firms, where multiple layers of oversight, tracking systems, professional project managers, and well-defined review procedures are in place. It is more likely to be found in small organizations with a high percentage of scientists newly arrived from academia.

The sociologist, Herbert Simon, noted that most complex problems are solved using an approach that he called "satisfycing" (Simon 1997a,b). Solving a problem by satisfycing means accepting the first solution that works, rather than seeking the best possible or optimal solution. Another way to put this is that "Organizations are happy to find a needle in a haystack, rather than searching for the sharpest needle in the haystack" (Allison 1971). Satisfycing offers the advantage of limiting the search for a solution to a subset of all possible solutions. Scientists in the private sector who attempt to manipulate all possible variables to ensure the best solution may be wasting their time. In most cases, the criteria for a good solution are not elegance or perfection, but how well the solution solves the problem within the constraints of available money, time, and manpower.

Because of their training and sometimes fastidious nature, scientists may be resistant to the notion of using the first good solution. Some will wish to find many possible solutions and choose among them. In some cases, this may be appropriate. For example, backup compounds for drugs are highly valued and appropriate. The art comes in knowing or judging when to stop. It is doubtful that there are rules to guide this process.

A company scientist needs to be comfortable making compromises between elegance and utility, full understanding and partial knowledge, and certainty and reasonable probability. Learning to choose between the best solution (which you may never find) and the first solution that works (which may not be the best) takes time. Many of the difficulties scientists experience in the private sector arise from the fact that the need to make such distinctions is rarely made explicit for them. Do not assume that because your group is called "discovery" or "research" that your mission is an open-ended one.

Get to the point

Our friend, Doug Kalish, a long-time veteran of the information technology world, recounted the following anecdote.

"In one of my companies, we hired many technical people right out of academia. I would assign them a project and ask that they come up with a solution that we could pass on to our development group and clients as soon as possible. I would then ask them to present their solution to the group. These presentations invariably contained dozens of slides that detailed every step, misstep, and approach that they took, and the conclusion was presented in the final one or two slides. I finally got so fed up that the next time they were scheduled for a project update I told them that I wanted to review their slides first. The next week, one of my staff showed me his presentation, which contained 50 slides. I looked at it for 30 seconds and said, 'See this last slide, the one that you have as your conclusion? Start there. Throw out all of the other slides and tell us where we go from here.'"

When presenting their work, scientists and other technical professionals typically feel the need to recount every experiment and every dead end that they encountered before reaching their solution. We have all sat through seemingly interminable seminars in which the speaker seemed compelled to show every experiment they did since graduate school.

We all want to get credit for how hard we worked and how clever we were to discover the right approach after slogging through all of the wrong approaches. But in the private sector, and maybe even everywhere, no one really cares. What is important is the end result and the data that support it. Your personal scientific journey is probably on no one's mind but your own. This is another difference between academia and the private sector about which no one will warn you.

This is not to say that there are never circumstances in which a recounting of each step in the experimental process is appropriate. In some instances, such as in a small, technically focused team meeting, this may be expected. The key is to adapt your content to the audience and the circumstance. If you are updating a project for senior management, or presenting an overview to the board, lose the history and the personal saga. You will get more credit for being succinct than for being overly inclusive.

You may need to discover whether you have this affliction on your own, because not everyone is as forthcoming or helpful as our friend, Doug. Alternatively, if you are in the habit of overpresenting, you may end up annoying your audience to the point that someone feels compelled to inform you in exasperation. Pay close attention to your audience when making a presentation. Listen and watch, not just for their reaction to your scientific content, but for their reaction to you and how you come across. Your audience's body language and nonverbal cues may be your best early indicator of how you come across. If you sense that they are impatient, it is perfectly acceptable to pause and say, "I just want to check whether I am giving too much detail or background here. Is this level of detail okay, or would you prefer that I move right to the conclusion?" You may discover that you get more credit for being sensitive to your audience's level of interest than for the content of your presentation.

SILO THINKING: THE BANE OF LARGE RESEARCH AND DEVELOPMENT ORGANIZATIONS

The following two quotes are representative of what we hear from executives about communication among different parts of their organizations.

> "Our information systems group spent the better part of two years developing a new knowledge management system, but for reasons that I'll never fathom, they only discussed it with about three users in the entire company during that time. As a result, we're now six months post launch of the system and it is not being used because no one can figure it out."
> "Members of the discovery group start foaming at the mouth every time I send one of my marketing directors to sit in on one of their quarterly meetings. For their part, marketing thinks that the scientists are a bunch of ivory tower idealists. As a result, one of our senior vice presidents has to serve as a go-between and this takes up half of his time."

The attitudes that lead to actions and reactions such as those above are often referred to as "silo thinking." This is a metaphor for the barriers between your group and others in the organization. The result is that each group functions in its own separate "silo" and communicates with other groups only as needed, or not at all. We have given silo thinking its own section in this chapter because its consequences to large science organizations are profound. Thankfully, application of improved self-awareness and improved interpersonal skills, especially in the realm of negotiation, can mitigate its negative consequences.

The principal reason for the creation of multifaceted science organizations is the need for contributions from those with widely different and complementary skills in creating a product or making a discovery. Silo thinking subverts this objective and results in absent or minimal communication among people or groups, delays, mistakes, and, in the worst cases, failure of products or projects. You do not need to understand the psychological origins of silo thinking to do something about it. In the following section, we show you how to recognize when you are succumbing to silo thinking and what to do about it.

Silos in scientific organizations

Perhaps the single most difficult conceptual barrier for scientists to cross in the private sector is that which separates basic or discovery research from the more applied or business-related functions. These include business development, clinical trials, marketing, regulatory affairs, and legal. The following table lists the various domains found in a typical life sciences company and one or more stereotypes (not entirely tongue-in-cheek) of that domain as it might be seen by those working outside of it.

Organizational domain	Example of cultural stereotype
Research	Impractical dreamers; no market sense. They fall in love with projects and never want to stop working on them.
Business development	Inundate research staff with half-baked proposals for in-licensing worthless programs. Conversely, company is unable to out-license its own programs.
Clinical research	Ossified clinicians whose notion of running clinical trials is stuck in the Neanderthal era.
Marketing	Slick salesmen who will try to sell anything to anyone, regardless of whether it works. They try to drive discovery based on market needs.
Regulatory affairs	Compulsive, humorless rule-followers whose job is to place hurdles and impediments in the way of research and development personnel.
Senior management	Have no clue about what's going on in the labs.
Fiscal	These are the people who really make the decisions. They are the reason that no one can do anything.
Human resources	Try to train scientists to learn what they have no interest in learning. Force managers to spend time doing personnel evaluations that no one reads.

The cultural and personal gulfs that separate the domains listed in the table are legend and exist in companies of all sizes and types. Research scientists vigorously object when they feel that representatives of any of these groups are attempting to influence the direction or focus of their work. From the researcher's perspective, the sole function of those outside research is to constrain them in their ability to do their job and shackle them with administrative requirements that make brilliant discoveries all but impossible. Conversely, business-related groups roll their eyes in disbelief when they talk about how strategic decisions are made by the research and discovery teams. In their view, these scientists are a bunch of ivory tower academics who, if left to their own devices, would develop products that have no intellectual property protection, require a 10,000-person, 12-year clinical trial, and address a nonexistent market. These are, of course, extreme views and neither universally nor consistently held. But they are representative of views that we have heard expressed at one time or another in every company with which we have been associated. Inasmuch as these attitudes hamper communication with those outside of one's own group, they are damaging to the organization and its progress.

The following case study illustrates some of these difficulties. It is a composite case derived from my own first-hand experiences and reports from others who were themselves involved in the experiences.

■ *Case Study: The Pet Project*

Alex is the program manager of a project at Mammoth Integrated Technologies to develop a new type of synthetic heart valve. This is a "pet" project of Shireen, the vice president for research and development, because she has a materials research background and knows the academic group that made the discoveries on which this program is based.

The chemistry group has synthesized modest amounts of several new polymers and the research team has been animal-testing valves made from these compounds. The results look promising, and Alex has petitioned the project review committee for authorization to move the program into the next phase of development, which would involve manufacturing more valves and getting the clinical groups involved to start planning for human testing if the animal studies progress.

This is the first meeting in which senior representatives from regulatory affairs, legal, and clinical trials are present. The chief executive officer (CEO) has also asked that a member from marketing attend.

The meeting opens with Shin, the lead scientist of the team, summarizing the progress of the program. Shin has been speaking for about ten minutes when the company's legal council, Sara, joins the meeting. After briefly listening to the discussion, she asks whether one of the valve manufacturing processes being used requires a license from the company that devised it. The head of manufacturing points out that they have used this method for more than nine months, reporting progress monthly. Sara replies, "This is the first project meeting that I have been informed of and it's not my fault that no one involved me earlier in the process." Alex tries to intervene, but then Sara's cell phone rings and she steps out of the meeting and does not return.

Alex recalls that he had asked the vice president for research whether to include a legal representative at these meetings. The vice president cautioned him that this was not necessary at this early stage because "Sara has a way of bogging down project meetings in unnecessary details."

Shin continues with his presentation but is immediately interrupted by Oren, a physician and manager in the clinical trials group. Oren asks Shin to define the specific patient population for which he anticipates the new valve will be appropriate. Several members of the research group jump in with different ideas but for each idea, Ira, the vice president for marketing, weighs in with skepticism in an offhand and dismissive manner. By now, Alex and the other researchers are exchanging "I told you so" looks. Shin sits down in frustration, feeling that the other groups were more interested in finding fault with the program than in hearing about his group's progress.

During this exchange, Alex resolves to himself that he will go to Shireen privately to argue for authorization to continue with the project as planned. He concludes that he will

never get this group to agree and it would be best to simply forge ahead and allow the group to deal with results as they unfold.

Indeed, the next day, Alex meets with Shireen and convinces her that the clinical and marketing representatives are nitpickers and cannot see that the project could lead to the most innovative product Mammoth has ever produced. Alex argues that the scientists are frustrated by the risk-averse stance of marketing, clinical, and legal and his group wants the chance to follow their scientific instincts. Given her personal interest in the project and her desire not to further frustrate the research team, Shireen agrees to allow the project to move forward.

We do not know whether the decision to move forward with the project was the right one. However, we do know that potentially valuable input from clinical, marketing, legal, and other groups was never properly integrated into the decision-making process. If the project moves forward without their participation, the team loses valuable input and may find itself with extra work to compensate for bad planning.

What happened here? Let us make a list of some of the events or circumstances that preceded this meeting.

1. Shireen is the scientific leader of the company but is intellectually bonded to the science of the project, a danger sign that we noted earlier. As a result, she encourages Alex to move ahead without seeking valuable input from others in the organization.

2. Alex and the other researchers have a long history of feeling alienated from and frustrated by marketing and clinical representatives. The researchers tried to keep them out of the loop for as long as possible, so that when they did become involved, they felt marginalized.

3. Because the nonresearch groups could give little prior input, they felt "blindsided" by Shin's presentation and the fact that the project has continued for so long without them.

4. Shin bears the brunt of the resulting frustration during his presentation. He is naïve about clinical trials and marketing issues, yet feels unfairly treated by those whom he views as outsiders with no understanding of the science.

5. Sara, the legal representative, has been actively excluded from any discussion of this project on the advice of Shireen. Shireen has, in effect, set Alex up for the confrontation that takes place in this meeting.

Although it may be unusual to find all of these problems in a single project, most projects manifest one or more of them at some point. Some will read the above account and focus on the various ways in which the case represents an example of failure to adhere to good project management principles. But project management, like science and technology, can be known and understood but poorly implemented.

Successful project management requires that the participants engage in face-to-face interactions, conversations, and meetings with those outside of their own areas of expertise. The research team needs input from legal to ensure that they are free to use the required methods and techniques. Conversely, legal needs input from research to plan their strategy for protecting the intellectual property that the project generates. The research team also needs to communicate with the clinical team so that they can decide on the appropriate patient population for their new valves. This decision is of great importance, because it will also influence the kinds of animal models the research team uses for testing. The clinicians need input from the researchers so that they can make the most appropriate recommendation and decide on other possible clinical indications for the new valves. Similar interactions must take place with marketing, regulatory affairs, manufacturing, engineering, and other groups within the company.

Each of these interactions has the potential to create discomfort for the participants. None of them fully understands the others' area of expertise, and each may use language or terminology with which the other is unfamiliar. A research scientist may not really understand the distinction between a phase 2a and phase 2b clinical trial, and a clinician may not appreciate the subtleties of why the valve that he is testing is unstable in acidic media. In addition, seemingly universal concepts such as time, which participants take for granted, may be viewed differently by each. In *Organizational Culture and Leadership,* Schein reviews studies showing that those in sales and marketing view time as having a very different (typically, orders-of-magnitude shorter) horizon than do scientists in research and development. The result is that when one group refers to completing a project "soon," the other group may have its own notion of what that means and neither recognizes the misunderstanding (Schein 1992, pp. 110–112).

As a result, conversations between members of different groups may be full of misunderstanding and false assumptions. If the participants have a preexisting skepticism about the competence, motivation, or importance of one another, the likelihood of a productive discussion diminishes further. Such difficulties are inevitable and even natural in large organizations. Your objective as a member is not to eliminate these difficult interactions, which is impossible, but rather to work through them productively.

Here is our analysis of the causes of the problems manifested in this case study.

1. As vice president for research and development, Shireen owns significant responsibility for the difficulties. She must become aware of her tendency to champion and protect projects in which she has a personal interest. Knowing this will sensitize her to the likelihood that she will yield to the research team's desire to be isolated from outside input. Shireen must also become aware of her attitudes toward members of other groups in the organization, including legal. It is almost certain that she is passing her views and attitudes on to her staff, thereby perpetuating her own biases and further solidifying the silo thinking of all involved. Shireen should address specific issues having to do with Sara's behavior at project meetings directly with Sara. Her reluctance to do this may stem from her inability to have what she expects will be a difficult conversation with Sara.

2. As project manager, Alex has a responsibility to build alliances with the various support and development groups and set an example for the members of the research project team. The ability to negotiate during uncomfortable or contentious situations is perhaps the single most important attribute of a successful project manager. Alex's inability to either anticipate or deal with the strong reactions during the project meeting suggests that he needs to work on acquiring these skills.

3. Both Alex and Shireen failed in their responsibility to educate the research team about the roles of the development and business groups. Moreover, they both encouraged the belief that these groups were mostly a pain in the neck. As a result, the research scientists' skeptical

and dismissive attitude is readily apparent during this and previous meetings, and exacerbates the atmosphere of mistrust and antagonism between company domains.

4. Despite the fact that he is a junior member of the team, Shin had a responsibility to understand the roles of the clinical and marketing groups in his organization. Had he been more aware of these, he might have anticipated the comments of these groups' representatives or, at the very least, taken them in stride.

Had even one of the participants been aware of what was happening from the perspective of silo thinking, and had they taken steps to bridge the differences among the various groups involved, the outcome of the meeting could have been quite different.

The proximate causes for the way the final decision was made are a combination of organizational, leadership, and individual behaviors that can and should be addressed by companies seeking to improve performance. The following section describes an approach to addressing silo-related problems.

Learn to identify silo thinking when it happens

The first step to overcoming silo thinking is to recognize it. Pay attention to your interactions with those from other groups in your organization. If you have difficulty concentrating on doing this during the interaction, make a point of debriefing yourself afterward. This is especially important when you have what feels like an uncomfortable or confrontational interaction. Refer to the first exercise at the end of this chapter.

Recognizing and avoiding silo thinking in science organizations

Ask yourself the following questions
• Do you or others in your group routinely criticize members of other groups in your organization?
• Do you notice a company or group mythology about misguided or inept behavior on the part of members in specific groups?
• Do you try to avoid having certain members from other groups or departments at meetings or research presentations?
• Should or must parts of your organization have input into what you do, but you have never understood their precise role, or you doubt their value?
• Do you find yourself regularly being dismissive or skeptical when someone from one of these other groups offers input?
• Do you find yourself routinely feeling annoyed or argumentative when you speak with particular people from other groups?
If you answered yes to one or more of these questions, it is likely that your work and your organization's progress are being hindered by silo thinking.

It is not necessary to like everyone in your organization. But you do need to be able to work with them. When feelings of animosity toward colleagues influence your professional interactions, your work, the advancement of your project, and the organization as a whole will suffer. However, the detrimental consequences of silo thinking need not result only from outright animosity. If you are ignorant of the function of another part of your organization or aren't on especially friendly terms with any of its members, you may simply ignore them out of habit or inertia. In either case, you have a responsibility to recognize and overcome silo thinking; a good place to start is to review the material in Chapter 2 that shows you how to become aware of your feelings and to anticipate how they might cause you to behave.

Work to build bridges to other parts of your organization

You have a responsibility to build bridges to other parts of your organization whose input and involvement you need in your project. Building bridges is also an effective way to distinguish yourself. It shows that you appreciate the need for input from and collaboration with those outside of your own area of expertise. In addition, it indicates that you are not reluctant to enter into discussions with those who may not fully understand what you or your group does. These are qualities that well-managed organizations recognize and reward.

It will not be possible to change an ingrained pattern of antagonistic interactions over-night. Building bridges involves taking small steps that change behavior and interaction patterns over time, and in a way that allows group members to become acclimated to the changes. Simple ways to start this process include

- asking a member of another group for information about the group's function, responsibilities, and organization
- actively seeking the point of view of a member of another group rather than waiting for them to offer it
- choosing one person within another group as a contact and using him to better understand the perspective of others in the group

The exercises at the end of the chapter list some ways in which you can modify your own behavior to begin mitigating the effects of existing silos. These exercises are based on the notion that as you reach out to other groups in a nonconfrontational, supportive, and open manner, they will eventually respond in kind.

Learn to negotiate in a way that fosters a fair and equitable outcome

Many interactions across organizational domains involve reaching agreements with others who seem to have very different viewpoints and agendas from your own. Circumventing the process by going to a higher authority, as Alex did in the previous case study, almost always results in increased antagonism and typically only postpones the inevitable showdown. Learn to resolve such issues on your own by identifying and focusing on your and the other party's underlying interests (as opposed to positions). By negotiating with equanimity in the face of hostility and stubbornness, you can distinguish yourself from those who either give up or become hostile.

It would be naive to believe that improved interpersonal or negotiation skills will break down every barrier to communication in science organizations. Some in other departments may resist your efforts to build bridges. Others may be unwilling to listen because they sincerely believe that they have no need for your input. Your organization may be led by someone with a pathological personality who intentionally sets people against one another. Fortunately, these are truly the exceptions rather than the rules. Most organizational members like their work and want to do a good job, but have just as much difficulty crossing cultural divides as you. Learning to build bridges across those divides is one of the most important skills that you can learn.

IF YOU ARE A MANAGER

A manager must take responsibility for educating new recruits from academia in the mores of the private sector. Scientists need to understand how decisions that affect their work are made including what scientific and economic factors affect the decision to create or terminate projects, how rewards are allocated, and how promotion decisions made. Review the sections in Mastering the Art of Scientific Research in a corporate setting, starting on page 268, to ensure that new recruits from academia understand the various ways in which science in the for-profit sector, and in your organization in particular, differs from academia.

Help your staff to avoid silo thinking. Pay attention to your own attitudes toward members of other divisions or groups. If you feel negatively about them, you will likely convey this to your employees unintentionally unless you maintain an awareness of these attitudes.

In summary, multiple opportunities exist for projects to succeed, fail, or head down unexpected, sometimes beneficial, paths. In many cases, it is the quality of the interactions among the participants that determines the outcome. Important observations by scientists involved with the performance of assays, manufacture of devices, or testing of new technologies can either be communicated openly to others or kept secret for use as self-promotional tools—it all depends on the quality of the interactions among members of the organization.

A willingness to share hunches, intuitions, and "crazy ideas" does not appear spontaneously upon the creation of interdisciplinary project teams (see the section on Psychological Safety on pp. 180–181 in Chapter 7). The ability of individuals to communicate with others outside of their own domain of expertise will determine the way and facility with which interdomain issues are discussed, explored, or resolved. The effectiveness of such communications can be influenced to some degree by the type of "organizational culture" found in a company. But it is most strongly influenced by the ability of the individuals involved to engage in productive interactions. As a manager, modeling such interactions for your team is part of your job.

REFERENCES

Allison GT. 1971. *Essence of decision: Explaining the Cuban missile crisis*, p. 72. Little, Brown, Boston.

Schein EH. 1992. *Organizational culture and leadership*. Jossey-Bass, San Francisco.

Simon HA. 1997a. *Administrative behavior: A study of decision-making processes in administrative organizations*, 4th ed., pp. 118–120. The Free Press, New York.

Simon HA. 1997b. *The sciences of the artificial*, pp. 28–30; 119–121. MIT Press, Cambridge, MA.

EXERCISES AND EXPERIMENTS

1 Identifying silo thinking

You probably already know who or which groups within your organization that you view with antagonism. Perhaps their values and focus are different from yours. One common divide within companies is between "research" and "development." An important step to becoming a skilled manager is identifying those whom you view as outside of your silo and examining the rationale and consequences of viewing them in this way. Who do you criticize? Who annoys you? The next time you react in one of these ways, try listening instead of judging. Put yourself in their place. Understand their mission. See yourself as they see you.

- Do you or others in your group routinely criticize members of other groups in your organization? If so, list the individuals and/or their groups and indicate how their function is connected to that of your group.

- Do you notice a company or group mythology about ineptness or incompetencies on the part of the members of specific groups? List one for each group.

- Do you try to avoid having certain members from other groups or departments at meetings or research presentations? Name them, their groups, and your reason for avoiding them.

- Should or must parts of your organization have an input into what you do, but you have never understood their precise role? List them and describe what you do not understand about their function.

- For one or more of the groups that you identified above, list three negative characteristics. For these same groups, list three negative characteristics that you suspect they might attribute to your group.

- For each of the groups, list one or two consequences to your organization of poor or strained communications among your group and the others.

2 Counteracting silo thinking

Open a line of communication. Choose an individual in each of one or more of the groups that you identified above. Ask this person two or three neutral information-gathering questions regarding their groups' functions. Listen to their answers with openness and curiosity.

Be supportive when you might normally be hostile. Choose one of the groups from the list above. The next time that you find yourself together in a meeting, say something supportive to someone from that group with whom you might normally disagree or ignore. Refer to Chapter 7 for guidance. You can always find something on which to agree, even if you disagree on the major point. Continue until you get a reciprocal response from that person or from someone in that group. Note whether your perception of that person or group changes over time and whether you notice an improvement in group interactions in general.

CHAPTER 12

Leading Science:
Empathy Rules

|f you have come this far in the book, it will come as no surprise to learn that we believe that the principal determinants of effective leadership are your ability to relate to and understand others (empathy) and your capacity for self-reflection. These are two key facets of what has been called emotional intelligence, discussed earlier in the book, and which many view as the single most important characteristic of effective leaders in all fields (see, e.g., Goleman 1998). Indeed, much of what we have already said in this book is a road map to acquiring emotional intelligence.

There are many books about leadership and it is probably worth your while to read as many as you can. Eventually, you will find common themes that make sense to you and that you can apply to your particular situa-

tion. Some leadership books that we have read and learned from include *Leaders, the Strategies for Taking Charge* (Bennis and Nanus 1985), *On Becoming a Leader* (Bennis 1989), *The Practice of Adaptive Leadership: Tools and Tactics for Changing Your Organization and the World* (Heifetz et al. 2009), and *Leadership Without Easy Answers* (Heifetz 1994).

This chapter is not meant to be a substitute for any of these books but, rather, a supplement. In our view, the type of leadership skills discussed in these and other popular books needs some calibration when applied to scientists and scientific organizations. The reasons for this need are threefold. First, unlike many types of workers, scientists are typically self-motivated. Second, as a scientist yourself, becoming an effective leader may require you to adopt behaviors that may not come naturally to you. And third, the characteristics that make scientific leaders effective may depend on the type of organization in which they find themselves (e.g., academic vs. for profit). So, although there is a lot to learn from the literature on leadership, our focus in this chapter is on those areas for which the processes of leading and acquiring the associated skills are different or unique for scientists.

There is a small subclass of literature relating to the study of leadership in the context of scientific or research and development organizations from which this chapter has benefited: Clarke (2002); Elkins and Keller (2003); Glen (2003); Sapienza (2005); Acton and Andrews (2006); Cuatrecasas (2006); Maccoby (2006); Lewi (2007).

SCIENTIFIC LEADERS: LEADERS OF SCIENCE OR LEADERS OF PEOPLE?

We often speak of a scientist as being a "leader in their field," meaning that they are a recognized authority or perhaps a founder of an area of investigation. In this sense, Watson and Crick were leaders in the field of genetics and Stephen Hawking in the field of cosmology. These scientific leaders helped to define new areas of research or new principles in their disciplines. Whether you are or will become a scientific leader is determined by your cumulative accomplishments, by the opinions of your peers if you are fortunate, and by history if you are not.

Another sense in which the term leader is used, which we call "leaders of people," refers to the type of leadership that is our focus in this chapter. Leaders of people inspire us to act in certain ways, move in certain directions, or achieve particular goals or objectives. Through their actions and words, they may also enable us to persevere in a difficult situation and overcome repeated setbacks and hardships. Examples of this type of leader in the social and political realms are Martin Luther King, Jr. and John F. Kennedy. Whereas these two are often cited as examples of inspiring leadership, we have all encountered people who were less well known, perhaps even unknown to most, who we have thought of as great leaders.

You don't have to be famous to be a leader and you don't have to be at the head of a large organization. The measure of leadership is in the quality of its impact on people, not necessarily on the number of people impacted. Sir Ernest Henry Shackleton, the great South Pole explorer, is often cited as an example of an outstanding leader, yet he commanded a ship carrying no more than 28 people on the voyage that made him famous.

Although authors on leadership often distinguish leadership from management, we think these distinctions, while semantically valid, are not terribly useful from an operational standpoint. For example, it is sometimes said that managers get people to do things right and leaders get them to do the right thing. In our view, good leaders need good management skills and vice versa. At times, good managers need to inspire and motivate and good leaders need to instruct and direct. A leader who focuses only on inspiration, vision, and the big picture, with no regard for implementation or tactics, risks being unrealistic, ungrounded, impractical, or all three. Conversely, a manager who focuses only on the nuts and bolts of getting things done with no regard for motivation, meaning, and empathy risks losing the engagement of those they manage. The following quote by Zia Khan captures this idea succinctly.

A vision without a plan is a fantasy. A plan without a vision is a forced march.
    – Zia Khan, Vice President of Strategy for The Rockefeller Foundation

For the purposes of this book, we define an effective leader of people as one who

- mobilizes and enables people to accomplish objectives or to address their toughest problems (Heifetz 1994, p. 15) and

- helps people to manage difficult situations or navigate challenging transitions.

In this chapter, we ask, "What is the place for people leadership skills in a scientific setting and what are those skills?" Is scientific leadership alone enough for you to weather the challenges of managing a team of scientists? We believe that the answer to this question is most definitely "no." Scientific leadership is certainly not enough when a project is dragging along and your team is demoralized. It is not enough when a project fails or a treasured hypothesis is cast in doubt, and you and others feel dispirited and question your own judgment. In addition, it is not enough when your scientific insight tells you to go in one direction and everyone else thinks that they ought to go in another direction.

In these cases, scientific argument and reasoning alone may not be enough to rekindle enthusiasm or overcome obstacles to progress. Rather, it is your people leadership skills that will enable you to move forward. These skills will be the ones that enable you to help people believe that they will overcome the seemingly insurmountable obstacles in the current project. They are the skills that will enable you to applaud the hard work and perseverance of an individual or team on a project that had to be abandoned for lack of progress. And they are the skills that will make the difference that will allow you to enlist allies in support of your scientific views when the facts alone are not enough. Outstanding scientific leaders are not necessarily successful leaders of people. The skills and behaviors required are distinct although not mutually exclusive. Science-based organizations often seek to elevate accomplished leaders of science to positions in which they are required to assume the mantle of leader of people—perhaps as the head of an institute or of a company. Thoughtful organizations make such choices carefully, cognizant of the fact that scientific leadership by itself is not enough to compensate for dysfunctional interpersonal skills. Often, but not always, great leaders of science acquire or learn people leadership skills as their careers advance and they are thrust into positions that demand more and more people leadership skills. The fact that such learning happens is central to the theme of this chapter and one to which we return shortly.

Sometimes, great scientific leaders never acquire good people leadership skills but continue to be advanced on the strength of their outstanding scientific accomplishments and vision. Indeed, it is the existence of just such people that we sometimes hear as a counterargument to the utility of the kinds of skills we espouse in this book.

Early in Carl's career of advocating for management and leadership training for scientists, he often heard comments such as, "Look at Professor X (a very famous and highly accomplished individual). They routinely pit their postdocs against one another, insult people in lab meetings, and alienate their collaborators. Yet, there's always a line of new postdocs outside their office willing to sign on for three years just to get trained there. Doesn't that prove that it's all about the science?"

Our response to this is often (half jokingly), "Well, if you're as famous as Professor X, you can behave that way too." The fact is that most of us are not as famous or accomplished as Professor X. We're not Nobel Laureates and we need to use every tool at our disposal to maximize our effectiveness and our productivity. Furthermore, we believe that Professor X could be even more productive were they a more effective leader of people. We have previously noted that scientists in training will endure considerable discomfort to be credentialed in the lab of a famous scientist, even one with an unpleasant management style. To use this fact of scientific life as a counterargument to the utility of learning more enlightened people skills is missing the point entirely.

Unenlightened leaders of people are less common in the private sector for reasons discussed in Chapter 11. Simply put, in the private sector too many people have a stake in what you do and how you manage to allow ineffective or clueless leaders to remain in positions of authority; not to say that this doesn't happen but that it happens with a low frequency.

Returning to the theme of the accomplished scientists who learn leadership skills as they are advanced to higher and higher leadership positions, we'll tell you at the outset that we do not believe that leadership is something with which one is born. Although we often hear the phrase "natural born leader," we think that this refers to people who have learned their leadership skills at an early age. This is great news for the rest of us, because even if we didn't learn those skills at an early age, we can always learn them later.

One of the themes of this book is that being effective in your job as a scientist or as a manager or leader of scientists involves adopting certain personal characteristics and practicing a set of behaviors. What distinguishes effective leaders from ineffective ones is whether they embody these characteristics and practice these behaviors and how well they do so. It is not a question of genetics; it is a question of intent and practice. You can decide to relate and behave in ways that make you an effective leader or not. From this perspective, it makes sense that we are less concerned about defining what leadership is than we are about enumerating these characteristics and behaviors. Once they are enumerated, you can go about recognizing and adopting them.

CHARACTERISTICS OF EFFECTIVE SCIENTIFIC LEADERS

What are the characteristics and behaviors associated with effective leaders of people in science? In February 2011, Carl, with the assistance of Danielle Kennedy, PhD, ran a workshop for scientists who had just assumed or were about to assume positions of leadership. The workshop, "Leadership in Bioscience," was held at Cold Spring Harbor Laboratory (CSHL). During the workshop, the participants engaged in an exercise in which they were asked to make two lists: one of characteristics of effective scientific leaders and another for ineffective scientific leaders (for similar lists, see also Clarke 2002; Sapienza 2005).

Below are some highlights from the list of effective leader characteristics. The characteristics or behaviors were generated in the workshop and the accompanying text is our interpretation or expansion. We have divided the list into two groups: personal characteristics, having mostly to do with personality and interpersonal skills, and professional characteristics, associated with the scientific workplace. The order of the list reflects how they were recorded in the workshop, not their relative importance.

Personal characteristics

Integrity, inspire respect

Your communications and interactions with others are honest and you only make commitments that you know you can keep. You make, rather than avoid or defer, difficult decisions and let people know why you made them. You exhibit behaviors that you expect others to emulate.

Empathy and emotional reliability

You have an appreciation for everyone's situation and needs. You convey this visibly by listening when people speak to you and using positive body language to convey your interest. Being empathetic to someone's situation or views doesn't mean that you can or will change their situation or agree with their views. Sometimes it's enough to just listen.

> ### ■ Case Study: Getting a Word In
>
> Pramod is the head of a large academic research group in a major teaching hospital. The group consists of a mix of MDs and PhDs all doing research in neurology. Pramod recently recruited Chen, a new postdoc who joined the lab six months ago. Chen is of Chinese nationality and has had a hard time feeling comfortable in the rough and tumble atmosphere of Pramod's weekly lab meetings. Chen asked to meet with Pramod about this. As she entered his office, Pramod got up from behind his desk, moved to a chair at a small round table in his office, and asked Chen to sit facing him. Chen said, "I don't feel like I can get a word in during lab meeting. All of these MDs talk so loudly and keep interrupting me when I talk about my work. I try to speak up, but they just keep talking over me. I'm worried that my work is not being understood."
>
> As Chen spoke, Pramod maintained eye contact with her and nodded his head in understanding. This encouraged Chen to talk about how she felt at a disadvantage in the face of the loud aggressive men in the lab and that she was concerned that this would affect Pramod's view of her work.
>
> Finally, Pramod said, "I know exactly what you're talking about. You know, some of these guys are very aggressive and they can get loud during a scientific discussion. I try to intervene only if I think we're getting off topic or if people are acting inappropriately. People like this are a fact of life here and I'm hoping that you will become comfortable interacting with them as you settle into the lab. I know it's uncomfortable for you but you shouldn't take their behavior personally. What you have to say is important and I encourage you to be persistent in what you need to say at these meetings. Don't take their behavior personally. Try saying "Please let me finish what I have to say." Can you try this, and we can get together again in a couple of weeks and see how it's going?"

Pramod demonstrated his interest in what Chen had to say by first removing the barrier of his desk and by sitting face to face with Chen. By nodding while Chen spoke, he encouraged her to be open about what was on her mind. Although Pramod didn't offer an easy solution to Chen, he did let her know that he heard her concerns and was willing to talk further.

Another leader might have told Chen that he would play a more active role in moderating the discussions during lab meetings. Pramod felt that this was both unnecessary and might actually work to Chen's disadvantage, depriving her of the opportunity to learn how to work in a lab culture different from that to which she was accustomed. Sometimes, leadership is simply about listening empathically.

Empathy also means that you have compassion for people's pain and discomfort when you present them with difficult choices, decisions, or changes that they must make. Empathy doesn't mean that you have to agree with or approve of what people do or say; but it does mean that you convey an understanding of what they did or why they said it.

Emotional reliability means that your affect or emotional demeanor is appropriate to the circumstance. Your reactions to people and circumstances reflect those people and those circumstances, rather than people and circumstances from your past.

Consistent management style: Neither micromanage nor undermanage

You have regularly scheduled meetings with your group and individually with those in it, and you do your utmost to be consistent in attendance. Your behavior is predictable; you don't micromanage one week and then not review anyone's data for the next month. You make decisions that are yours to make and do not avoid them or hand them off to others. As noted in Benis and Nanus (1985, p. 44), "The truth is that we trust people who are predictable, whose positions are known and who keep at it; leaders who are trusted make themselves known, make their positions known."

Build positive culture and values

Through your own behavior, you set an example of positive and respectful interactions with others, of impeccable scientific integrity, and of a strong work ethic balanced by a respect for your team's personal and home lives.

Communicative, enthusiastic, optimistic

You understand that communication is a two-way process. Not only are you clear in what you say, but you are attuned to, and inquisitive of, those with whom you speak to make sure that you are being understood. You are receptive to others who wish to communicate with you and work at listening to ensure that you understand what was meant. You provide appropriate feedback that is both honest and considerate. You exhibit realistic optimism about the value and outcome of your team's work in a way that gives the team strength and encouragement, and you understand that they depend on you for that. You act as cheerleader for individuals and groups in a way that helps people through difficult or uncertain times. You recognize and manage conflict.

You can identify tensions between yourself and others as well as among others in your group. You intervene in a manner appropriate to your role and position and don't tell people to work out for themselves difficult situations that are legitimately your responsibility.

Professional characteristics

Developer of teams: Receptive to giving and receiving feedback

Your decisions about what projects people work on are based in part on the learning and training needs of those individuals, distinct from your own needs for publication or advancement. You take seriously your role of mentor and teacher and act as an advocate for people when they need your help or guidance.

The following anecdote was related to Carl by a scientist in a biotechnology company.

"We were at an all-company event last month, where some of the senior management were supposed to update us on the company's progress in several areas. At the end of his talk, one of the vice presidents, I won't say which one, said something like, 'We all need to pull together on this project—it's very important for the company and also for me—my bonus depends on it—ha ha—so I'll be paying particular attention to it.' His comment got a laugh, but most of us thought that it was in very bad taste, especially because we knew that this guy really did only care about his bonus. It was kind of demoralizing."

There's nothing wrong with a healthy desire for personal gain, but as a leader you also have an obligation to the advancement and welfare of the people you lead. If your team senses that this is lacking, their motivation may be blunted. At times, this may mean allowing a scientist to pursue a path that you view as risky or ill-advised in the service of their learning from the experience.

Clear focus, vision, and expectations

In this category are such traits as clarity of expectations for those doing the research and reporting to the leader, a clearly focused research program, and a vision for the field of research in which the lab works.

Scientific expertise

Although it is not surprising that this is on the list, it will perhaps be surprising to some that it was neither the first thing mentioned nor especially prominent during the discussion of leadership during the workshop (see also Sapienza 2005 for a similar conclusion). We suspect that scientific expertise is almost a given—no scientist wants to be led in a scientific setting by someone whose expertise and experience doesn't command respect.

Yet, under other circumstances scientific expertise assumes greater prominence. For example, when scientists talk among themselves about which lab to choose for their training, the discussion is frequently heavily weighted toward scientific focus, eminence in the field, productivity (publications, grants, etc.), and the like rather than on the other characteristics in our list.

How can we explain that, on one hand, when asked to make a list of desired leadership characteristics our CSHL group came up with this list, whereas on the other hand, we don't often hear scientists discussing most or any of these qualities when evaluating possible mentors? One explanation is that scientists may be willing to overlook or deemphasize people leadership skills in favor of scientific productivity or eminence to get credentialed in their chosen field. Another explanation is that scientists simply don't discuss leadership (except when complaining about its absence) in the way we did at the CSHL workshop.

But the fact that we don't discuss it explicitly doesn't mean that it plays no role in guiding our decision about which person to choose as a mentor. Here is an example of how this might work, in the form of an anecdote relayed by Carlos, now a professor at a large research university.

■ **Case Study: Junior or Senior?**

When I was finishing my graduate work, I started looking for a place to do a postdoc. There were two people with whom I thought I might like to work, and I set up interviews with each of them. One was a highly regarded senior faculty member at a nearby university whom I'll call Dr. Senior. He had pioneered a new approach to studying membrane structure and had a large lab with maybe six or seven postdocs at that time. Many people thought of him as an imposing and rather stern individual. The other person was a new faculty member at another institution who had just arrived there and had come with a stellar reputation and an impressive publication list—let's call him Dr. Junior. Dr. Junior was just setting up a new lab focused on an area of research that interested me.

There were definite pros and cons to working with each of these people. The pros for Dr. Senior were easy to see—he was already established in the field and very productive. The cons were that the lab was large and it was unclear how much guidance or attention I would get. It was also not clear from our discussions the project on which I might get to work. The pros for Dr. Junior were that I would get a lot of individual attention and could learn a lot in that way. I also thought that I would have a much greater say on what I worked on and the direction of the lab. Because Dr. Junior was younger, I also believed that I might get on better with him. The cons for him were that he was still relatively new to the field.

These were all the (mostly) scientific considerations. But when I met the two of them, other considerations came into play. After meeting Dr. Junior, I had vague uncomfortable feelings about him. He seemed too eager to get me on board. I began to wonder whether he was unsure of himself. I worried that this eagerness might betray other underlying insecurities that might make working for him unpleasant. In the end, this was what swayed my decision. Even though Dr. Senior wasn't that excited about my joining his lab, he seemed to be a bit more stable in some subtle way that I don't think I could have defined at that time. Maybe it was some feeling of predictability and stability that made me more comfortable. In the end, that was how I made my decision.

Now, decades later, I see that I made the right choice. Watching Dr. Junior's career, I have seen that he has been unable to attract the best people to his lab and those he did attract were managed (maybe even micromanaged) with a heavy hand. My gut instincts served me well.

The point of this story is that when making decisions about the course of our professional lives, we factor in considerations other than scientific and technical, whether or not we realize it. The personal characteristics that Carlos found attractive in Dr. Senior (predictability, stability) and those that made him uncomfortable with Dr. Junior (overeagerness, insecurity) impacted his decision on a deeply emotional level. He got a "feeling" about these people that helped him make up his mind. That Carlos was attentive to these feelings indicates that he had good emotional intelligence.

We view our list of leadership characteristics as an operational definition of leadership that answers the following questions: Would you like to work for this person? Do you find this person inspiring? Does this person generate enthusiasm for their work? Do you trust this person's scientific judgment?

Cold Spring Harbor meets Google

A few weeks after this list was generated, we were reading the business section of the Sunday *New York Times* and encountered an article called, "The quest to build a better boss" (Bryant 2011). The article described how Google had undertaken an extensive internal study to define the characteristics of effective managers and leaders at their organization. They emphasized that this wasn't a study to define leadership in general but to determine specifically what works at Google, the implication being that Google was somehow unique. Many organizations like to highlight their unique "culture"—the way in which people interact with one another as well as their beliefs, values, and mutual expectations. There may indeed be characteristic ways in which people interact and behave in some organizations, just as different cultures around the world have their own mores and patterns of interaction. Was there evidence for a unique culture in the list of desired leadership characteristics generated at Google? As we read Google's list, we were struck by the fact that it had, with some leeway for word usage, a remarkable concordance with the Cold Spring Harbor list. The following table shows Google's leadership characteristics (in order of importance) next to the CSHL characteristics that most closely correspond.

	Google's list of desired leadership characteristics	CSHL list of desired leadership characteristics
1	Be a good coach—provide feedback, meet frequently	Be a developer of people and teams. Be receptive to giving and receiving feedback
2	Empower and don't micromanage	Use a consistent management style; neither micromanage nor undermanage
3	Express personal interest in people	Have empathy and emotional reliability
4	Be productive and results oriented. Focus on what employees want the team to achieve; help the team be effective and help with prioritization	
5	Be a good communicator, listen, encourage dialogue	Communicate; be enthusiastic and optimistic
6	Mentor career development	See entry 1, above
7	Communicate vision; get team involved in setting vision	Have clear focus, vision, and expectations
8	Be technically proficient	Practice scientific expertise
9		Build positive culture and values
10		Recognize and manage conflict
11		Have integrity; inspire respect

Thus, 23 young scientists at our CSHL workshop generated in 90 minutes much (and perhaps much more) of what a gaggle of Google consultants pouring over 10,000 observations and 400 pages of interview notes spent one year doing.

Because we imagine that the people sampled in the Google survey were heavily weighted to the technical professional category, it was not surprising to us that the list generated by our CSHL scientists and the list generated at Google looked alike. There are, of course, some differences. One that is especially interesting is Google's item 4, referring to team effectiveness, which is absent from the CSHL list. The Google list was compiled by people in the corporate sector, where a high premium is placed on working as a member of a team. The CSHL list, on the other hand, was compiled by academic scientists who, more often than not, work and are evaluated independently (see Chapter 5).

Also worth noting, and consistent with our discussion above, was that while technical proficiency was also on the Goggle list, the Googlers rated it as last in importance. We don't think this means that technical expertise was not important but rather that it was considered as almost a given in choosing or evaluating a leader.

Recognizing and dealing with conflict

We have discussed the prevalence of conflict avoidance at several places in this book and have reviewed specific examples of how feeling uncomfortable with conflict or how managing it can be a detriment when managing a team or leading a meeting. This theme also assumed prominence during the CSHL workshop as we explored the characteristics of effective and ineffective leaders.

Just as we noted that there were elements of the Google list that were absent from the CSHL list, the opposite was true as well. The most interesting absence from the Google list was the ability to manage conflict (item 10). There is a bit of a backstory to how this entry came to appear on the CSHL list.

The CSHL list was generated during the course of two workshop sessions—one held on the very first day of the workshop and another on the last (fourth) day, to edit and amend the initial list. The list generated on the first day did not contain any reference to conflict management. However, during the next three days, as part of the workshop curriculum, we discussed techniques for recognizing and managing conflict in the context of negotiations and difficult conversations. In addition, before the workshop, we had asked the participants to prepare a brief case study relating to a difficult leadership situation with which they were dealing in their lab. On the third day of the workshop, we reviewed these case studies and attempted to ascertain the *fundamental* issues and themes that they represented. Interestingly, the avoidance of conflict by mentors or leaders was a theme underlying the majority of the individual case studies brought for discussion by the participants. Our informal review of these cases after the workshop suggests that 17 of the 23 cases presented had conflict avoidance by a lab director or leader in one form or another as a contributing theme.

Here is a condensed example from one of the cases.

> Currently in the lab, we have a postdoc who thinks it's OK to grab and put away for their own use common lab supplies such as tubes, buffer, and media. They also never contribute to the preparation of common solutions or media and never reorder supplies that run low. When given a polite request to change this behavior, they respond by screaming and yelling. Our lab director does nothing about this.

Other examples included themes such as

- Mentors failing to take responsibility for making or guiding decisions about authorship on multiauthored papers.

- Mentors failing to intervene or adjudicate in open conflicts among lab members.

- Mentors placing junior scientists in de facto positions of responsibility without providing them with either authority or guidance.

- Mentors giving the message (either explicitly or implicitly) that they do not want to hear about problems or interpersonal conflicts in the lab.

- Mentors failing to be clear about responsibilities in the lab and being unwilling or unreceptive to discussing this.

These are all examples of leaders who have either abdicated their responsibility or failed to recognize that direction and leadership were required. In most instances, the circumstance resulted from or led to a conflict in the lab that had gone unrecognized or unaddressed.

Yet not until these cases were discussed with others and with the workshop facilitators on the third day of the workshop, and we had performed our final review of the list of leadership characteristics on the last day of the workshop, did it become evident that this common theme existed. That is, without the heightened awareness provided by the workshop discussions, neither the workshop participants nor even the instructors recognized that the avoidance of conflict on the part of mentors or leaders was an underlying theme and was in many cases the cause of the dilemma presented in these cases. Nor did any of us notice that the ability to effectively manage or deal with conflict was absent from the list of desired leadership characteristics until the last day of the workshop.

The point here is that one has to be attuned to the existence of conflict and how it is managed to recognize whether it is being managed effectively or even at all. The workshop did this for our participants, so that by the last session, the theme took on a visibility and importance that was absent earlier in the workshop.

What is most interesting about this isn't the fact that conflict avoidance by leaders is problematic or that it is common, because it is both. What is most interesting is that none of the participants knew that this avoidance was central to the problems with which they were dealing in their case studies. When faced with a situation in which a leader abdicated responsibility, either explicitly ("You two work it out for yourselves") or implicitly by ignoring egregious or disruptive behavior, the scientists assumed that it was their responsibility to fix or manage the difficult situation themselves.

When we reviewed the list of desired leadership characteristics on the last day of the workshop, the ability to recognize and manage conflict was added. We feel that it is one of the most important elements on this list.

For our workshop participants, knowing that their leaders and mentors were ignoring or not managing conflict wasn't always a completely satisfying outcome. Some expressed a desire for approaches by which they could solve the problems generated by their leaders' abdication of responsibility. We discussed the related matter of dealing with a passive or absent mentor in Chapter 10 and cautioned trainees against feeling like it is their responsibility to step into the void and take matters into their own hands. In some cases, we could offer the use of the tools of negotiation described in Chapter 3. If lab members find themselves in situations in which their leaders are ignoring a conflict or foisting the responsibility for its management onto lab members, having these tools at their disposal is better than not having them. Indeed, even one person with good negotiation skills in a group can make the difference between resolving a dispute from which the leader has absented herself and a protracted battle of wills among peers.

On the other hand, employees or lab members can feel helpless when their leaders abdicate responsibility for lab difficulties. Even with the best negotiation skills, lab members may find themselves in situations that require the intervention of a leader or authority figure. The primary lesson in this area that came out of our workshop was that when the participants became leaders themselves, they would almost certainly be less likely to ignore or mismanage conflict than their own leaders or mentors.

What your organization needs from you as a scientific leader:
Get with the program

We were mostly pleased with the list of leadership characteristics generated by the CSHL group. However, when we reviewed it further, weeks after the workshop, we noticed that there were in fact quite a few characteristics of what we thought an effective leader should be or do that were on neither the CSHL or Google lists. Both of these lists were generated by people who for the most part were not, or at least not yet, in positions of leadership. That is, the lists were in essence "wish lists" for what people wanted their leaders to be, not what they wanted to be as leaders, or, more important, what the organizations in which they worked needed from them as leaders. This last point is key, because your definition of leadership will depend on where you sit in the organization. If you are the one being led, you may desire one set of traits. If you represent the organization in which the leader functions, you may want other traits.

We have reorganized our CSHL leadership wish list and added a few items representing what your organization needs from you as a leader. But what your organization needs from you will depend to some degree on the type of organization in which you find yourself.

What your people need from you	What your organization needs from you	What both your people and organization need from you
• Have empathy and emotional reliability • Communicate; be enthusiastic and optimistic • Recognize and manage conflict • Use a consistent management style; neither micromanage nor undermanage	• Align work with and focus on company or organization's goals • Deliver organization's goals on time and on budget • Hold people accountable for their work and its quality	• Build positive culture and values • Have scientific expertise • Have clear focus, vision, and expectations • Have integrity; inspire respect • Develop people and teams

The expectations of a for-profit organization such as a pharmaceutical company for its scientific leaders are relatively straightforward. A company might want its leaders to keep its members focused on the mission and goals of the organization. It also might want its leaders to hold its members accountable for performing their jobs according to certain standards of excellence and for performing them within specified times and within budget. A good leader will make these corporate objectives clear to employees, provide feedback to them on how well they are meeting them, and help them find ways to meet them if they are not doing so. A good leader will also inspire employees with a vision of why meeting these expectations is central to their mission and will create in employees a desire to meet these objectives that is as strong as their desire to satisfy their scientific curiosity.

On the other hand, the expectations of a university for its scientific leaders are more subtle. As noted earlier, at some level, the expectations of a university for a research professor are identical to the individual goals and expectations of each professor individually: scientific productivity and funding. Beyond this, the university might expect that a professor follow and promulgate the rules and bylaws of the institution, but that is less about leadership than about good citizenship. So, although we have included a column heading in the table above entitled "What your organization needs from you," the entries therein relate mostly to non-academic enterprises.

Perhaps it is not surprising that these characteristics in the second column didn't show up on either the CSHL or Google lists. After all, although most people would understand the need for holding people accountable, that's probably not the first thing on their wish list for the one who leads them. It's especially not surprising that these items didn't show up on the CSHL list, because the people who generated that list were all scientists working in an academic setting.

Meaning and context

We have one final addition to the list of leadership characteristics. Certain leaders have the ability to frame the task or mission of their group in a way that calls attention to what might be called "higher meaning." This characteristic or behavior didn't show up on either the Cold Spring Harbor Laboratory or Google lists, yet it is a hallmark of leaders to which we respond positively when it is there.

We have all heard leaders who speak of their company's mission in terms of improving the health or lives of those with a specific illness or their university's mission in terms of advancing basic knowledge. When this is spoken with sincerity, it can serve several purposes. First, we think that people feel good about themselves and what they do when they can relate what they do on a daily basis to something greater than themselves. Second, hearing a leader speak about higher meaning can create a community bond among team members who may otherwise feel that all they have in common with their lab mates is the space and equipment that they share.

In a lab in which everyone is working on their own project, it can feel forced or artificial when the lab head exhorts the group to think and behave like a team. But by calling attention to shared higher aims such as curing illness or even advancing basic knowledge in a challenging area of research, the leader creates an implicit bond among group members that can have beneficial effects on cooperation and communication.

Third, we believe that people appreciate hearing it explicitly said that what they do has a meaning beyond data, publications, or advancement. Unlike some other characteristics of leaders that we have discussed, providing such meaning is uniquely the job of the leader. When you speak of the larger context and meaning of your work to your group, you define yourself as a leader, both in your own mind as well as in that of your listeners, in a way that almost no other behavior that we have discussed does.

Bennis and Nanus (1985, p. 39) describe this in somewhat loftier language: "In short, an *essential* factor in leadership is the capacity to influence and *organize meaning* for the members of the organization." As a science or technical professional, speaking about meaning, mission, and higher aims may not come naturally to you. The good news is that even a small amount of such behavior goes a long way. On occasion, saying even a few words about the higher goals and mission that bind your group together may be all that is needed.

Here are a couple of simple examples.

"When I gave my seminar last week at Central Hospital, I got a quick tour of the new children's oncology unit they opened recently. If any of you have ever been to a place like that, you know how important what we're doing is—we may be many steps removed from helping the kids I saw in the waiting rooms there, but just the fact that we're in the game somewhere along the line should make us all feel good about what we do."

Or in a lab even further removed from human applications,

"Personally, I have to tell you all that I'm really proud of what comes out of this lab. What you all do here provides the basis for understanding the fundamentals of living systems. That's enough in and of itself, but you are also providing the basis for future breakthroughs in understanding and treating human disease."

If you observe the response of your group and discern that they are receptive and appreciative, you'll be encouraged to say a little more the next time. We have therefore added the entry "meaning and context" to the column heading "What your people need from you" in our table of leadership characteristics and behaviors in recognition of its importance

What your people need from you	What your organization needs from you	What both your people and organization need from you
• Have empathy and emotional reliability • Communicate; be enthusiastic and optimistic • Recognize and manage conflict • Use a consistent management style; neither micromanage nor undermanage • Provide meaning and context	• Align work with and focus on company or organization's goals • Deliver organization's goals on time and on budget • Hold people accountable for their work and its quality	• Build positive culture and values • Practice scientific expertise • Have clear focus, vision, and expectations • Have integrity; inspire respect • Develop people and teams

Motivation

We often think of leaders as having the ability to inspire and motivate people. Motivation in this context means getting people to focus on or want to achieve something that they might not otherwise want to focus on or achieve on their own. In a work setting, that might involve a leader finding ways to get people excited about a project or a set of objectives or goals.

We believe that most scientists have a high degree of inner motivation to do their work and to understand and unravel the natural world through conceptualization and experiments (see Lewi 2007 for a discussion in the context of drug discovery). At his workshops Carl is often asked by scientists in management roles, "How can I get someone in my group to be more motivated?" His response is often, "I can't tell you how to motivate a scientist but I can tell you how to demotivate them—ignore them, keep changing their goals, never give them any feedback, and keep them in the dark about your plans." Scientists get tremendous satisfaction from doing the work of science, from designing experiments to interpreting data, and the (rare) insights that arise from these activities. Most scientists that we know are deeply motivated to do their work and rarely need the kind of cheerleading that is required of leaders in some other areas of endeavor where intrinsic motivation may be low or nonexistent.

So we might ask, is there a need for motivation from a leader in the realm of science? The answer, as is so often the case, is "it depends." Just because scientists do not need to be given the basic motivation to do the work of science does not mean that leaders cannot influence their enthusiasm and commitment.

There are times when leaders see the need to inspire scientists to contribute more openly to a team or company effort, for example. Glen (2003), in his insightful book *Leading Geeks*, notes that creative technical professionals are motivated more by challenge than by extrinsic rewards such as salary and notes that leaders can build on intrinsic motivation by talking about the significance and meaning of the work. Rather than creating motivation, the leader is building on and taking advantage of the intrinsic motivation of the people they manage. Being empathic helps the leader to know what is important to those they manage.

When faced with a string of failures, young scientists may become discouraged and disillusioned about either their abilities or the scientific course that they are on. Under these circumstances, it's hard to be motivated to do one more experiment, manipulating one more in a seemingly infinite number of variables to get something to work or to answer a question.

Here, a leader can be a source of inspiration by helping scientists to accept that scientific progress is made slowly and that the process can be tedious and repetitive. Leaders often use stories of their own experiences persevering on projects that seemed to drag on forever with no results as a way of inspiring or motivating team members to keep plugging away at a thorny problem. In fact, this approach, called "story editing," can be an effective and powerful tool for helping people reframe or normalize challenging situations (Wilson 2011). In the case of a scientist who is doubting that they have what it takes to succeed on a tough project, the mentor can tell a story about themselves or others in a similar situation, feeling the same way. The point is not to instill a false or unrealistic hope but rather to show that self-doubt and flagging morale are common and natural occurrences for scientists at all career stages and can be dealt with. Such encouragement also falls into the "empathy" category of leadership traits, because it shows the team that the leader understands and cares about the frustration that they may be feeling. Often, such expressions of empathy are all that is needed to remotivate a frustrated scientist.

Whereas most scientists get deep personal satisfaction from their work, they also have the expectation that they will be recognized for their contributions to a project. When this is not forthcoming, scientists may become demotivated despite their affinity for the project itself. An important role of the leader is therefore *not to demotivate*, but rather to ensure that work is structured to allow for individual as well as team recognition and to make it a point to provide that recognition regularly.

Recognition doesn't mean providing positive feedback only when projects or experiments succeed. It is perhaps even more important to provide positive feedback when they fail, because failure often happens. It is highly motivating, especially to a young scientist, to hear from a mentor that despite a notable failure or lack of experimental progress, that they are seen to be making a diligent effort. The following case study illustrates this point.

▪ *Case Study: Motivated or Demotivated?*

When I was running an academic research lab, we started work on a membrane protein whose function was quite unknown. As a result of a collaboration with another group, we had some reason to believe, based on genetic mapping studies, that the protein might be involved in type 2 diabetes. Based on this hypothesis, we launched a major effort in the lab involving three postdocs. We investigated the problem from multiple perspectives, using tools of cell biology, mouse genetics, and comprehensive genomic sequencing. However, after about a year of intense work, it become clear to us that the hypothesis was unsupportable, if not incorrect. The results leading up to this conclusion built up over time and eventually got to the point where the conclusion was clear—we had been working for over a year on a dead-end project. The feeling of disappointment in the group was palpable during lab meetings. Eventually, there came a time when I knew I needed to call a halt to this line of investigation once and for all.

I called a special meeting of the entire lab and we reviewed together where we were in the project and the reasons for stopping this particular line of investigation. We reviewed the approaches we had taken and the logic behind them. I went out of my way to praise each member of the group for their effort and stated truthfully that if I had it to do over again, I would still have taken the same approaches. I told the group that despite the failure to confirm our hypothesis, I thought that this was some of the best science we had ever performed in the lab and that we should all be proud of having done it. I always had the feeling that as a result of this meeting, lab morale actually improved rather than the opposite, which you might expect when a big project needs to be abandoned.

This case shows how a well-timed intervention and interpretation by a leader can forestall the possible demotivating effects of a disappointing outcome. It is also a good illustration of the importance of a leader's empathy. In this case, the leader understood the likely disappointment on the part of the team after cancellation of a project on which they had all worked intensely and that it was not just personal disappointment, but practical as well, because it was certain that hope for publication was no longer possible.

Another leader, lacking empathy for team members' disappointment, might simply have called a meeting announcing the next big project, leaving the team to fantasize about what was really on the leader's mind (that perhaps they were "a bunch of losers") and to deal with their disappointment individually. Although it might be questionable that what the empathic leader did was "motivating" to the team, there can be little doubt that the actions of the unempathic leader would be demotivating. We also discussed this earlier, in Chapter 5, in the context of managing uncertainty in a scientific setting.

Finally, leaders can build on scientists' intrinsic motivation by expressing realistic confidence in the abilities of team members. Leaders who challenge their team members to excel show that they have high expectations for their team. Such leaders can enhance people's confidence and self-assurance and thereby improve their performance. Various terms including "multipliers" (Wiseman and McKeown 2010) have been used to characterize leaders who manage to bring out the best in their team in these ways. The opposite type of leader ("diminishers") may make people feel inadequate by demeaning or humiliating them, not expressing confidence in them, or not giving them opportunities to improve or excel. Savvy leaders monitor their interactions to make sure they are showing the right level of confidence and providing appropriate challenges to motivate the individuals in their team.

ADAPTIVE VERSUS TECHNICAL PROBLEMS: MISS THIS DISTINCTION AT YOUR PERIL

In their book *The Practice of Adaptive Leadership*, Heifetz et al. (2009) draw an important distinction between two very different types of challenges that leaders face. They suggest that some challenges are principally technical in nature—for example, finding a candidate gene or genes associated with a particular disease or teasing out the components of a novel cell-signaling pathway. Technical challenges are best met using the scientific tools at hand or ones that are developed for that purpose. There are known ways to associate genes with medical conditions and to systematically enumerate elements of signaling pathways. This is not to say that these methods always work or that better methods don't need to be developed for specific problems, but even in those cases, the work of new method development is a technical challenge involving hypothesis generation, testing, and evaluation.

Adaptive problems, on the other hand, are fundamentally different. They relate to people, the organizations that they have created, and their attitudes, beliefs, and loyalties. Solutions to adaptive problems often require people to change not just what they do but the ways in which they do them. Adaptive problems almost always require changes or adjustments in how we interact with others, how people behave, or in what they believe. If you are a scientist and not inclined to think about or deal with such matters, recognizing and addressing adaptive problems may not come naturally to you.

If you have a disinclination to think about or an inability to discern problems that are interpersonal or attitudinal in nature, you may mistake an adaptive problem for a technical one. For example, you may believe that the reason your experiments aren't working is that the experimental variables aren't being controlled tightly enough, whereas the real, or at least a contributory, reason is that the technician and postdoc aren't communicating.

If you're blind to the communication problem, you will focus on finding technical solutions—perhaps you will spend weeks having your lab prepare updated and more detailed protocols that everyone must follow. But if you are attuned to the behaviors of and relationships among the people in your group, you will inquire about how the technician and postdoc are getting along and interact, how the technician responds to instructions or guidance from the postdoc, and how the postdoc responds to questions or looks of confusion from the technician.

The solution to this problem may be that the postdoc needs guidance in how to better involve the technician in the planning and execution of an experiment, rather than just handing him a cryptic protocol and expecting the technician to read his mind. Solving that problem requires conversations with people about behaviors, interactions, attitudes, and possibly feelings. Making the experimental protocols clearer and more explicit may also be a big help, but without the first intervention, the second will be only a half measure.

If the underlying problem is really an adaptive one, a technical solution will either not work or will work for a time, whereupon the original problem, or a similar one, will recur. Until the underlying adaptive problem is addressed, you may be spinning your technical wheels to no effect. The following case study illustrates this point.

▪ *Case Study: Damitol Pharmaceuticals*

Damitol is a manufacturer of custom monoclonal antibodies for scientific and medical applications. It is a medium-sized company with 350 employees located in San Francisco. The company is working on a lucrative manufacturing project involving production of a single antibody, slated to be used in a clinical trial commencing in 24 months by a major pharmaceutical company, Optimal Pharmaceuticals. Damitol had signed a contract that requires them to produce the antibodies according to a very strict time schedule so that the trial can begin as planned. There are monetary penalties to Damitol if the production schedule cannot be met.

For the past two months, Damitol has been experiencing intermittent by-product contamination of purified antibody lots in their manufacturing facility. At first, each incident was explained as being an isolated case and ascribed to operator error, even though specific causative errors or "deviations" in the manufacturing procedure could not be identified.

Although Damitol had built into the schedule sufficient downtime to accommodate occasional manufacturing failures, as failures continued, senior management began to get nervous about the possibility of financial penalties to Damitol. Indeed, Damitol was so reliant on this one manufacturing contract that even the delay in a single payment from Optimal would require them to seek interim financial support, something Damitol's board would be reluctant to do.

As the year wore on, contamination increased in frequency to the point at which it was now inevitable that the delivery schedule could not be met and that Damitol would suffer a cash flow problem. Maria, the CEO of the company met frequently with the vice president of manufacturing, Axel, who assured her that everything possible was being done to identify the root cause of the contamination. Despite this assurance, Maria was unconvinced and insisted that Axel bring in consultants to help him solve the problem. Axel was at first resistant, claiming that they were working on the problem vigorously, but under pressure from Maria, agreed to look for someone with the requisite skills.

Axel looked for a consultant who had a track record in the manufacture of monoclonal antibodies for pharmaceutical purposes, and who had experience in dealing with contamination issues. Through his network of colleagues, he identified Shin, a pharmaceutical manufacturing expert who was willing to visit Damitol and investigate their problem.

When Shin showed up, he was assigned by Axel to meet with the five directors of the various manufacturing teams involved with the Optimal product. These meetings went on for days during which time Shin was bombarded with technical reports, spreadsheets, and PowerPoint presentations all relating to the manufacturing process. Shin felt that he was getting nowhere with this approach, so he asked Axel if he could attend some of the group meetings that were focused on solving the contamination problem. To his surprise, Shin was told that there were no specific meetings to do this and that the issue was discussed at the company's routine weekly meetings, along with other business.

Shin attended the next of such meetings and was shocked at what he saw. Axel was in complete control of the agenda and discussion at the meetings. When it came time to discuss the contamination problem, Axel dominated the discussion, dismissing ideas that seemed irrelevant to him and forcing the discussion in directions he favored. When Shin tried to interrupt to ask a question, he was met by a hostile look from Axel who commented, "Look, we're under a lot of pressure here and we need to make some decisions. Why don't you hold your questions until after the meeting?"

During the next several days, Shin attended more meetings and met in confidence with people who reported to Axel individually. He began to see that although there were many capable people working for Axel, their opinions and viewpoints were rarely taken into consideration; they had learned over time not to disagree with him. The result was that Axel was making the decisions in almost every area of the manufacturing process including the contamination troubleshooting.

Shin met with Maria and told her that although he didn't have a magic solution to their problems, he thought that the company's ability to solve the problem was being hampered by the way that Axel was running the manufacturing operation. Shin thought that the people within manufacturing were fully capable of solving the problem, but that their enthusiasm and creativity were being sapped by the authoritarian structure created by Axel. Shin told Maria, "I know you hired me to find a technical solution to your contamination problem, but I've found instead a much bigger problem. Once you solve that problem, I believe we can tackle the technical problem. But unless you solve your organizational problem, an army of technical consultants won't be able to help you."

Maria was at first resistant to this idea. Axel was essentially the second in command at Damitol and she was reluctant to make a change in such an important position. What would the board think? What would they say to their investors? Was Shin sure that there wasn't some other consultant they could bring in to troubleshoot their manufacturing problem? Shin's reply was, "Look, Maria, of course you could bring in more consultants, but frankly none of them will know your manufacturing process as well as the people you have working right here in your company. Whatever the problem is, you have a much greater chance of figuring it out if you can just mobilize and motivate your people to do it. That's not being done now. Axel's management style is demotivating people, and no one wants to risk disagreeing with him."

Maria reluctantly agreed with Shin and put together a plan to replace Axel with Colin, a promising director from the process development group. Colin immediately created a "contamination swat team" from the most capable people in the manufacturing group and within two weeks, the team had developed a list of actions to be taken to eliminate the contamination problem. As these actions were implemented, the frequency of contamination decreased, until after eight weeks, it disappeared entirely.

This may seem like a complex case, but it is actually a very simple one when viewed from the perspective of technical versus adaptive challenges. Maria assumed that she was dealing solely with a technical problem. This may have been because that was how Axel presented the problem to her. Because Maria was a "hands-off" leader, she gave Axel wide authority to manage the manufacturing operation. Unfortunately, Axel did not do the same with the people who reported to him. Axel had his own ideas about what was causing the contamination and, under the guise of time pressure, didn't allow divergent views to be aired or discussed, resulting in a demotivated team. Even when Maria was made aware of this systemic problem, she was reluctant to intervene.

In pointing out the adaptive challenge to Maria—the need to change the leadership of the manufacturing group—Shin recognized that without this change, the technical problem might never be solved. The adaptive challenge was multifaceted; it required that Maria own up to the fact that she may have made a poor choice in appointing Axel vice president of manufacturing, she didn't know how he was managing the manufacturing team, and the people on that team felt marginalized. Once she was able to face and accept these realities, she was able to rise to the challenge and do what Shin, as an outsider, could clearly see needed to be done.

This case also illustrates the distinction between authority and leadership that Heifetz et al. (2009) discuss at length. Axel had authority—that is, he was vested by Maria with the power to make decisions and implement plans—and he used that authority to the detriment of the company. One might argue that he could have, but did not, exercise leadership in that he chose to focus exclusively on finding a technical solution. The fact that Axel himself was an impediment to the solution begs the question of whether he could have, in fact, exercised leadership in this case. But we shouldn't be too hard on Axel. After all, he had done well by the company for several years, when all or most of the problems were in fact technical in nature.

Maria exercised leadership in confronting the adaptive challenge faced by the organization. Recognizing and facing the adaptive challenge was not easy and might not have happened at all had she not taken the advice of Shin. But Maria gets credit for recognizing the truth in Shin's analysis and then acting decisively on it.

It is not hard to see why exercising the type of leadership that Maria exhibited is difficult. In this case, there were likely to be multiple repercussions. First, Maria had to confront her board of directors with a major change in the organization at a time when the company was under intense external scrutiny by the investment community. She also had to risk her own credibility—after all, she had hired Axel. Second, Maria had to have faith that the other managers in Damitol would be able to rise to the challenge and step in to fill the void in authority left by Axel's departure. Despite his sometimes overbearing nature, Axel was well liked by Damitol employees. Thus, Maria risked a backlash to her unexpected dismissal of Axel. Finally, dismissing Axel was no guarantee that the technical problem would be solved. Maria faced the possibility that not only would she have radically shaken up the organization, but that the technical problem might still persist. All of these factors might have been enough to make a hesitant or insecure leader do nothing and simply hope for the best.

Of course, not every technical challenge has an underlying adaptive component. Often in science, the technical challenges are multiple and significant, and although there is an appropriate organization and leadership structure in place, the problems can persist. However, it is always a good idea to pose the question of whether there is an adaptive component to the problem. The table below suggests some questions to ask.

Technical problem or adaptive problem?

- Is the technical problem one that has persisted for longer than usual and for which all the "standard" approaches have failed?
- Have you solved the technical problem but have a persistent sense that you haven't seen the last of it, or do other similar problems keep cropping up after you've solved this one?
- Do the problem-solvers feel that only certain types of solutions can be discussed or that there are unspoken constraints on how the problem can be addressed? An example from the previous case study might be that no one in Damitol's leadership team felt comfortable bringing up Axel's management style as a possible factor in the company's inability to solve the contamination problem.
- Does the group have the resources it needs to address the problem? If not, why not? Is the organization withholding resources as a way of limiting the kinds of solutions that can be implemented?
- Is there a sense that certain aspects of the problem cannot be discussed during problem-solving sessions? You can often tell that this is that case if certain questions or topics "fall flat" in the room when asked. The leader may be sending subtle messages that certain topics (perhaps the most contentious ones relating to behaviors or interactions among people) are "off the table" for discussion.

Some examples of adaptive problems disguised as technical problems in a scientific setting include the following.

Technical problem/solution: You've been working on a scientific problem for a long time and have not made headway. You feel as though you have exhausted all possible approaches but must continue plugging away because the solution may be just around the corner. The technical solution is to find new ways to approach the problem.

Embedded adaptive problem/solution: You have invested a lot of money and time in this project and many people know you're working on it. There's a part of you that thinks you ought to abandon it and put your efforts into other projects but you don't want to risk being viewed as a quitter or being seen as having made a bad choice in investing in this project. The adaptive solution is to address your underlying need and motivation to continue working on the problem as well your resistance to putting it aside and moving on to something that might be more productive.

Technical problem/solution: You continue to have disagreements with your collaborator about the focus and data to be included in a paper that you are writing together. The technical solution is to focus your discussions on the merits of each figure and data table and how the data are to be represented and hope that you will reach an agreement about each element one by one.

Embedded adaptive problem/solution: You each have had a different idea about what the real focus of the paper should be but you have avoided talking about that because the last time you discussed it, you ended in an argument. The two of you now focus exclusively on the data while avoiding the larger underlying problem. The adaptive solution is to revisit the contentious discussion about the focus of the paper in a way the addresses the underlying interests of you and your collaborator (see Chapter 3 for negotiation strategies).

Technical problem/solution: Google is hiring new managers at a rapid pace to keep up with its expanding business. Google's senior managers and executives feel that because of this expansion, many new managers have poor or limited managerial and leadership skills. Therefore, Google commissions a survey to define what traits are most desired by their workforce for their leaders in the hopes that new managers will adopt these traits.

Embedded adaptive problem/solution: Google's technically focused senior leaders and managers feel much more comfortable reviewing code, developing market strategies, and reading consultant's reports than acting as mentors for new managers. The adaptive solution might be to coach senior managers and leaders so that they behave in ways that new managers can observe and emulate.

We suspect that anyone with a technical bent, who is more comfortable dealing with technical issues than interpersonal or organizational ones, might be especially prone to mistaking adaptive problems for technical ones. It is not always easy to know when you are avoiding an uncomfortable adaptive challenge by focusing on a technical fix.

I recall a young female investigator, let's call her Sandrine, who was approached by her department chairman, Harold, at a medical center with the possibility of collaborating on a project. The chairman's line of work was quite different from hers, but Sandrine had access to powerful biological tools that, if adapted, could be applied to the chairman's problem. Sandrine felt torn by the request. On one hand, she wanted to be seen by the chairman as helpful and collaborative, but on the other hand, she had very limited staff and resources to devote to anything but her own project. She spent weeks agonizing over how she could adapt her lab's work schedule and plans to incorporate the chairman's project but to no avail; she couldn't solve the technical problem of doing more work than she had the resources or time for. The only technical solution she could envision was if the chairperson would fund a postdoc researcher to do the work in her lab, but even then, the additional time it would require for her to train the postdoc to do the work seemed daunting.

When Sandrine discussed the problem with a close friend, the friend said, "Your real problem is that you're just starting out in this department and you need to focus all of your efforts on getting your research program up and running. At the same time, you feel obligated to give the chairman what he wants but you can't tell him the reality of your situation—perhaps because you think he will be angry with you."

Sandrine's friend had identified the adaptive problem—that Sandrine didn't want to have a difficult conversation with the chairman. It may also have been that Sandrine felt angry at the chairman for even making this request—after all, she was in a vulnerable situation as a new person in the department and may have felt some degree of implicit pressure to accede to the request. All of these factors made having a realistic conversation with the chairman unpleasant for Sandrine to contemplate, causing her to focus her energy on trying to find a technical way to work his project into her schedule.

Without the intervention of her friend, Sandrine might never have taken the logical step of having a conversation with the chairman about the realities of her situation. Sometimes, perhaps the majority of times, it takes an outsider, such as Shin in the Damitol Pharmaceuticals case, or Sandrine's friend, to identify the underlying adaptive problem.

LEADERSHIP IS A ROLE

The test of a first-rate intelligence is the ability to hold two opposing ideas in mind at the same time and still retain the ability to function.
— F. Scott Fitzgerald

If you are a scientist leading other scientists in the private sector, you may not always agree with corporate decisions about which programs get support and which get dropped. It may be that a program in which you strongly believe is slated for a reduction in funds or even termination. However, in your role as a manager or leader, you have a responsibility to support the mission and decisions of the organization as a whole, although you may have personal doubts about the wisdom of those decisions. The time and place to express these doubts are either in private with the person to whom you report or in a meeting in which the merits of specific programs are being debated. The following comment was made to Carl by Lisa, a team leader in a top-10 pharmaceutical company:

"Everyone in my team was really disappointed when our program on inhibitors for the c-ruf signaling pathway (name changed for this story) was terminated by the executive committee. All week, my people were complaining to me about how much time we all had put into that program and about all the promising leads that we had. The problem was that I agreed with them. I was really torn about what to say—should I tell them, 'Yes, it was a stupid decision by people at the top who don't understand the science, but there's nothing we can do about it,' which is what I really felt?"

As a leader, Lisa is playing multiple roles at once. She seems most concerned about keeping her team motivated in the face of disappointment. In this role, Lisa wanted to empathize with her team (one of the traits that individual scientists value most highly in their leaders). By showing the team how much she cared about their disappointment, Lisa could strengthen her bond with the team and possibly help them to deal with their disappointment. The characteristics of this role show up in the "What your people need from you" column of our leadership characteristics table (see p. 296).

However, Lisa is also playing another role, that of a leader in the organization. She is required to ensure that the organization's mission is accomplished (see the second column in the leadership characteristics table), even if she doesn't agree with its every detail. Characterizing the decision as inane and maligning others in the organization is not an appropriate nor productive way to deal with disagreement. It sends a divisive message to her team, reinforcing the "us versus them" mind-set that is characteristic of silo thinking discussed in Chapter 11. Ultimately, the effect of such pronouncements will be to diminish trust within the organization. Here is what Lisa might say instead:

> "I realize that you are all disappointed in this decision. I was part of the group that reviewed this project and argued strongly in its favor. However, the group felt there were other, more pressing, projects that needed attention. I am disappointed as well, but I understand the basis on which the decision was made. Do any of you have any questions or comments about this decision? [She waits for people to speak and answers a few questions.]
>
> Now that it's decided, we all need to put this behind us and move on to the next project. I am personally aware of the hard work that each one of you put into this project and think that you should all be proud of what we accomplished despite this disappointment."

By making a statement like the above, Lisa meets the needs of her team for empathy and recognition and also fulfills her role as a leader in the organization by helping her team to focus on a new project. Importantly, Lisa allowed time for her team members to express their thoughts and disappointment and ask questions, further demonstrating to them that she cares about what they are experiencing and feeling.

Here is an anecdote from Alan, another team leader, about a different situation but with the same underlying theme.

> "My lab has eight people in it, ranging from grad students to postdocs. We work with miniscule amounts of radioactivity and the typical garden variety of toxic substances that you would find in any biochemistry lab. It's really hard for me to make my people pay attention to proper biosafety protocol. They just run into the lab to check on a gel or a reaction with a coffee cup or can of soda in their hand. Personally, I don't think it's a big deal. It's not like they are laying a pizza out on the lab bench or anything like that. The problem is that our department safety officer is a real pill. If the officer sees someone walking out of the lab with a soda, they immediately makes a major case out of it. I tell my people that they need to humor the safety officer and to peek out into the corridor if they're going to exit the lab carrying a coffee or something. There's no point in provoking the safety officer."

In this example, Alan, like Lisa above, wants to play the "good guy" and be empathic with his lab members, who don't see the need to follow all the lab safety guidelines. However, just as with Lisa, Alan has a responsibility to his department and institution. If Alan disagrees with how safety guidelines are being enforced, he needs to bring that up with the safety officer or the institution. Encouraging lab members to circumvent the guidelines is inconsistent with Alan's role as a leader in the organization and creates an atmosphere of cynicism. By doing so, Alan is avoiding his responsibility to the organization, an essential aspect of his role as a leader.

The notion that as a leader you are playing a role is an important one. Bennis and Nanus (1985, p. 56) use the phrase "creative deployment of the self" to describe the way in which leaders use themselves, their behavior, and their interactions with people to inspire and get things done. Leading is not about acting or behaving according to how you feel; it is about acting according to what your team and organization need from you. As the examples above illustrate, being a leader sometimes means having to do or say things that don't come naturally to you.

On occasion, you may feel that your various roles are in conflict. In your role as manager, you have a responsibility to your organization that at times may seem to conflict with your role as mentor. Although there are no simple solutions to such dilemmas, you will always do better to think through your various roles and responsibilities and what they require of you. Without such analysis, you may end up ignoring key responsibilities, be they to your organization or its people, that come with your role as leader. As the quote from F. Scott Fitzgerald at the beginning of this section indicates, you will inevitably need to learn how to live with and manage conflicting ideas and responsibilities as a leader. Even if your organization requires you to do something that you feel goes against your responsibilities as a mentor, you can implement it in a way that helps those affected to understand the underlying rationale and that demonstrates your empathy.

WHAT MAKES A LEADER EFFECTIVE?

Emotional awareness is what separates good leaders from good managers.
– Zia Khan, Vice President of Strategy for The Rockefeller Foundation

Leadership is perhaps the most complex and subtle topic that we have covered in this book. It inevitably means different things to different people. In the context of science, leadership has multiple meanings. At one end of the spectrum are the largely technical or scientific elements, for example, charting a new or challenging path for a research project in a way that engages and inspires others. At the other end are the more interpersonal elements, such as helping a team to recognize and overcome the psychological consequences of a stagnant research project. The leader's skill lies in knowing what type of leadership is required for each circumstance and, at times, for each person in the team. It will be the leader's empathy—the ability to read and respond to the emotional needs of the members of the team—that will guide the choice of leadership behavior in each case. Being empathic does not mean being swayed by emotion. You can be attuned to the feelings of others, and let them know that, and still make decisions based on objective criteria, reason, and fairness. Some team members may be less inclined to experience feelings of self-doubt during difficult times and may need only technical guidance and support from the leader. Others may need periodic encouragement and positive talk from the leader to help them rekindle enthusiasm and maintain feelings of self-worth. However, it is not just your team members who need this kind of support.

The self-aware leader knows when they themselves need guidance, mentoring, or simply a trusted friend or colleague to use as a sounding board when a tough decision needs to be made. Do not fall into the trap of thinking that just because you are the leader you don't or shouldn't need advice, guidance, or a sympathetic ear when you have questions, doubts or insecurities. You show strength when you can admit you do not have all of the answers and humanity when you can acknowledge your own uncertainties. Find a peer or mentor who can support you in your role as leader and cultivate that relationship as you would a rare orchid.

Although listing the characteristics of effective leaders, as we have done in this chapter and as Google did in its survey, is useful, the list does not tell you how to go about acquiring these traits or how to know if you are leading effectively. Use the Exercise at the end of this chapter as a tool to help you ascertain where your leadership skills need improvement and whether they are improving over time.

TEN ESSENTIAL CHARACTERISTICS OF SCIENTIFIC TEAM LEADERS

In the following we summarize some of the key elements of leadership that derive from the themes and techniques we have developed in this book. Although this list isn't unique to the scientific setting, it contains elements of leadership that we think are especially important in that setting. Following each element, we have given two examples, the first in an academic setting and the second in a company (biotech or pharma) setting. The elements fall into three categories: Managing Yourself, Managing Others, and Integrating with Your Organization. This last set of elements is especially important in the business world where company goals play a much more influential role in the work of scientists than they do in academia or in private research foundations.

Manage Yourself

1. Leaders are self- and other-aware.
2. Leaders model behavior.
3. Leaders hold and manage uncertainty.

Manage Others

4. Leaders create a safe environment for innovative thinking.
5. Leaders ensure that team members know the criteria for success.
6. Leaders share plans for the future.
7. Leaders help teams celebrate success and learn from failure.

Integrate with Your Organization

8. Leaders provide interpretation and context for team members.
9. Leaders keep team and management in the loop.
10. Leaders provide liaison with the organization.

1. **Leaders are self- and other-aware.**

   a. Aware of own feelings in real time

   b. Able to anticipate and intercept their own counterproductive behaviors (passivity, insulting or dismissive behavior)

    c. Aware of own hot buttons

    d. Aware of other's reactions through words and body language

Situation	Behavior	Example	Impact
Self-aware. A postdoc stubbornly resists taking the leader's suggestion regarding an experimental approach.	Be aware of your own reaction and feelings ("I'm angry. We've had this conversation three times now"). Take a brief time-out if needed to let anger subside. Focus on the problem not the person.	"I respect your viewpoint. How about trying it my way and seeing how it goes? I'll commit to revisiting the approach once we have some preliminary data. We can change track if necessary."	Leader doesn't react to frustration and focuses on a path that gets the work started.
Other-aware. A project team member isn't participating in the discussion and looks disengaged.	During interactions the leader watches participants for out-of-team behaviors.	"Alice, you haven't said anything yet. What's your perspective on this?" (open invitation rather than demand).	By being sensitive to a team member's body language and affect, the leader facilitates engagement and participation.

2. **Leaders model behavior.**

    a. You get the behavior you exhibit.

    b. Set the standards for your group.

Situation	Behavior	Example	Impact
A collaborator is trying to get a much larger share of a joint project budget than you think is appropriate.	Go into win–win negotiation mode. Don't get caught in turf battles. Model collaborative attitude.	"Ian, I agree that resources are tight. What if we start by mapping out our respective manpower needs based on the project plan and see where that gets us?"	By showing the team how to be collaborative in a difficult situation the leader sets behavioral standards for team interactions.
A company team member says something derogatory about the way senior management makes decisions.	Empathize if appropriate and clarify if possible, never join. Model positive attitude.	"I know this decision-making process is frustrating but typically senior management has information that we don't, and that influences their thinking in ways that it's hard for us to appreciate."	By displaying a positive attitude, the leader helps members work through situations where they have little control leading to feelings of powerlessness.

3. **Leaders hold and manage uncertainty.**

   a. Your job is to manage ambiguity without denying it.

   b. Acknowledge uncertainty when it exists and develop plans for dealing with it.

   c. Share plans and contingencies with team members.

Situation	Behavior	Example	Impact
The team has heard that a renewal application for the big grant that funds your research got a low priority score. You need to keep them motivated.	Acknowledge the disappointment and uncertainty, even if there's nothing you can do about it. Help team maintain focus.	"I know this is a big disappointment for all of us. I read through the critique, and I'm certain we can address the concerns raised. We need to continue to build on our data so when we resubmit the proposal it will be that much stronger."	By acknowledging the uncertainty and validating members' concerns the leader can help the team to refocus on its work.
A partnership with another company that was supposed to fund your project has fallen through.	Acknowledge frustration; don't look for easy answers. Share information and contingency plans.	"This is really disappointing for all of us. We did put some contingency funds in this year's budget for this project, although not at the level it might have been with the added funding. Let me get the exact numbers and we can meet tomorrow to map out our revised work plan. We're still committed to this project."	Acknowledging disappointment, sharing budget information, planning for contingencies, and involving the team in new plans help them manage uncertainty.

4. **Leaders create a safe environment for innovative thinking.**

   a. Demonstrate respect for all ideas.

   b. Delegate, don't micromanage.

   c. Release control to teams when appropriate.

   d. Encourage risk, tolerate failure.

   e. Spread decision-making around.

Situation	Behavior	Example	Impact
You and your postdoc cannot agree on which cell lines to use to test a CRISPR-mediated gene knockout.	Ask clarifying questions rather than being dismissive. Be willing to believe that their point of view may have merit.	"Jennifer, I'm not sure I see how the CHO cells would yield useful data. Can you explain your rationale one more time?"	By not dismissing ideas that seem irrelevant or incorrect out of hand, the leader shows the team that it's OK to take chances and think out of the box.
You have asked your team to make a decision on which model to use for a new antiangiogenesis compound screen. They make a decision that you disagree with.	Probe the team for their rationale. Don't authorize team to make a decision unless you're willing to try it.	"This isn't the approach I would have come up with, but I respect the process we agreed to. Let's give it six months and see how it goes."	Team members will feel empowered and will have ownership of the outcome.

5. **Leaders ensure that team members know the criteria for success.**

    a. Project advancement

    b. Individual advancement, promotion, and recognition

Situation	Behavior	Example	Impact
You bring two postdocs together to discuss a joint research project. You want to clarify the ground rules for authorship when the work gets published.	By having this conversation before the fact, you ensure that people know what to expect from joint projects and collaborations.	"Alan and Sharon, let's agree from the outset on how we will determine authorship of any work that comes out of this collaboration."	People have clarity about how their work will be recognized and rewarded.
You have been approached by a team member who says, "I've been here two years—where's my promotion?"	Know and communicate the criteria for advancement in your organization.	"Fran, just being here two years doesn't automatically qualify an employee for a promotion. Let's review the guidelines... ."	When team members know how they are being evaluated they know what kind of behavior and performance is expected of them if they wish to advance.

6. **Leaders share their plans for the future.**

    a. Scientific plans for the lab or group

    b. Plans relating to the department or organization

Situation	Behavior	Example	Impact
You're thinking about starting work in an experimental area that's new to your lab. It may require changing some people's research focus.	Be honest about your plans and the reasons for them, including their impact on team members. If possible, seek volunteers for the new project.	"I've been thinking about moving into neuronal stem-cell identification for a while and I think now is the time for reasons I will share with you. I need one or two of you to begin the preliminary work that will probably take time away from your main project. If you're interested, let me know and we can review the pros and cons for your work and your future plans."	By being up front about your plans and by giving team members choices on whether or how they get involved, you create trust and a sense of agency in the team.
You're about to start work on a joint project with another company.	Advise the team on the expected longevity of the project and possible outcomes.	"Management has given us two years to show preclinical proof of concept in this program. I've got a good feeling about this project, but if it doesn't pan out we've got a long list of other targets to focus on."	By sharing what you know and being honest about what you don't know you build trust in the team.

7. **Leaders help teams celebrate success and learn from failure.**

    a. Recognize and applaud failed projects just as you would successful ones.

    b. Hold postmortems to learn from your organizational mistakes.

    c. Focus on lessons for the future not on "what we should have done."

Situation	Behavior	Example	Impact
The team has been unable to link a mutation in a target gene with a phenotype they've been focused on.	Review what the team did and how it was done. Tell the team what it did well. Review whether the process could have been improved.	"I saw each of you working hard on this project. Juan, you did a great job setting up the assay. In retrospect, would we do anything different in the future?"	Team members get credit for their work whether or not the project succeeds. Team members probe to learn lessons.

(Continued)

Situation	Behavior	Example	Impact
The executive committee has agreed to advance your lead molecule to the preclinical stage of development. This is a big win for your group.	Take the time to review how you got here. If possible, highlight what each person did and how that advanced the effort. Don't let it go at "Good work!" be specific with your feedback.	"We knew we were taking a risk with this molecule, but your hard work proves us right. Sven, your work with the Medicinal Chemistry group really paid off. Amanda, the screens you set up for us really paid off."	By giving timely and specific feedback to your team you demonstrate that you know and care about what each one is doing and help overcome the challenges some scientists face when working in a team.

8. **Leaders provide interpretation and context for team members.**

   a. If you do not explain and interpret events someone else will and they will likely get it wrong.

   b. The leader puts the group's work in context and thus provides meaning.

   c. The leader validates the importance of the team's work.

Situation	Behavior	Example	Impact
You just got a new grant funded that will allow you to hire three new postdocs.	Provide contextual information for the team about what this means for them and the group as a whole.	"This grant will allow us to expand into a new area of research and add to our team. I'll be asking all of you to help me chart the path forward and to help select our new team members. There will be opportunities for all of you to learn this new field if you so choose but I'm still deeply committed to our existing projects."	Providing information and context limits rumors and fears and provides reassurance to the team.
One of your company's most promising new drug applications is rejected by the FDA. Employees are nervous.	Review the company's strategy to deal with the situation. Educate the team on regulatory procedures even if it's outside of their domain.	"I've spoken to the leadership team. Like me they believe our filing was solid and that you all did a great job putting it together. I want to share with you what our strategy is for responding to this ruling."	Knowing the plan will reduce anxiety and help the team maintain focus on their own domain.

9. **Leaders keep the team and management in the loop.**

   a. The leader communicates up (senior management, department or institutional agents, the outside world) and down (to team).

   b. Through the leader, the team knows what upper management is focused on, and management knows what the team is doing.

c. In this way, the team is assured of visibility to management and management is assured of the team's focus and progress.

d. The leader ties day-to-day scientific activities with corporate/ department or institutional objectives.

Situation	Behavior	Example	Impact
The leader has participated in a three-day conference on the lab's scientific focus.	Recognize the legitimate interest of team members to know what's going on in the field in which they work.	"I'll start today's meeting by giving you a summary of what was discussed at the conference I attended."	Team members will have a sense of involvement in the lab's focus and will better understand the context of their work.
The leader reviews the team's progress at the company's therapeutic area retreat attended by senior management.	During the report on the team's progress the leader highlights accomplishments or contributions by specific members.	"Our team validated three new angiogenesis targets this year. Samantha was especially helpful with the RNAi studies, and Zhawei did a great job coordinating the histology work."	Team members feel less invisible to those outside of their group. Those outside the group identify and can connect with specific team members, reinforcing their role and accomplishments.

10. **Leaders provide liaison with other parts of the organization.**

a. Work to promote good communication between teams and between science and other parts of organization (development, clinical, commercial, etc.).

b. Be alert to "cultural" differences within organization that can lead to silos.

Situation	Behavior	Example	Impact
A lab member makes a dismissive comment about the Department's Health and Safety Officer who has just inspected your lab.	Intervene to prevent stereotyping and scapegoating.	"I understand that some of you feel that these regulations are burdensome, but Harriet is just doing her job and it's actually an important one. Let me explain... ."	Team members need the leader's help in putting their work in the context of the organization as a whole and in understanding their responsibility to the organization. If you don't take this seriously, how can you expect them to?

(Continued)

Situation	Behavior	Example	Impact
The team is trying to decide on the appropriate preclinical model to use for testing its new anti-inflammatory compound.	Involve other parts of the organization sooner rather than later.	"I suggest we ask someone from Clinical for input on this before we make a decision. The earlier they weigh in, the lower is the chance that we'll have to change models later."	By acknowledging the expertise of those in other parts of the organization the leader brings in appropriate and necessary input.

REFERENCES

Acton GA, Andrews T. 2006. Impeding scientists' productivity. *Res Technol Management* **49:** 59–60.

Bennis W, Nanus B. 1985. *Leaders: The strategies for taking charge.* Harper and Row, New York.

Bennis W. 1989. *On becoming a leader.* Addison Wesley, Reading, MA.

Bryant A. 2011. The quest to build a better boss. *The New York Times,* March 13, p. B1.

Clarke T. 2002. Why do we still not apply what we know about managing R&D personnel? *Res Technol Manage* **45:** 9–11.

Cuatrecasas P. 2006. Drug discovery in jeopardy. *J Clin Investig* **116:** 2837–2842.

Elkins T, Keller R. 2003. Leadership in research and development organizations: A literature review and conceptual framework. *Leadership Quarter* **14:** 587–606.

Glen P. 2003. *Leading geeks.* Josey Bass, San Francisco.

Goleman D. 1998. What makes a leader? *Harv Bus Rev* **76:** 93–104.

Heifetz RA. 1994. *Leadership without easy answers.* Belknap Press of Harvard University Press, Cambridge, MA.

Heifetz RA, Grashow A, Linsky M. 2009. *The practice of adaptive leadership: Tools and tactics for changing your organization and the world.* Harvard Business Press, Boston.

Lewi PJ. 2007. Successful pharmaceutical discovery: Paul Janssen's concept of drug research. *R&D Manage* **37:** 355–362.

Maccoby M. 2006. Is there a best way to lead scientists and engineers? *Res Technol Manage* **49:** 60–61.

Sapienza A. 2005. From the inside: Scientists' own experience of good (and bad) management. *R&D Manage* **35:** 473–482.

Wiseman L, McKeown G. 2010. *Multipliers: How the best leaders make everyone smarter.* Harper Business, New York.

Wilson TD. 2011. *Redirect. The surprising new science of psychological change.* Little Brown, New York.

EXERCISES

Your leadership characteristics: Self-evaluation checklist

Here is a list of questions to answer that can help you to monitor and improve your leadership skills. Copy the checklist so you can use it repeatedly. Circle the number that comes closest to your agreement with each statement. Perform the evaluation and start working on those areas where you scored as needing most improvement (scores of 1, 2, or 3). Retake the evaluation after six months to document your improvement.

	Agree 5	4	3	2	Disagree 1
1. People in my group feel free to express views that differ significantly from what the rest of the group is saying or thinking.					
2. I play an active role in lab meetings without dominating the discussion.					
3. I consciously spend time in team meetings observing the flow of the meeting to ensure that we accomplish what we need to accomplish. I also monitor the temperature of the discussion and make sure that the group isn't falling prey to group behavioral traps.					
4. I periodically tell my team what's going on in our institution that they may not know about (especially things that might impact them) and what I hear and learn at scientific meetings.					
5. I talk to my group about the "big picture"—how what we are doing fits into the larger scientific field in which we work and why what they are doing in the lab is important.					
6. I share plans for the future with my team so that they know what projects are coming or being planned, and what our fallback plans are if a current project doesn't work out as hoped.					
7. I regularly talk to my team about lab matters and policies that affect all of them (i.e., equipment or reagent sharing, cleanliness, etc.) and do not rely solely on my lab manager or the members themselves to implement or enforce policies.					
8. I have realistic optimism about what we do and I share this openly with my team.					
9. I show people through my own behavior how to critique results and ideas while being respectful of the person presenting them.					
10. I avoid saying or doing things to others that are intended to be, or likely to be, felt as humiliating or embarrassing, either in private or (especially) in public.					

(Continued)

	Agree 5	4	3	2	Disagree 1
11. I provide regular feedback on performance to my team members, focusing on observable actions and behaviors rather than on personality characteristics. When providing such feedback, I offer suggestions for improvement when needed.					
12. I let people know publicly when they have done a good job.					
13. I can think of at least one recent instance in which I identified a conflicting relationship among people in my lab and tried to explore its origin and what could be done to improve it.					
14. I regularly talk to my team about the overall progress of our projects.					
15. I regularly talk to the individual members of my team about the overall progress of their individual projects.					
16. I provide appropriate guidance and direction to my team members but also allow them the freedom to make their own decisions and mistakes.					
17. I go out of my way to help my team and its members learn from failures and mistakes rather than just showing my disappointment and moving on to another project.					
18. When I interact with people in a professional setting, I work to maintain an awareness of my own feelings and reactions so that I can anticipate and minimize negative or unintended messages that I might convey (e.g., through body language).					
19. I am aware when I start to feel angry or upset and am able to moderate my behavior and maintain a professional demeanor.					
20. I let people know that I care how they feel about how decisions that I, or my organization, make affects them, even when I cannot change those decisions.					
21. I really *do* care about the people I work with and how they feel.					
22. I am attentive to the learning needs of my team members. I provide both encouragement and help in acquiring new skills and learning new techniques.					
23. Whenever possible, I seek to resolve disagreements using the principles of collaborative negotiation, focusing on the underlying interests of both the other person and myself.					

(Continued)

		Agree 5	4	3	2	Disagree 1
24.	I show people that I am neither perfect nor invulnerable by having a sense of humor about myself and by admitting mistakes as quickly as possible.					
25.	I am not afraid to admit when I don't know something or when I am uncertain about a decision I have made, nor do I think admitting this leads people to disrespect me.					
26.	I understand what my own boss or my organization expects and needs of me as a leader.					
27.	When I am uncertain about the appropriate course of action or about a decision that I need to make, I have at least one person who I respect that I can turn to for advice.					

CHAPTER 13

Shape the Future of Science and Technology

A few years ago, the editors of a scientific magazine with a large national distribution asked Carl to write a brief response to two questions: How should senior scientists and principal investigators manage their labs to avoid conflict between lab members? What is the best way to handle strong personalities? When he told them that he would be happy to write an article on these topics, they said that they were only looking for a few bullet points to place in the magazine. He respectfully declined and advised them to read this book.

Learning to be more self-aware and work productively with others is no simple task and can hardly be captured in a few bullet points. Although we have introduced tools and concepts to help you along this path, we do not for a moment believe that it is an easy one. First, changing how you see yourself and interact with others may require you to alter lifelong and deeply ingrained traits and habits. Second, science and technology communities and the organizations and members within them are resistant to change. Yet change can happen, even if only by one person at a time.

PUTTING IT ALL TOGETHER

We present one final case study that illustrates the possibilities for change. The case is presented in more detail than our previous cases because we wish to depict the complexity of real-life situations. It is based on a true story and real people, whose identities we will reveal at the end.

As you read through the case, note the problems or potential problems that you see. By problems, we do not mean simply "who wasn't being nice to whom?" because there's plenty of that going on. Rather, we mean what behaviors do you detect that are likely to limit communication and productive interaction between the characters? What patterns of interaction, management, and leadership (or lack thereof) do you detect that will limit or reduce progress in this project? After all, this is the bottom line. Without putting too fine a point on it, whereas much of what we have said in this book has to do with improving how you relate to others in the science workplace, the underlying objective was never to help you make friends, although that might be a beneficial side effect. The objective was to provide a set of behaviors and tools that increase the likelihood that you will be successful in your research or scientific work. Sometimes those behaviors will seem friendly in nature. Other times—for example, when you have to make an unpopular decision—your behavior or actions will anger and frustrate people. In either case, the underlying principle is that your behavior is in the service of your role as a manager or leader and your task as a science professional, with consideration for those you work with.

You may find it helpful in the following case to categorize the various ineffective behaviors and interactions you encounter in terms of the themes of the previous chapters. Do you find issues related to self-awareness, self-control, peer relationships, mentoring, group management, leadership, negotiation, etc.? In addition, write down how you might have handled those specific situations differently from the participants in the case.

At the end of the case, we present an alternative hypothetical scenario to illustrate how several aspects of the case could have been handled differently by one or more of the protagonists.

▪ *Case Study: Rewriting History*

Susan was a senior investigator in the structural biology unit of USA Pharma. She worked in a group loosely organized under Ed, the head of structural biology. Her project focused on the three-dimensional structure of a growth-factor receptor called PAN that had been implicated in pancreatic cancer. Susan was an ace at X-ray crystallography and was recognized as an expert in receptor structure even before she joined USA. But Susan had problems. Ed noticed from day one that she was secretive and reluctant to share her insights and data with others. Ed was dismayed at this, but took a rather laissez-faire attitude. Her work was superb and he was reluctant to make waves for fear of alienating or losing her. In addition, every time he broached the topic of openness with Susan, she became defensive and angry. Ed and most of the other scientists at USA felt at a loss in these situations and decided to simply ignore the problem and hope for the best. After all, Susan was producing good results, so better leave well enough alone. He was reluctant to bring the problem to human resources because he did not have faith in them and thought that they might just make matters worse.

Athan, another investigator at USA who was several years senior to Susan, had been working on PAN as well. In fact, he was taken by surprise when he learned from Ed that Susan would be working on PAN, but he was reluctant to say anything about it. Ed had managed another project in a similar way once before, but when Athan queried his decision, he denied having any ulterior motive beyond moving the science forward, and he cut the discussion short by saying that members of the lab had to resolve such matters on their own.

Athan's relationship with Susan had been tense. Shortly after Susan arrived at USA, she noted that Athan used a sample preparation technique that she believed was totally inappropriate for PAN-type proteins. She told him this in such a condescending manner—referring to his approach as "amateurish"—that Athan was both hurt and angry for days.

Like others, Athan was uncomfortable with Susan's style: She was direct, argumentative, and to the point, whereas he was much more circumspect and reserved. As a result, Athan and Susan rarely spoke to each other. Although they did not get on too well, they were well aware of each other's data. At one point, after seeing some of Susan's especially exciting data, and despite their earlier clashes, Athan suggested that he and Susan collaborate. Susan turned red in the face and practically shouted that she did not need him to help her interpret her own data.

When Ed got wind of this confrontation, he called them both into his office. He said nothing about their interpersonal clashes. Instead, he told them that to avoid overlap in their work, they must each confine their efforts to different aspects of the PAN problem, which he delineated for them. They agreed, but their relationship never improved.

In October, two investigators in the cancer biology division of USA, Mohan and Ralph, told Athan that they believed PAN was related to another receptor called MBC that they had implicated in certain malignant brain tumors. In fact, they had already developed lead compounds that inhibited the activity of MBC. Unfortunately, these compounds had no effect on tumors in animal models, and Mohan and Ralph were being pressured by Ruth, their supervisor, to abandon the project. They argued against this and asked if they could determine whether any of their lead compounds might be effective against pancreatic cancer before they put the project aside. Ruth would not authorize synthesis of more test compounds, or new animal studies, without more definitive data to support the similarities between MBC and PAN. So Mohan and Ralph wanted to take a closer look at the PAN structural data.

Mohan and Ralph had previously met with Susan on two occasions during the past 12–18 months. Mohan, in particular, was put off by her abruptness and abrasiveness. After each meeting, Mohan came away feeling that a give-and-take discussion would be impossible. More than once, Mohan felt insulted by Susan's comments, and became angry in return, exacerbating an already difficult situation. Eventually, Mohan decided that dealing with Susan was just too uncomfortable and, because of other commitments, temporarily shelved the PAN problem.

For her part, Susan felt that Mohan was a naïve dilettante when it came to receptor structure. Because she had never felt the need to suffer fools gladly, she was dismissive of him. Sharing data was not an issue, but having to deal with someone who tried to tell you your own business was quite another.

Despite the discomfort, at the second of their meetings, Mohan and Ralph were able to ask Susan about the similarity of the PAN receptor to MBC. Susan replied that she had suspected for months that the receptors were related, but that this "didn't mean a damn thing" and she needed more data to establish this definitively.

Now, 18 months later, during a casual conversation about the PAN project, Athan showed Mohan and Ralph both his own data as well as some of Susan's new data. He showed them Susan's slides from a recent team meeting that contained her close-to-final structural model for PAN. After a brief and excited examination by Ralph, they all concluded without question that PAN was of the same class of receptor as MBC.

In a burst of enthusiasm, they met with Ruth who immediately approved the synthesis of more of their lead compounds and sent them off to one of their academic collaborators, who was an expert in pancreatic cancer. Within two months, the results showed that one of their compounds was highly effective at slowing or stopping the growth of this cancer in mice. Based on this information, USA decided to move the candidate drug into human testing. Within a year they had obtained promising positive results.

Athan felt very uncomfortable about this whole process. He knew that if Mohan, Ralph, and Susan had been able to forge a collaboration sooner, or at least had a discussion that did not result in them all going away insulted or angry, a therapy for this deadly cancer might have been available to patients at least two years sooner than it would be otherwise. Moreover, he also knew that the financial implications for his company were huge. During the next six months, Mohan and Ralph were widely credited for their roles in discovering the PAN therapy, and Athan was given credit for assisting them. Susan's contribution was downplayed and Mohan, Ralph, and Athan did not go out of their way to correct this impression. No one seemed to remember that Susan herself had seen the similarity between PAN and MBC before anyone else. Within the year, Susan left USA for another job.

In our version of this story, the two-year delay in the development of the pancreatic cancer therapy would have been unfortunate for USA and possibly fatal for many cancer patients. What could have been done better?

Here are some of the interactions or behaviors that we have addressed earlier in this book.

1. Neither Ed nor anyone else at USA tried to help Susan see how her behavior was affecting her work and USA's progress. Ed was likely conflict-averse and lacked the skills to manage a difficult person such as Susan, who did not have a lot of self-awareness.

2. By putting Susan on the same project as Athan without discussing it with both of them, Ed had set the stage for a confrontation. Even when asked by Athan about the advisability of doing this, Ed refused to discuss the matter. We need to wonder whether the resulting tensions in the group were invisible to Ed or whether he consciously chose to ignore them.

3. Athan responded emotionally to Susan's insults, rather than using the agree, empathize, and inquire or assure approach, which might have enabled him to open a dialogue with Susan (e.g., Agree: "Look, Susan, I readily admit I don't know everything, and I suspect there's a lot I could learn from you." Empathize: "I'm also frustrated that Ed has given us the same project." Inquire and assure: "Is there some way we could work together on this? My only goal is to advance this program.")

4. Hearing of a confrontation between Athan and Susan, Ed focused only on the technical problem of how to divide up the work on the receptor and ignored the adaptive problem of Athan and Susan working on the same receptor and not being able to communicate or collaborate. Predictably, things got worse.

5. Athan shared Susan's data with Mohan and Ralph in Susan's absence, further marginalizing her. Ed was either not aware of this or choose to do nothing to try to get Susan more involved with the expanding project. Perhaps he was being passive–aggressive toward Susan. By now we might also wonder whether Susan was being scapegoated for delaying the project. But as the above analysis suggests, she was not the only one with responsibility for the delay.

Could it have turned out otherwise? In many ways, the outcome to this story could have been different. Below we examine several alternative scenarios, focusing first on Ed's role. Let us take Ed's perspective, and pick up the story just before Mohan and Ralph approach Athan.

■ *Case Study: Rewriting History (revised version)*

For some time, Ed was concerned about both Susan's protectiveness of her data and her hostility. He felt that it was detrimental to the group and would certainly not benefit Susan's career, either at USA or elsewhere. After some thought, Ed realized that his own reluctance to deal with Susan was a big part of the problem. He could not handle conflict. His tendency was to avoid angry people, especially women, at all costs. He realized that in the past, he had let hostilities in the lab simmer and hoped that they would just disappear. He had behaved this way with Susan for two years and concluded that he was doing a disservice both to Susan and the company.

Ed also suspected that he knew the reason for Susan's paranoia about her data. When he hired her, he had deliberately assigned her to the same project on which Athan had already been working for two years. He had tried this approach before with the idea that either the synergy or the competition would accelerate the project. In this case, some benefit resulted, but the tense atmosphere created by the situation had made the whole department uncomfortable. Thus, Ed detected multiple problems, including his poor judgment in assigning projects, Susan's abrasive manner in the lab, and his own unwillingness to intervene.

For her part, Susan believed that others in the group—mostly men—were given advantages that she was denied. She felt that she had to prove herself every time she joined a new company. The only way that she could ever get any respect at USA, she believed, was to solve the PAN problem herself and prove her worth.

After much thought and preparation, Ed called Susan into his office and had the following conversation with her:

"Susan, I know how hard you've worked on the PAN project and I believe I know how important the project is to you. Although I can empathize with you, I'm frustrated by the divisive atmosphere in the department. I know that you get along with many people, but are you aware that you have a tendency to be confrontational with others? For example, in the group meeting last week you told Athan that his sample preparation technique was naïve and amateurish.

This behavior is not in the best interest of the company or you. Frankly, I even had difficulty deciding to discuss this with you because I was afraid that you would get angry. It's not easy for me to talk to people who get angry so I've avoided being honest with you. I'm sorry that I haven't discussed this with you before, but I'm doing it now because you're a very important member of the department and we need your talent and skills.

I suspect that you were frustrated because of the way your project was assigned. Is that true? If so, I take responsibility for that. Moreover, as head of the department, I feel that I haven't been honest with you and I haven't given you the opportunity to grow professionally. I'd like to correct that now. I need to give you some honest feedback about how your behavior impacts your professional relationships. Then I'd like to work with you to help you change that behavior. If you're frustrated with the way in which you have been treated, we can deal with that to see if we can fix it. I'd also like to help you deal with Athan, Mohan, and Ralph in a way that allows you to address scientific differences of opinion without anger or hostility. Do you agree that there is a problem?

I know that you're a dedicated scientist and that getting the science right is your primary motivation. I'd like to try to help you get your scientific points across without others feeling that you are disrespectful of them.

This is not to say that you don't have reason to be angry at times and we can and will discuss the substance of those issues. But right now, the priority is learning how to best express your concerns in ways that give you the greatest opportunity to be heard and have a positive impact on the project."

Ed concluded, "Look, we're all human. If we're angry or sarcastic, we're bound to generate a negative response even when what we say is demonstrably and scientifically true. People may hear what we say, but they may not take it in because they're reacting to their own feelings of being insulted, hurt, or angry. It doesn't have to be that way.

Finally, having said this, I need to tell you that you do need to improve your interactions with others in the company. I am convinced that the difficulties you have in working productively with others slow the progress of our projects, and I can't let that continue. I need your commitment that you will work to improve your professional relationships, and we need to set some specific goals for improvement. I'm willing to work with you on this because I value your contributions. Is this something that you're willing to work on with me?"

We will never know what the effect of such a conversation might have been because in the real case it never took place. But we can imagine. We can imagine that Ed's nonconfrontational discussion with Susan, his honesty about his own feelings, and his taking responsibility for his own role in the situation might have had some impact on Susan. After some time, with guidance and feedback, she might have learned to control her anger and focus her energies on building alliances with some and working collaboratively with others.

In the following table we show on the left what Ed said to Susan and on the right what principles or tools he used.

Ed's comments to Susan	Principles used
	Prepare. Ed has prepared for this discussion by reviewing his interests. He has concluded that his strongest interest is in maintaining the productivity of his group. Therefore, he needs to keep Susan and find a way to improve her working relationships with others.
"Susan, I know how hard that you've worked on the PAN project and I believe I know how important this project is to you. Although I can empathize with you, I'm frustrated by the divisive atmosphere in the department.	**Empathize to defuse or forestall anger.** Ed starts by empathizing with Susan, recognizing the importance of her work as well as her frustration.
I know that you get along with many people, but are you aware that you have a tendency to be confrontational with others?	**Address the behavior.** "You have a tendency to be confrontational" rather than attacking the person (e.g., "You can't get along with people"). **Ask questions.** Ed asks questions to ensure that Susan sees the problem and understands what he is saying.
For example, in the group meeting last week, you told Athan that his sample preparation technique was naïve and amateurish. This behavior isn't in the best interest of the company or you. Frankly, I even had difficulty deciding to discuss this with you because I was afraid that you would get angry. It's not easy for me to talk to people who get angry so I've avoided being honest with you. I'm sorry that I have not discussed this with you before, but I'm doing it now because you're a very important member of the department and we need your talent and skills.	**Be specific.** Ed gives Susan a specific example of the behavior he is talking about. **Be willing to examine your own complicity in the situation.** Ed shows that he is willing to examine his own behavior in the same way he is asking Susan to examine hers. **Develop self-awareness.** Ed has thought through his interactions with employees and peers and knows that he tends to avoid conflict. This alerts him to the fact that he might have done the same with Susan.
I suspect that you were frustrated because of the way your project was assigned. Is that true? If so, I take responsibility for that. Moreover, as head of the department, I feel that I haven't been honest with you and I haven't given you the opportunity to grow professionally. I'd like to correct that now. I need to give you some honest feedback about how your anger impacts your professional relationships. Then I'd like to work with you to help you change that behavior. If you're frustrated with the way in which you have been treated, we can deal with that to see if we can fix it.	**Accept responsibility.** By showing his willingness to take responsibility for his own role in the situation, Ed creates an atmosphere of trust. **Be attentive to reactions.** Ed carefully watches Susan's facial expressions and body language to be sure that she is willing to discuss her behavior. This is a critical point in the interaction: Ed must ascertain whether Susan is both willing and capable of discussing her own behavior and is prepared to take responsibility for it. If at this point she becomes angry and denies that the problem is entirely caused by how others treat her, Ed will need to tread carefully and either terminate the discussion or jump right to the last paragraph to discuss his expectations for improvement.

(Continued)

Ed's comments to Susan	Principles used
I'd also like to help you deal with Athan, Mohan, and Ralph in a way that allows you to address scientific differences of opinion without anger or hostility. Do you agree that there's a problem? I know that you're a dedicated scientist and getting the science right is your primary motivation. I think that I can help you to deal with and control your anger in certain situations so that you get your scientific points across without others feeling that you are disrespectful of them.	**Check for understanding and agreement.** If Susan doesn't see a problem with her behavior, Ed should be prepared to provide specific examples for her. **Notice body language.** Ed watches Susan's body language and notices that she begins to nod almost imperceptibly as he speaks. Picking up on this cue, he gives her an opportunity to speak.
This is not to say that you don't have reason to be angry at times and we can and will discuss the substance of those issues. But right now, the priority is learning to best express your concerns in ways that give you the greatest opportunity to be heard and have a positive impact on the project."	**Focus on the problem, not the person.** Ed frames the problem in terms of its consequences to the company. He avoids attacking Susan.
Ed concluded, "Look, we're all human. If we're angry or sarcastic, we're bound to generate a negative response even when what we say is demonstrably and scientifically true. People may hear what we say, but they may not take it in because they're reacting to their own feelings of being insulted, hurt, or angry. It doesn't have to be that way.	**Focus on the problem, not the person.** Rather than being accusatory, Ed focuses on how Susan's behavior affects her work and her professional relationships. He maintains the focus on her behavior, not her personality, and offers support.
Finally, having said this, I need to tell you that you do need to improve your interactions with others in the company. I'm convinced that the difficulties you have in working productively with others slow the progress of our projects, and I can't let that continue.	Ed makes clear that the problem is decreased productivity attributable to poor communication and interactions within the company. **Use your authority.** Ed makes it clear to Susan that he will not let the situation continue along its present course.
I'm willing to work with you on this because I value your contributions. I need your commitment that you will work to improve your professional relationships, and we need to set some specific goals for improvement.	**Assure.** Ed assures Susan of her value and his willingness to help her.
Is this something that you're willing to work on with me?"	**Get a commitment.** Ed concludes by asking for Susan's commitment to work with him.

An important element in this interaction is Ed's ability to observe how Susan responds to his analysis of the problem and to his suggestions for improvement. As we show later in this chapter, Ed's assessment of whether Susan is willing and able to engage in this type of discussion, and whether she accepts responsibility for her own behavior, is key to the success of his attempted intervention. If, early in the discussion, Susan becomes indignant, denies any responsibility for the situation, and blames the strained relationships on others, Ed may conclude that this approach will not be successful. In that case, he has to decide either to let things continue or find a solution that alleviates the problem but does not require that Susan have insight into herself, which of course is out of Ed's control in any case. One possibility might be to appoint another scientist to serve as liaison between Susan and the others. However, if Susan's insights, personal knowledge, and active participation in a joint project are what is needed, there may not be an easy solution.

In the above scenario, we focused on how Ed could have used his role as mentor and leader to improve interactions and collaboration between Susan and others in the company. The situation could also have been improved in other ways. Let us list a few of them.

1. Susan might have read this book and started taking notice of her own behavior during interactions with her peers. She might have concluded on her own that some people, Athan and Mohan especially, pushed her hot buttons. She might have discovered that she was especially sensitive to having her work questioned in the kind of glib manner that Athan and Mohan seemed to favor. She felt that they were arrogant and dismissive of her, and in some cases they actually were. Had Susan come to these conclusions, she might have decided to ignore their behavior and to observe, rather than react, to the feelings engendered by those behaviors. Getting some distance on her feelings might have given her the opportunity to choose how to behave rather than simply to act in response to her feelings of anger or annoyance. Finally, she might have applied the 95% rule, after observing that Athan and Mohan interacted this way with everyone, not just her.

   This is not to suggest that Susan should have stood quietly by if she was insulted, mistreated, or ignored. Rather, our suggestions might have allowed her to deal with these problems and satisfy her underlying interests (doing good work and advancing the project) without resorting to either withdrawal or hostility.

2. Had Athan read the section on conflict avoidance in Chapter 5, he might have noticed that Ed was conflict-averse and that this was affecting the project. Athan might have concluded that Ed was either blissfully unaware of or consciously ignoring the difficulties that he and Susan were having because he did not know what to do about them. If Athan had been a bit more interpersonally savvy, he might have felt comfortable approaching Ed and outlining the difficulties as he saw them. Taking negotiation cues from Chapter 3, Athan might have appealed to Ed's underlying interests in ensuring that his projects progressed, getting Ed to focus on the several problems needing his attention.

3. Had Mohan read this book, he might have been successful at developing a collaborative relationship with Susan. By ignoring or deflecting her insults (Chapter 3) and using techniques to defuse and neutralize her anger, he might have been able to overlook Susan's tough exterior and maintain his focus on developing a relationship that would have furthered the interests of the company. By framing discussions with Susan in the context of the problem (the company's need for diffusion of information across departments) rather than the person (accusing Susan of being obstructionist or hostile) as described in Chapter 5, Mohan might have been able to appeal to Susan's highly developed professional standards.

It is always tempting to pinpoint specific individuals as causing the problem because individual behavior is so easy to see. In some cases, the behavior of one individual may be so aberrant that it seems logical to identify it as the cause of a group problem.

Was the problem that Susan behaved in a hostile and noncollegial manner? Was it that Ed seemed to lack many of the important skills we associated with effective leaders in the previous chapter? Or was it that Athan, Mohan, and Ralph couldn't deal with an argumentative colleague and then used her data without her knowledge?

The alternative scenarios listed above suggest that it may have been possible for any one of the people involved to initiate a resolution of the problem by behaving differently. But this should not be taken to imply that any one of the individuals was either at fault or solely responsible for the genesis of the problem or its outcome.

It is difficult to see the web of relationships and interactions that inevitably underlie all problems in groups. In our experience, focusing attention on one group member as the singular cause of a group problem is almost always a mistake. Problems between people in the workplace happen in the context of relationships, and as such, all parties own some responsibility for their origins and outcomes.

Having said this, however, as a leader Ed owns some special responsibility for the situation. His tolerance of Susan's hostility and of the strained relations between Susan, Athan, and others sent a message that dysfunctional behaviors and interactions were acceptable. His reluctance or inability to deal with the conflict in the group and to coach Susan, Athan, and others in more effective ways of interacting, signaled that it was acceptable to display animosity and hostility in the workplace. In effect, he was modeling the ineffective behaviors the group members were displaying.

Because this case takes place in a company responsible for a lifesaving and financially lucrative product, the benefits of any of the interventions suggested above to Susan personally, the company, and sick patients could have been huge. If the two-year delay in the delivery of the product did not occur, it might have saved the lives of many patients. Susan might have received just credit for her role in the discovery, rather than being marginalized and ignored. (Note to the reader: The scientific details of this case are purely fictional. The PAN receptor does not exist, and unfortunately, as of this writing, nor does an effective treatment for pancreatic cancer.)

But even if this case had taken place in an academic setting, an alternative outcome could have been equally important, at the very least for Susan. Rather than being treated as a pariah or an obstructionist, she could have been recognized as a key contributor to important work, with significant, positive implications for both her science and her career.

The astute reader may have recognized some similarities between this case and the well-known historical account of Rosalind Franklin and the structure of DNA, from whose story this case has been created. Replace Ed with J.T. Randall, head of King's College biophysics unit, Athan with Maurice Wilkins, and Mohan and Ralph with James Watson and Francis Crick, and you have what for many is a familiar story (Maddox 2002; Sayre 1975; Watson 1968).

As illustrated by our alternative scenarios, if any or all of the protagonists in that story had more highly developed self-awareness and interpersonal savvy, Franklin's involvement and recognition might have been very different. In our revisionist version, Randall managed to face his own unwillingness (here imagined, but not inconsistent with the historical facts) to deal with controversy and conflict in a way that might have helped Franklin face and manage her own frustrations. He might have also helped Franklin to recognize the impact of her behavior and act in ways that advanced the science as well as her own career at the same time.

Recognizing Franklin's role in the outcome of this case is in no way meant to excuse any mistreatment Franklin may have endured, either by the various participants or in historical accounts of the events. But human interactions are always of a reciprocal nature. We can only guess how the outcome might have differed if Franklin had an ally such as the hypothetical Randall in the modified case.

Whereas Franklin's data helped reveal the structure of DNA, Susan's data helped reveal a promising cancer therapy. In both cases, the discoveries or breakthroughs almost certainly would have occurred eventually. In the cancer story, the principal tragedy was that an important and possibly lifesaving discovery was delayed. In the DNA story, an excellent scientist who could have been a full collaborator and participant in the discovery of a lifetime was marginalized by Wilkins, Watson, and Crick because of their inability to interact productively with her, and her with them.

Our alternative version suggests that it did not have to turn out this way. Scientists can learn to deal with anger and frustration without damaging their professional interactions, mentors can recognize and address conflict before it becomes explosive, and colleagues can develop collaborations that fairly satisfy the interests of all participants. None of this comes naturally to anyone, least of all to those whose professional lives are spent focused on concepts, data, and technical matters. Just as science can be and is learned by those who are so inclined, so too can self-awareness and interpersonal skills. Just as science is taught and required as part of the education of a science professional, so too can self-awareness and interpersonal skills.

TRANSFORMING SCIENCE ORGANIZATIONS

Technical professionals in training are hungry for extra-scientific skills that will increase their effectiveness and success. Educators and funding agencies are becoming increasingly aware of the need for this sort of training. A recent report from the National Academies of Sciences, Engineering, and Medicine, *Graduate STEM Education for the 21st Century* (Leshner and Scherer 2018), provides a comprehensive roadmap to making graduate science education more responsive to the needs of students and to preparing graduates for the multiple career options available to them.[1] But the challenges to making such changes are many. Let us examine a few.

How do science organizations differ from other organizations?

In the following, we review two classes of challenges that face the science organization: those that are unique to science-based organizations and those that are shared with other types of organizations.

It is naïve to assume that scientists and science organizations can be managed in the same way as other professionals and other organizations. It is equally naïve to assume that because of their profession, scientists are devoid of the same blind spots and frailties as the rest of humanity. This latter belief can lead to the mistaken notion that science organizations operate with a degree of rationality that is absent from other organizations.

Here is our list of challenges that distinguish science organizations, especially those focused on discovery, from other types of organizations.

- Science organizations engaging in discovery research present unique managerial challenges. Discoveries cannot be made according to a predetermined timetable. This creates an inherent tension between scientists and managers. Managers must produce budgets and timelines based on expectations of when certain events are most likely to occur. Because of the impossibility of scheduling discoveries, a rigid bureaucratic approach to management will result in a disjunction between what scientists can produce and what management espouses they should or will produce.

---

[1] Available as a free download at https://www.nap.edu/download/25038.

- Because some scientists are disinclined to deal with interpersonal issues, "people" problems may be ignored and, as a result, escalate. Problems that could have been resolved by line or lab managers often end up in human resources or the chairperson's or vice president's office.

- Questioning and skepticism play a central role in the scientific process. However, skepticism expressed by scientists with poor interpersonal skills can lead to divisive confrontations, reducing productivity.

- Scientists are trained in an environment in which rewards and reinforcement derive from individual achievement. Despite declarations to the contrary, real rewards, promotions, and advancement in nearly all organizations are based on individual performance. Cynicism results when scientists are asked to behave as team members by sharing knowledge, insights, and results but observe rewards going solely to those who impress management with individual achievements.

- Science organizations are often led by executives who have little or no science training. It is not uncommon to find individuals with financial, legal, and business development backgrounds heading such organizations. Such leaders suffer from definite disadvantages if they do not understand the nature of the scientific process, and especially if they believe that research and discovery can be managed by conventional "command and control" approaches.

- During the past two decades, companies in many sectors of science and technology have suffered from financial instability, short lifetimes, and major organizational changes such as mergers and acquisitions. As a result, the stability required to train employees and create a positive corporate culture is frequently missing. Model programs for organizational change, team building, and management that have been developed based on Fortune 500 companies (as most have) need to be carefully examined for relevance, especially in the science sector.

How are science organizations similar to other organizations?

The following is a list of challenges that scientists and science organizations have in common with their nonscience counterparts.

- Like professionals in other fields, scientists have a strong need to take ownership of projects and ideas as a way of defining themselves and gaining recognition. Managing scientists without understanding this produces a frustrated and unmotivated workforce.

- Scientists have as much of a tendency as anyone else to rely on what they know. It is dangerous to believe that because someone is a scientist, they will dispassionately evaluate all possible approaches or alternatives. A pharmaceutical company manager we knew, when asked why a division in his company was using *Drosophila* (fruit flies) as the model organism for a certain study, replied, "Oh, that group is run by Jim Hess. He did his last postdoc in a fly lab."

- Every scientist has a ready willingness to believe that their latest discovery is the most important breakthrough of the century.

- Eminent scientists join advisory boards of science companies for dozens of reasons. A belief in the company's technology may be among these reasons, but not necessarily. Do not judge the value of a company's science, or the likelihood that the company will succeed, based on the number of Nobel laureates on its scientific advisory board. As we demonstrate in this book, science-based companies need more than a stellar scientific advisory board to succeed.

Focusing on people

Science and technical professionals can begin transforming the organizations in which they work, from the ground up, by taking the following steps.

Hire the right people

When he was an executive in biotechnology companies, Carl divided the qualifications of candidates for scientist and science management positions into three categories: (1) academic or technical skills, (2) professional and managerial experience, and (3) personal characteristics and interpersonal skills. The first two categories could be captured on a resume and were typically the basis for the first round of selection. During a candidate's interview, however, it was mostly the skills in the third category that were of interest; these skills could not be captured on a resume. Here is a list of key qualifications and characteristics that you might look for in a potential employee.

Group 1. Academic and technical

- Specific scientific or technical skills, academic degrees, publications

Group 2. Professional and managerial

- Previous experience in similar organizations or jobs
- Managerial/supervisory experience as appropriate
- Strategic experience/vision as appropriate

Group 3. Behavioral and psychological

- Good self-awareness and ability to step back, observe, and comment on own behavior
- Willingness to take personal ownership for behavior
- Ability to describe feelings experienced in a difficult situation
- Openness to constructive criticism and observations about behavior; absence of defensiveness during such conversations
- Ability to control actions under tense or stressful conditions
- Ability to constructively provide feedback to others
- Ability to read others well enough to sense which people will work well together
- Absence of need to dominate or belittle
- Ability to accept ambiguity without seeking quick and possibly inappropriate solutions to difficult problems

- Ability to deal productively with implied or direct affronts in a group setting
- Ability to negotiate equitable agreements when parties have differences of opinion
- Understanding of organizational objectives and how to help meet them
- Ability to remain focused and avoid tangents
- Willingness to seek advice of others when stuck
- Willingness to admit ignorance
- Ability to handle disagreements directly but collegially
- Willingness to share knowledge

Whereas the first two categories might identify technically proficient and experienced candidates, technical proficiency and experience alone are not necessarily the most important criteria for choosing a leader, manager, or even a scientist. We wonder how many of the scientists in a typical company meet more than one-third of the Group 3 criteria.

In Chapter 4 we introduced an interviewing and hiring process that allows you to assess candidates for nearly all of the elements that we just discussed. If you are hiring candidates who will have managing or supervising others as a major part of their job, use Scoresheet 5 and the Guide to Scoresheet 5 in Chapter 4 to evaluate and compare them. If you do that you will have taken a giant step toward transforming your group or organization. If you are evaluating someone who is already part of your organization or group and for whom a formal interview is either inappropriate or impossible, here is a list of questions to ask yourself based on the above characteristics.

- How self-aware is this person? Can they step back, observe, and comment on their own behavior?
- Do they take personal ownership for their behavior, or do they explain it as being the inevitable consequence of external circumstances or of things that others have done?
- Can they describe their feelings or experiences in difficult situations?
- Are they open to constructive criticism and observations about their behavior? Do they become defensive or uneasy during such conversations or receptive and curious?
- Can they control their actions under tense or stressful conditions?
- Can they provide constructive feedback?
- Can they read others well enough to sense who someone will work best or collaborate with?
- Do they dominate or belittle others?
- Are they comfortable with ambiguity or do they seek quick and possibly inappropriate solutions to difficult problems?
- Can they deal productively with implied or direct affronts in a group?
- Can they negotiate equitable agreements when parties have differences of opinion?
- Do they understand the organization's objectives and how to help meet them?
- Do they tend to go off on tangents or get so bogged down in detail that they cannot see the big picture?

- Will they seek the advice of others when stuck?
- Do they have difficulty admitting ignorance or are they overly confident of what they do know?
- How do they handle disagreement? Do they remain silent? Do they attack?
- How readily do they share their knowledge?

Sometimes we can gain insights by asking candidates for anecdotes or stories about previous jobs. In many cases, insights come not from the actual content but from other information. Here are three ways to get answers to some of the questions listed above.

1. Ask the candidate to describe their last position and the projects on which they worked. One candidate might be obsessively technical and neglect to mention others who worked on the project, whereas another might discuss the science as well as the group and the way in which the project fit into the company's goals. All other things being equal, we would surmise that the second candidate would be a better candidate for a managerial or leadership position.

2. Ask the candidate to describe an especially difficult problem (both technical and managerial, if appropriate) and how they approached it. Listen for evidence of interactions or collaborations: Was input or advice from others sought and welcomed or were projects bogged down because of unwillingness to seek such input or advice? As above, pay attention to the balance between technical detail and contextual, interpersonal, and organizational references.

3. Pose a hypothetical situation for the candidate. One our favorites is seeing how they would deal with a scientist in a group situation, who is very smart and capable but who hoards data and is abrasive to others. If you have read this book, you know that there is no single right answer. The kinds of responses you get will tell you much about how the scientist is attuned to the complexities of managing other scientists. One candidate responded by saying that he would give the hypothetical scientist an ultimatum to either "shape up or ship out." Another suggested that he did not see a problem as long as the scientist shared the data with him. Neither of these candidates would be high on our list of new hires.

Finally, note that no matter how assiduous you are about hiring the right people, their effectiveness will depend strongly on your organization and how it is managed. In dysfunctional organizations effective people become less effective, and ineffective people become incapacitated. Individuals can make behavioral decisions and interact with others in ways that reduce the impact of organizational dysfunction but they probably cannot neutralize it.

Train the employees that you hire

Scientists who have more than a few of the qualities identified by the questions above are few and far between. Consequently, the most effective way to transform your organization is to train employees to be better managers and colleagues. Of course, this is easier said than done. By far, the best way for scientists to learn the skills presented in this book is to observe and emulate those who already have them. If you have these skills yourself, model them for your lab, team, or company. You may wish to take advantage of specific workshops in conflict resolution and negotiation, as well as individual coaching (see Other Resources on pp. 345–346). Unfortunately, few groups or individuals specialize in providing such training in a way that addresses the specific needs of scientists.

The exercises at the end of our chapters can be used by individuals or groups as "continuing education" material. But if you choose to discuss some of these issues in a group format, do so with caution. Many people are uncomfortable or embarrassed to discuss themselves, their sensitivities, and their feelings. Never force participants to engage in a discussion if you sense that this may be the case. Forcing someone with poor self-awareness into a discussion about themself in the presence of others can be at best uncomfortable for that person and at worst traumatic, especially if you are not trained in group facilitation.

Finally, remember that for some, interpersonal skills are far more difficult to learn or to change than are technical skills. Unlike technical skills, which are learned anew, how we interact with others is deeply ingrained in our personality. Thus, depending on the individual, there may be a significant amount of unlearning that needs to take place before new learning can be effective. This is no mean feat for most and may be impossible for some.

A manager or leader has only so much time to devote to training scientists in these skills and therefore must judge who is trainable and who is not. Answering the questions in the previous section is one way to sense someone's potential. Training someone with limited self-awareness and little sign of being able to acquire it is not the best use of a manager's time. If you supervise such employees who are valuable from a technical perspective, you may need to find a way to take advantage of their technical skills while ensuring that their interpersonal deficits do as little damage as possible. The case study involving Raju in Chapter 2 illustrates one way to accomplish this. The next section provides a framework within which you can identify those in your organization who, with mentoring, have the greatest potential to become good managers and leaders.

The myths of scientific research

- Research is done in mental and social isolation.
- The most important determinant of scientific success is how much science you know.
- Success in science requires a single-minded devotion to science and nothing else.
- Hire the most technically qualified people and do not worry about anything else.

Identify and mentor your future leaders

The single most important element of a transformed science organization is its people. Once you have begun training your science staff in the skills of self-awareness and interpersonal relations, you have begun to create future leaders of people, not just of science. Just as you can identify scientists who have or can acquire the requisite technical skills, so too can you select the ones with the greatest promise to become the future managers and leaders of your organization.

We have found it useful to identify future leaders by assigning scientists to one of three categories.

Category 1. Scientists in this category are technically proficient and already have good inter-personal skills. A subset of these will also have good self-awareness, and the remainder may be able to improve this skill with a modest amount of mentoring. Category 1 is likely the smallest group of scientists in your organization, perhaps fewer than 5%. These scientists need only a modest amount of your mentoring time, largely to help them attain the skills needed to manage and to continue to develop their self-awareness and relational skills. Give these scientists managerial or leadership responsibilities early in their career so that they can benefit from feedback as soon as possible.

For some readers, identifying Category 1 scientists will be intuitive. They are the ones who interact smoothly with others, take feedback in stride, and avoid getting involved in laboratory cliques or in power struggles. These scientists can also be identified by answering the evaluation questions listed earlier in the chapter.

Category 2. These scientists are technically proficient, have modest or indeterminate inter-personal skills, but sufficient self-awareness to be candidates for improvement. Although they may at times behave in an inconsiderate or insensitive manner, they are nonetheless open to discussion and feedback about their behavior. This group may be the one with which the savvy manager can have the most impact. If a scientist is open to observing their behavior and reflecting on their underlying motivations, there is an excellent chance that their behavior can be improved.

One way to identify scientists in this group is to give them some feedback on their behavior and observe how they react. (Remember to use "I" phrases and to focus on specific behaviors, i.e., "I noticed that you seem to be having a hard time working with Farid" not "You seem to have difficulty getting along with people.") Denial, defensiveness, and excuses are all red flags. On another occasion you might ask why they behaved in a way that was counterproductive in a group setting or that was insulting or devaluing to someone. Gently probe for an indication that they are willing to examine their motivation and can understand and empathize with the victim of their disfavor. Positive signs in either of these circumstances indicate that you may continue such discussions and help the employee use their self-awareness to gain better control of their behavior and improve rapport with others. Helping a scientist in this category can be one of the most satisfying things you do as a manager or leader and can have the greatest impact on your organization.

Category 3. These are scientists who are technically proficient but have poor interpersonal skills, low self-awareness, and a limited capability to improve or limited interest in improving. This is usually the largest of the three groups. These employees

- become defensive or evasive or seem to be totally baffled when observations are made about their behavior or performance;
- find it difficult or impossible to discuss their feelings or motivations;
- have a history of being confrontational, argumentative, and devaluing of others; or
- have a history of being passive, withdrawn, and uncommunicative.

Scientists in this category are sometimes the most knowledgeable, technically adept, and scientifically productive. But they are unlikely to see the utility of the skills offered in this book, and they may be resistant to efforts aimed at helping them to acquire such skills.

If Category 3 scientists have superior, unique, or essential technical or scientific skills, place them in a position from which they can do the most good scientifically and the least damage interpersonally.

Insisting that Category 3 scientists work in a highly structured or interactive team setting can be a recipe for disaster. But there are ways in which valuable intermediate-level Category 3 scientists can be incorporated into teams that allow them to be productive. Making them responsible for specific tasks or deliverables that can be accomplished independently or with limited group interaction is one approach. The key is to maintain a formal relationship between the scientist and the team, with well-defined deliverables, expectations, and responsibilities, while at the same time giving the scientist enough freedom to accomplish these with some degree of independence.

Often, those in Category 3 are very concrete thinkers who function best with clear and unambiguous assignments. Be certain that the responsibilities and deliverables that you assign are defined in as clear a manner as possible. It goes without saying that you should think twice before giving such employees line management responsibilities. If you must place them in a managerial position or have others reporting to them, be alert to signs of stress in those individuals, and be prepared to intervene.

There are always exceptions. We have seen cases in which scientists who were viewed as having weak interpersonal and managerial skills managed to be effective team leaders. In one case, a senior scientist, who under most circumstances would never intentionally have been given managerial responsibility, had attracted over time a small group of dedicated scientists very much like himself in personality and temperament. This group was remarkably effective and productive. Although they worked in virtual isolation, their role was such that this was not a major detriment. Their work output was integrated into those projects that required it, and feedback was provided through ad hoc but effective communications. This kind of group would be almost impossible to create intentionally, but if you have one in your organization, think carefully before you disband it.

RECOGNIZING AND ACKNOWLEDGING ATTITUDES TOWARD GENDER, RACE, ETHNICITY, AND MINORITY STATUS

In proposing ways in which science organizations can be improved upon or transformed we cannot ignore the now well-documented disparities in funding, hiring, and promotion associated with gender and minority status. The recent National Academies of Sciences, Engineering, and Medicine report "Sexual Harassment of Women. Climate, Culture and Consequences in Academic Sciences, Engineering, and Medicine" (2018)[2] thoroughly documents the personal and professional challenges faced by women and minorities in the sciences. Throughout this book we have espoused a pragmatic and empathic toolkit-based approach to managing science, one in which we are "hard on the problem and soft on the person." Yet we know from

[2]Available as a free PDF download at https://www.nap.edu/catalog/24994/sexual-harassment-of-women-climate-culture-and-consequences-in-academic.

experience that we all harbor both conscious and unconscious attitudes and biases that affect how we interact with, manage, and judge others. In Chapter 4 we reviewed cognitive biases that can influence our judgments of people in the context of hiring. These same biases can affect all of our interactions. A good place to start exploring your biases is the website Project Implicit (https://implicit.harvard.edu/implicit/), which provides a suite of tools that can give you a glimpse into your own unconscious or implicit attitudes toward a variety of people.

Gender-based bias is well documented in science, resulting in lower funding, hiring, and promotion rates for women. Compounding these barriers are a wide spectrum of overt harassment behaviors experienced by women in both training and professional settings in science. As noted in the National Academies report, the academic setting is a fertile environment for such harassment, being characterized by perceived tolerance for harassment, male-dominated work settings, hierarchical power structure, and symbolic rather than substantive compliance with regulations designed to limit or sanction unwanted or frankly illegal behaviors.

The glimmer of good news in the report is the idea that discriminatory and harassing behaviors may be reduced or stanched by clear organizational policies proscribing them. "Organizational climate is, by far, the greatest predictor of sexual harassment, and ameliorating it can prevent people from sexually harassing others." So, although it may not be possible to eliminate harassers, it may nevertheless be possible to eliminate or reduce acts of harassment by creating an organizational culture that has no tolerance for such behaviors. Creating this culture is the role of those in leadership roles, whether it be as head of a lab or research group or as President or CEO of the organization. As we have noted throughout this book, people emulate the behaviors and attitudes of those in leadership roles and of those who they admire and trust. By showing through your actions that gender, race, or ethnicity have no roles in how you behave toward others in your professional and personal life, you are establishing a norm. This is perhaps the most important transformation you can bring about in your own organization.

TRANSFORMED SCIENCE ORGANIZATIONS

Transformed science organizations consist of employees and leaders who share the belief that the way in which scientists do science has an impact on the science that they do. Such organizations

- consist of members who think about and manage how they interact with one another, deal with conflicts productively, and iron out animosities before they become feuds;
- perform both internal and external negotiations and interactions respectfully, directly, and honestly;
- have leaders who pay attention to their own behavior and interactions, and encourage others to do the same.
- have clear policies proscribing behaviors that discriminate against or disadvantage any individual or group based on gender, ethnicity, or minority status and have a culture that supports these policies.

None of this is to suggest that transformed science organizations are populated by scientists with saccharine smiles behaving like lobotomized drones. The underlying emotions that ignite nonproductive behaviors are universal. They cannot, and probably should not, be extinguished. Human feelings of jealousy, anger, fear, competition, and more are natural. Some individuals will, at times, feel ostracized, misunderstood, and isolated.

A transformed science organization does not eliminate the spark that ignites creativity, or the healthy disagreements that often stimulate discovery. Rather, it provides science and technical professionals with the tools to deal with the inevitable human problems they will encounter in the scientific workplace. Successful application of these tools enables them to direct more of their energy and excitement toward solving scientific problems and accomplishing their organization's objectives.

Transforming a science organization cannot be accomplished without the support and attention of its most senior scientific leaders. "People" skills must be taught to students, trainees, and employees in the same way that technical skills are taught. We must stop referring to such skills as "soft" and stop characterizing those who espouse them as "touchy-feely." Moreover, we must either stop relegating the training for such skills to human resource departments or imbue these departments with an importance and focus that they rarely have in most organizations. This will not happen unless we agree to classify these skills as "must have" and not "nice to have."

We have identified several categories of "lab dynamics" skills that should be taught to scientists. First are skills that help scientists become better aware of their own feelings and behavior. These skills are prerequisites for productive interactions in difficult or tense circumstances. Another set of skills helps scientists to become better attuned to others and to how others respond to them. Next are skills that help scientists to become good negotiators and to reach agreements that are equitable and satisfy the interests of all parties. We showed how such skills can improve the way scientists interact with peers, supervisors, and team members. We have suggested that it is possible to expand the scope of science education so that scientists are as knowledgeable about how to interact with people as they are about science. We have outlined what scientists need to know and the skills that they need to use to thrive in the private sector. Finally, we have suggested ways in which science organizations can use our recommendations to select the best candidates for scientific management and leadership positions.

Lab dynamics can be taught through workshops, instructional materials, and case studies. We believe that workshops are especially useful for teaching negotiation and associated skills because of the immediacy of the learning experience. Case studies (such as "Rewriting History" on pp. 322–323) can be used to present complex science-management problems. They can also be used to show how scientific progress is affected by the quality of the interactions of the people involved. These studies may be used as didactic tools on their own or to catalyze discussion in formal training settings.

If you are a human resources professional, be cautious about trying to introduce training in interpersonal skills for scientists to your organization. If your efforts do not have the full and vocal support of your scientific leaders they will either fail or have limited success. It is best if the impetus for such training comes from the scientists or scientific leaders themselves. Try giving your scientific leaders a copy of this book as a way to kindle their interest. Next, suggest having a workshop in negotiation skills. Negotiation skills are a great place to start because they are concrete, the basics can be communicated in a one-day workshop and scientists readily see the utility of being good negotiators. As we noted in Chapter 3, learning principled negotiation develops many of the characteristics we associate with high emotional intelligence.

We believe that with research and experimentation, we can develop additional approaches to help scientists improve communications, interactions, and collaborations. This book offers a starting point. But vision and support are needed to develop other options and approaches. We call on educational institutions, as well as agencies and foundations that support scientific research, to follow our lead.

REFERENCES

Leshner A, Scherer L, eds. 2018. *Graduate STEM education for the 21st century.* National Academies Press, Washington, DC.

Maddox B. 2002. *Rosalind Franklin: The dark lady of DNA.* Harper-Collins, New York.

Sayre A. 1975. *Rosalind Franklin and DNA.* W.W. Norton, New York.

The National Academies of Sciences, Engineering, and Medicine. 2018. *Sexual harassment of women. Climate, culture, and consequences in academic sciences, engineering, and medicine.* National Academies Press, Washington, DC.

Watson JD. 1968. *The double helix: A personal account of the discovery of the structure of DNA.* Atheneum, New York.

EXERCISES AND EXPERIMENTS

1 Identifying potential leaders

Becoming aware of who among your scientific staff are good candidates for leadership and managerial responsibility will enable you to mentor and advise your leaders-in-training. Answering the following questions can help you to identify candidates for Category 1, 2, or 3 (for a description of scientists in Categories 1-3, see pp. 336–337). Check the box under a number for each person you are evaluating in the column to the right of the trait listed: Check 5 if you strongly agree with the statement as it applies to the person or a smaller number if you agree less. Add them up for a measure of your views of those you are rating. The higher the number, the better the potential of the candidate. This exercise is not meant to be used to rank your staff in any absolute way, but rather as a tool to help you ask the appropriate questions when you are considering the leadership potential of your staff. Remember that the rankings are your own views, which may themselves be skewed or biased.

	Strongly agree			Disagree	
	5	4	3	2	1
1. Will spontaneously step back, observe, and comment on their own behavior.					
2. When prompted to observe and comment on their own behavior is open and willing.					
3. Takes personal ownership for behavior.					
4. Does not seek to blame others or external circumstances for mistakes or failings.					
5. Able to describe feelings or experiences in difficult situations.					
6. Open to constructive criticism and observations.					
7. Does not become defensive or uneasy during such conversations.					
8. Able to control behavior under tense or stressful circumstances.					
9. Can deal with implied or direct affronts in a group.					
10. Can provide constructive feedback.					
11. Does not dominate or belittle others.					
12. Able to negotiate and arrive at equitable agreement when parties have differences of opinion.					
13. Understands organizational objectives and how to help meet them.					
14. Can see the big picture associated with a project.					
15. Willingly seeks the advice of others.					
16. Has little difficulty admitting ignorance.					
17. Is not overly confident of own knowledge.					
18. Does not avoid or withdraw from disagreements or difficult conversations.					
19. Willing to share knowledge.					

Of these 19 items, traits 1, 2, 5, 6, and 7 relate to the ability of the individual to listen and respond productively to observations about their behavior and to have insight into it. Low scores in these traits should not be viewed as having absolute psychological validity. These are your assessments of the person's ability to be open and receptive to discussing their own behavior and motivations with you. If your evaluations for each of these traits are consistently low, then you may have difficulty establishing sufficient rapport to help these individuals observe and improve their behavior. Because others in your organization may have a better rapport with specific individuals than you, always validate your perceptions about leadership or management potential with one or more of your colleagues or a contact in human resources.

2 Rewriting your own history

Recall a situation from your own organization or group, in which you were involved, that had a negative impact on a project or the individuals.

- Summarize the situation, focusing on its impact on a project, task, or objective, and on the individuals involved, as well as the consequences for the team or organization.

- Refer to the skills presented in this book. Choose one skill and consider how acquiring or exercising that skill by one of the participants in the incident might have changed the outcome for the better. Try this for as many of the individual skills and participants as possible. Refer to the skills used by Ed in the case study "Rewriting History" to get started.

- Focus on your own involvement in the incident and do the same for yourself as you did above for the others. Identify as many ways as possible that exercising specific "Lab Dynamics" skills could have helped you to effect a better outcome.

- Use the results of this exercise to anticipate ways in which you can avert or mitigate future incidents that may impact your work.

Resources

BOOKS

At the Helm: Leading Your Laboratory. 2nd ed. Kathy Barker (2010), Cold Spring Harbor
  Laboratory Press, Cold Spring Harbor, New York.

A "must have" for every new manager of a science group. Written by a working scientist, it
provides concrete and useful answers to questions about organizing and administering a
scientific laboratory and group. Topics also include psychological issues such as lab culture,
conflict, and morale.

Dealing with People You Can't Stand: How to Bring Out the Best in People at Their Worst.
  Rick Brinkman and Rick Kirschner (1994), McGraw-Hill, New York.

Builds on some of the themes developed in *The Feeling Good Handbook* (see below) and
provides extended illustrations and amusing examples.

**Developing Managerial Skills in Engineers and Scientists: Succeeding as a Technical Man-
ager.** M.K. Badawy (1982), Van Nostrand Reinhold, New York.

Provides guidance on choosing science management as a career and preparing for it. One of the
few books included in this list that contains material on planning, decision-making, and
project management.

Emotional Intelligence. Daniel Goleman (1995), Bantam Books, New York.

Excellent guide to this important topic. Illustrates the nature of emotional intelligence and
argues persuasively for its importance in our daily lives and interactions.

The Feeling Good Handbook. David D. Burns (1990), Plume/Penguin, New York.

Excellent and highly readable introduction to the concepts and applications of cognitive behavior therapy, on which our approach to self-management is based. An outstanding book for those
seeking further guidance in dealing with difficult people (see Chapters 20–22). Our use of the
"agree, empathize, inquire" approach in Chapter 6 is taken from these chapters of Burns' book.
Also contains useful insights into understanding and managing your moods and interactions.

Getting Past No: Negotiating Your Way from Confrontation to Cooperation. William Ury
  (1993), Bantam Books, New York.

This short paperback is probably the single most useful book written on the topic of negotiation and is the foundation of Chapter 3 of our book. Should be required reading for every
scientist, regardless of whether they aspire to be a manager.

How to Manage the R&D Staff: A Looking Glass World. James E. Tingstad (1991), AMACOM, New York.

Written primarily for the experienced manager. Provides guidance on many of the practical aspects of running a research laboratory from a psychological perspective and covers such topics as group cohesion, building self-confidence in scientists, and task delegation.

The Human Side of Managing Technological Innovation: A Collection of Readings, 2nd Edition. Ralph Katz, Ed. (2004), Oxford University Press, New York.

Fascinating and wide-ranging collection of readings on managing in science and technology organizations. Topics include teamwork and creativity.

Leadership Without Easy Answers. Ronald A. Heifetz (1994), Belknap Press/Harvard University Press, Cambridge, Massachusetts.

An insightful study of what leaders do and why. Focuses on why those in leadership positions feel pressured to provide quick and easy answers to complex issues and how to recognize and avoid this. Introduces the important concept of "managing ambiguity." Contains many useful lessons for those who lead science groups.

Leading Geeks: How To Manage and Lead People Who Deliver Technology. Paul Glen (2003), Jossey-Bass, San Francisco.

Insightful book about managing technical professionals. Although it gives the impression of having been written primarily for the manager with a nontechnical background, it is full of insights and ideas that scientists will find enlightening.

Making the Right Moves: A Practical Guide to Scientific Management for Postdocs and New Faculty, 2nd Edition (2006).

http://www.hhmi.org/resources/labmanagement/downloads/moves2.pdf

This is a free, downloadable book based on the Burroughs Welcome, Howard Hughes Medical Institute's 2002 course in laboratory management. Offers advice and guidance from working scientists on everything from negotiating a position offer to the meaning and practice of leadership in science. Highly recommended.

Managing Scientists: Leadership Strategies in Scientific Research, 2nd Edition. Alice M. Sapienza (2004), Wiley-Liss, Hoboken, NJ.

Excellent introduction to various approaches that may be taken by scientists who manage other scientists. Addresses psychological and cultural issues in ways that are very accessible to most scientists and science managers.

Organizational Culture and Leadership. Edgar H. Schein (1992), Jossey-Bass, San Francisco.

Classic study of organizations and organizational culture. Despite its rather academic-sounding title, it contains case studies and insights that are illuminating for scientists seeking to understand the psychological underpinnings of the behavior of those in organizations and groups.

The Practice of Adaptive Leadership. Ronald Heifetz, Alexander Grashow, and Marty Linsky (2009), Harvard University Press, Boston.

In our view, a "must-read" especially for the more experienced leader seeking to fine-tune their skills at understanding how people and groups function in an organizational setting. The book highlights the distinctions between technical and adaptive leadership challenges and provides valuable lessons in dealing with adaptive challenges.

Scientists in Organizations: Productive Climates for Research and Development. Donald C. Pelz and Frank M. Andrews (1976), Institute for Social Research, University of Michigan, Ann Arbor.

Somewhat dated but interesting use of quantitative techniques to uncover the organizational and individual correlates of creativity in scientists and managers of science.

OTHER RESOURCES

The A.K. Rice Institute for the Study of Social Systems
http://akriceinstitute.org/

A.K. Rice Institute, through its local affiliates, sponsors experiential workshops that focus on providing insight into the way in which people interact and work in groups. The Institute "... seeks to deepen the understanding and the analysis of complex systemic psychodynamic and covert processes which give rise to nonrational behavior in individuals, groups, organizations, communities and nations." Do not be put off by the jargon. We recommend these workshops for those who are interested in and committed to learning about their own behavior and reactions in groups in a work setting.

American Group Psychotherapy Association (AGPA)
http://www.AGPA.org

A professional organization that trains human services professionals in group therapy and group dynamics. Provides a national referral service for those seeking certified group psychotherapists. Participation in group psychotherapy is an excellent way to acquire emotional intelligence skills for those who have difficulty learning these skills on their own.

Cold Spring Harbor Laboratory courses and workshops
http://meetings.cshl.edu/courses.html

Each year since 2011 Cold Spring Harbor Laboratory has offered a three-and-a-half-day workshop entitled "Leadership in Bioscience" (http://meetings.cshl.edu/courses/html) at the Banbury Conference Center. The course, which is based heavily on the lessons in this book, is designed and facilitated by Carl M. Cohen with the assistance of Suzanne L. Cohen and is geared specifically to scientists who had just or were about to assume positions of leadership. Visit the Cold Spring Harbor Laboratory courses website for future offerings of this workshop.

European Molecular Biology Organization (EMBO) Lab Management workshops and courses
http://lab-management.embo.org

EMBO has taken the lead in providing comprehensive multiday courses in managing and leading scientific enterprises. There are separate courses for postdocs and for independent investigators. To our knowledge, as of September 2011, there were no comparable routinely offered workshops of this type in the United States.

Emotional Intelligence (EI)

Dozens of books have been written on this topic, including applications of EI to the workplace and leadership development (see, e.g., *The Emotionally Intelligent Manager*, by David R. Caruso and Peter Salovey [2004], Jossey-Bass, San Francisco; *Primal Leadership: Realizing the Power of Emotional Intelligence*, by Daniel Goleman, Richard E. Boyatzis, and Annie McKee [2002], Harvard Business School Press, Boston). Books on improving your EI are also available (see, e. g., *Raising Your Emotional Intelligence: A Practical Guide*, by Jeanne Segal [1997], Henry Holt, New York; *The Emotional Intelligence Activity Book: 50 Activities for Developing EQ at Work*, by Adele B. Lynn [2002], AMACOM, HRD Press, New York). In addition, a growing number of online resources offer information, assessment quizzes, and links on emotional intelligence. One reputable commercial organization with which we are familiar is the Hay Group (http://ei.hay group.com). A Google search done in September 2004 returned more than 1,000,000 hits to the entry, "emotional intelligence," and one done in 2012 returned nearly 9,000,000. *Caveat emptor.*

National Training Labs (NTL)
http://www.ntl.org/

A highly regarded nonprofit organization that originated the "T-group" concept for experiential learning. NTL specializes in experiential training in leadership, organizational, professional, and personal development. Readers of our book may be particularly interested in their "Human Interaction Laboratory": "… an introduction to interpersonal relations, group dynamics, with a focus on developing and practicing effective interpersonal skills and giving and receiving feedback responsibly."

The Program on Negotiation of the Harvard Law School
https://www.pon.harvard.edu

Offers numerous lectures and workshops throughout the year focused on education, techniques, and applications of negotiation and mediation. Website contains links to a variety of resources, seminars, courses, and materials that the program makes available for purchase.

Science Management Associates (SMA)
www.sciencema.com

Science Management Associates provides workshops, training, and coaching based on the principles of this book for scientists, science managers, and executives who wish to improve their interpersonal, management, and leadership skills. These workshops are presented at host institutions specifically for their employees or staff and can be customized for specific audiences. Clients have included "top 10" pharmaceutical and biotechnology companies in the United States and the European Union as well as educational institutions and government agencies. The SMA website contains useful management tools for scientists including forms for individual development plans (IDPs), performance reviews, and links to other resources at www.sciencema.com/resources.

Women in Science and Engineering Leadership Institute
www.wiseli.engr.wisc.edu

This group is based at the University of Wisconsin, Madison. Their website contains many excellent and useful links to resources, reports, and references focused on women in science and engineering. They also offer workshops addressing many facets of bias in STEM including gender.

Index

About the Authors

Carl M. Cohen, Ph.D., is President of Science Management Associates in Newton, Massachusetts, providing training, coaching, and consulting to the biomedical research community. Carl has been a senior executive in Boston area biotechnology companies since 1996 and was a Professor of Medicine and of Anatomy and Cell Biology at Tufts University School of Medicine. Carl received his Ph.D. in Physics (Biophysics Research) from Harvard University.

Suzanne L. Cohen, Ed.D., CGP, is a psychologist and Partner in Science Management Associates. Her many roles have included Clinical Instructor in Psychiatry at Harvard Medical School, President of the Northeastern Society for Group Psychotherapy, and Fellow of the American Group Psychotherapy Association. Suzanne is in private practice in Newton, Massachusetts.